COMPUTER ORIENTED NUMERICAL METHODS

Theory, Problems, Algorithms and Implementation
Using C, C++ and Python Languages

"We want that education by which character is formed, strength of mind is increased, the intellect is expanded, and by which one can stand on one's own feet..."

... Swami Vivekananda

"The meaning of education is to encourage in people a sense of self-introspection, the ability to grasp and to develop the capacity for constructive works..."

... Dr. S. Radhakrishnan

"Always surround yourself with the kind of people who can make you believe that you can do it..."

... R. S. Salaria

"Treat others the way you wish to be treated..."

... R. S. Salaria

"May the blessings of the almighty go with this book to one and all..."

... R. S. Salaria

COMPUTER ORIENTED NUMERICAL METHODS

Theory, Problems, Algorithms and Implementation Using C, C++ and Python Languages

SIXTH EDITION

R. S. SALARIA

Alumnus of IIT, Delhi.
Certified Software Quality Professional, Govt. of India.
SUN Certified Programmer, SUN Microsystems, USA.
SUN Certified Trainer, SUN Microsystems, USA.

Former Head

Department of Computer Science & Engineering
Thapar University, PATIALA-147004.

Assisted by

Anita Salaria

KHANNA BOOK PUBLISHING CO. (P) LTD.

Publisher of Engineering and Computer Books

4C/4344, Ansari Road, Darya Ganj, New Delhi-110002
Phone : 011-23244447-48 **Mobile:** +91-9910909320
E-mail : contact@khannabooks.com
Website : www.khannabooks.com

Dear Readers,

To prevent the piracy, this book is secured with HIGH SECURITY HOLOGRAM on the front title cover. In case you don't find the hologram on the front cover title, please write us to at contact@khannabooks.com or whatsapp us at +91-99109 09320 and avail special gift voucher for yourself.

Specimen of Hologram on front Cover title:

Moreover, there is a SPECIAL DISCOUNT COUPON for you with EVERY HOLOGRAM.

How to avail this SPECIAL DISCOUNT:

Step 1: Scratch the hologram
Step 2: Under the scratch area, your "coupon code" is available
Step 3: Logon to www.khannabooks.com
Step 4: Use your "coupon code" in the shopping cart and get your copy at a special discount
Step 5: Enjoy your reading!

Price: ₹ 425.00

ISBN: 978-93-89139-00-6

Computer Oriented Numerical Methods *by* R.S. Salaria

First Edition: 1999
Sixth Edition: 2023

To view complete list of books, Please scan the QR Code:

Published by:
Khanna Book Publishing Co. (P) Ltd.
CIN: U22110DL1998PTC095547

Visit us at: **www.khannabooks.com**
Write us at: **contact@khannabooks.com**

Printed in India by: Vaibhav Products, Delhi.

Dedication

Our Dearest Parents,
Daughters Surbhi & Mehak,
Beloved Readers,

AND

Our Country Men/Women at Large.

My Slogan for My Fellow Citizens

India . . .

. . . is a Great Country

AND

Indians . . .

. . . are Outstanding.

Preface

The study of numerical methods has become essential for scientists and engineers with the widespread use of computers in the field of scientific research and engineering design. To prepare young scientists and engineers to meet this challenge, most of the universities have introduced a course on numerical methods as an integral part of the course curriculum in BCA, BSc (IT), BSc (Comp. Sc.), BE/BTech (All Branches), PGDCA, MCA, and M. Sc. in physical sciences.

This book has been designed to serve as a textbook for these courses. In addition, the practicing engineers, scientists and researchers will find it useful as a reference book.

Throughout the book, the emphasis is laid to ensure the conceptual understanding of various numerical methods. The formulae for these methods have been derived in a simplest manner. In addition, the formulae for estimating the accuracy of the computed results are also worked out wherever possible. Further the use of these methods is illustrated with the help of well-designed examples.

The formulation of algorithms for these methods is illustrated using pseudo language, an algorithmic language based on English. It is hoped that this approach will enable readers to write programs on their own with considerable ease. This is illustrated by implementing the algorithms in Turbo C version 2.01, Turbo C++ version 3.0, and Python 3.7.2.

I earnestly hope that this book would suit your needs and you will enjoy reading this book. Please do write my publishers or me and let us know your reactions. We will enjoy hearing from our readers and welcome any comments, suggestions and criticism that will enable us to improve the book.

I am certain that you will enjoy reading this book. I do need a little favour from your side in the form of your response about the book. Your free and frank comments, suggestions, and positive criticism will help me to improve the book further to meet your expectations.

. . . R. S. Salaria

Email : *rssalaria@yahoo.com*

Acknowledgments

It is my pleasure to take this opportunity to thank all those who have helped me directly or indirectly in the preparation of this book.

I would like to place on record and acknowledge the works of all those great authors whose works I have referred in the preparation of this book.

My very special thanks are reserved for honourable Dr. APJ Kalam, former President (popularly known as people's president), Republic of India, my role model, whose teachings fascinate me, make us believe that India is a great country, and we should take pride in the brand "Made in India".

I also thank Late Prof. C. M. Bahl, Managing Director, Rayat-Bahra Group of Institutes (having three campuses at - Mohali, Ropar and Hoshiarpur, in the state of Punjab), and former Director, Computer Centre, Panjab University, Chandigarh, for his continuous advice and encouragement that always motivates me to believe in greatness and doing great things.

I also owe thanks to my daughter, Surbhi Salaria and Mehak Salaria, who always have my driving force.

Finally, I must thank my publisher, Mr. Mukul Seth, Khanna Book Publishing Co. (P) Ltd., New Delhi, for publishing and establishing this book as one of the preferred book on the subject.

. . . R. S. Salaria

A Word of Wisdom for Students

My beloved students, we would like to remind you that we can categorize people as:

People who watch the things happening,
People who look into how things are happening, and
People who make the things happen.

Now you are to find out where you fit-in and where you want to be.

We firmly believe that youth of our country is second to none, and this fact is recognized world wide.

Our honorable former president Late Dr. APJ Kalam (first and may be the last president known as *people's president*) has been saying at all the platform that in India there is no dearth of resources and talent, only thing we need to change is *the mindset* (we have to shed *chalta hai* attitude). He has set a vision for us to see India a developed country by 2020, and the basis for his vision is the talent of our youth.

Take some time to discover yourself and try to assess how much talent you possess. Utilize your period of study to harness that talent. I can assure you that you will have a great future.

We want to share a popular game with you which all of us used to play in our childhood. The game is of burning a peace of paper using a lens under the sun. When the lens and the paper are held stationery, it starts burning in a few seconds. However, if you keep moving either the paper or the lens, the paper will never burn. *Why it happens so?* The answer is simple – the light energy emitted by the sun is not focused. The same is with us. We have enough energy in ourselves, but we don't focus it, rather we waste it in meaningless acts. The next question arises is – *where to focus our energy?* The natural answer of this question should be – what you want to become in your life (that basically defines your goal). So you need to focus your entire energy on your goal, and surely you will achieve your goal.

We would also like to share with you a scene from the popular epic – *Mahabharta*. When Dronacharya, Guru of Pandvas, asked his disciples to aim at the eye of the sparrow placed on a tree. All his disciples, except Arjuna, replied in a same manner – *Guru ji, we see a sparrow*. However, when the turn of Arjuna came, his reply was – *Guru ji, I see only the eye*. That is the kind of the focus you also need to develop in your life.

The next point is that when you have identified your goal, you need to identify the steps that you need to follow in order to accomplish your goal. Once the goal and the steps to accomplish your goal are identified, write them nicely on a piece of a paper and paste that paper at a place which you visit regularly. Whenever you will visit that place, the writing on the paper will remind you about your goal and what you have to do to achieve your goal.

The final word, you may set any goal for yourself and that may be of becoming a successful doctor, an engineer, an administrator or a businessman, but you must also try to be a *nice human being*. You know a nice human being is like a fruit of all seasons, and is welcome every where, anytime. We believe, the basic mantra to become a nice human being is too simple and can be stated as

"Treat the others the way you want to be treated,
in other words,
don't do anything with others which you wish others shouldn't do with you."

Love yourself & enjoy your student life and make best use of it . . .

Wish you good luck and a nice time ahead.

. . . R. S. Salaria

Contents at a Glance

Table of Contents

APPENDICES

Preamble

Learning Outcomes

After reading this chapter, students will have the ability to

- explain the scope of numerical analysis differentiate between analytical (closed form) solution and numerical solution
- explain the type of problem that can be solved using numerical analysis
- explain how computers can be used effectively to find numerical solutions of variety of problems

1.1 INTRODUCTION

Numerical analysis is the study of the appropriate methods for solving mathematical, scientific and engineering problems by means of basic arithmetic operations such as addition, subtraction, multiplication and division. Scientists and engineers are widely using numerical analysis to solve their problems. A major advantage of numerical analysis is that a numerical solution can be obtained even when a problem has no *analytical solution,* also known as *closed form solution.*

For example, the integral

$$\int_0^{\pi} (\sqrt{1+\cos^2 x})\,dx$$

gives the length of the arc of the curve $y = \sin(x)$. This integral has no analytical solution. However, with the help of numerical analysis, using standard methods, we can compute the length of the arc of this curve. These methods, of course, apply to any integrand. There is no need to make substitutions or to do the integration by parts to obtain the result. Further, in numerical analysis, the only mathematical operations required are addition, subtraction, multiplication and division plus the comparison operations. Since these simple operations are the only operations that computers can perform, thus numerical analysis and computers make a perfect combination.

1.2 SOME OBSERVATIONS ABOUT NUMERICAL SOLUTIONS

It is important to note that the following points about numerical solutions:

1. The results obtained using numerical analysis is purely numerical in nature, whereas analytical methods give a solution in terms of mathematical function that can be evaluated for specific instances. The behaviour and properties of the function are clearly visible and this is not the case with numerical solutions. In this sense, the analytical solution has an advantage over numerical solution. However, the numerical solution can be plotted to show some of the behaviour and properties of the solution.

2. The numerical solution is always an approximation to the true solution, but the approximation can be made more accurate to some extent. There is a limitation to achieve a certain level of accuracy because of the way the computer stores numbers and performs arithmetic. These limitations are described in *Chapter* 2. Even to evaluate an analytical solution to get the numerical solution for a specific application incurs similar errors as in obtaining numerical solution directly.

The analysis of computer errors and other sources of errors in numerical methods are an important part of the study of numerical analysis. This issue will occur throughout this book.

Here are some of the problems that numerical analysis can do and are described in this book:

- Find the roots of a non-linear equation.
- Solve a system of non-linear equations.
- Solve a system of linear equations.
- Interpolate to find the intermediate values within a table of values.
- Find efficient and effective approximations of the functions.
- Fit an appropriate curve through a given set of points.
- Find derivatives, when function is known only as table of values.
- Integrate function, when the function is known only as table of values. Multiple integrals can also be solved.
- Solve ordinary differential equation, when the initial values of the variables are given.

1.3 NUMERICAL ANALYSIS USING COMPUTERS

Numerical analysis requires a large number of repetitive arithmetic operations. Handling of these many operations is possible only if we use a computer to carry out these operations. Though these operations are very basic, a human would make so many mistakes and it will be difficult to rely on these results. In addition, the cost of manpower and the time required would be much more.

Recall that computer is a dumb machine (i.e. not an intelligent machine which can think and act of its own) it must be given detailed and complete instructions for each step it is supposed to perform. In other words, a computer program must be written so that the computer can do the numerical analysis.

As you study this book, you can learn a lot about many numerical methods and the way you can write programs to implement them.

You can use a computer language such as BASIC, FORTRAN, Pascal, C/C++, Java, Python, etc. to write programs for these numerical methods.

As a user of FORTRAN since 1985, I must say that FORTRAN is most suitable language to write programs for numerical methods. In particular, the Microsoft FORTRAN 77 version 5.1, a version of FORTRAN developed by Microsoft Corporation and implemented on PC's, provides all sort of facilities to write programs for numerical analysis.

In this book, various numerical methods are described through a pseudocode. This pseudocode is written in such a form that translating it into a program is relatively straightforward.

Number of example programs, as per the demand of the curricullum of various universities/institutions, written using C, C++, and Python languages are given in Appendices.

These programs have been written for better understanding rather than showing professional proficiency.

❑ ❑ ❑

2 Data Representation and Computer Arithmetic

Learning Outcomes

After reading this chapter, students will have the ability to

- differentiate between exact numbers and approximate numbers
- the concept of significant digits
- explain how numbers are represented in memory
- explain the concept of number system
- convert numbers from decimal number system to binary number system and vice-versa
- explain the concept of complements for representing negative integer numbers
- explain the possible schemes for storing negative integer numbers
- explain the possible schemes for storing real numbers
- perform arithmetic operations on floating-point numbers
- explain the consequences (pitfalls) of floating-point representation
- explain the variety of errors that occur in numerical computations
- explain the various measures of accuracy
- explain the relationship between relative error and significant digits
- explain how error propagates in various arithmetic operations

2.1 INTRODUCTION

In many applications in the field of science and engineering, one would require numerical results that involve the use of all kinds of numbers and formulas. In such situations, one must be able to use these numbers and solve the formulas to obtain the best possible results. But in most cases, the data being used is not always exact; therefore the numbers used to express such data cannot be exact. They are approximations correct to certain number of significant digits. These approximations introduce errors in the computed results. In addition, there may be instances when the formula or method used to obtain the solution is not exact. As an example, consider the evaluation of e^x, which is expressed as infinite power series. This series has to be truncated to finite number of terms. This truncation introduces an error in computed value of e^x. The amount of error will depend on number of terms retained in the series.

In general, we can say that the formulas or methods used to obtain numerical solutions can be approximate. Therefore, the use of approximate numbers and/or approximate methods gives us a solution that is not exact but approximate. However, our main objective is to obtain an approximate solution as close to the exact solution as possible.

This chapter describes the concept of approximate numbers, significant digits, the way numbers are expressed and operations are performed on them, types of error and their sources, propagation of errors in successive operations, etc.

The clear understanding of these concepts will enable the reader to apply various methods described in the book to obtain numerical solution of real life problems with considerable ease.

2.2 CONCEPT OF APPROXIMATE NUMBERS

It is very important to make a distinction between numbers that are exact and those that express approximate values. For example, numbers 2, 1/2, 11, etc. are exact numbers because there is no approximation or uncertainty associated with them. On the other side, the numbers π, $\sqrt{2}$, e, ..., appears to be exact, but they cannot be expressed exactly by finite number of digits. These numbers can be expressed as 3.1414, 1.4142, 2.7183, ..., respectively, which are approximation to the true values and hence are called *approximate numbers* or *approximate values* of the numbers. Therefore, an approximate number is defined as the number that is used as an approximation to an exact number and differs slightly from the exact number for which it stands.

2.3 CONCEPT OF SIGNIFICANT DIGITS

Any digit is a significant digit except when it is used to fix the decimal point or to fill the places of unknown or discarded digits. For example, in the number 0.00263, the

significant digits are 2, 6, and 3, whereas zeros are used merely to fix the decimal point, and therefore are not significant. Similarly, in the number 438000, the significant digits are 4, 3, and 8. However, in the number 3809.025, all the digits including the zeros are significant.
333
Generally, the best way to find the significant digits is to write the number in the scientific notation with the first digit being non-zero. Using scientific notation, the above numbers can be expressed as 0.263×10^{-2}, 0.438×10^{6}, and 0.3809025×10^{4}, respectively. Now just read the digits in the mantissa except the trailing zeros, if any.

2.4 REPRESENTATION OF NUMBERS IN MEMORY

The smallest element of information is the binary digit, abbreviated as *bit,* which represents digit 0 or 1. The instruction to the computer and the data to be processed is internally organised in groups of bits. The group of eight bits, known as *byte,* has become standard. One byte may represent a letter, a digit or a special character such as +, *, -, ?, $, #, @, etc..

In the present text, we are concerned only with numbers that can be integer numbers or real numbers. As these are represented internally as a sequence of bits, operations on them follow the laws of binary arithmetic.

The computer's memory, in case of microcomputers (PCs) based on Intel processors, is organised as a long sequence of bytes. Therefore in order to store a number, one or more bytes are used. The way of storing an integer number is quite different from that of a real number. We will discuss this aspect separately.

2.4.1 Number Systems

A *number system* is formally defined as a set of

1. Symbols or digits.
2. Rules describing the ways numbers are formed from these digits.
3. Rules specifying the operations performed on these numbers.

The total number of digits in a number system constitute *base* of the number system. The base of a number system is also referred to as *radix.*

We all are familiar with decimal number system, which is defined on digits 0, 1, 2, 3, 4, 5, 6, 7, 8, 9, and has a base of 10. Each digit has a certain *face value* and a *place value* (also known as *weight*) based on its positions in the number, which is always a power of 10. And any decimal number can be expressed as a sum of products of face values and place values.

For example, the decimal number 1537.25 can be expressed as

$$1\times10^3+5\times10^2+3\times10^1+7\times10^0+2\times10^{-1}+5\times10^{-2}$$

The binary number system is defined on digits 0 and 1 only and has a base of 2. The *place value* is always a power of 2. Like a decimal number, a binary number can also be expressed as a sum of products of face values and place values.

For example, the decimal number 1010.01 can be expressed as

$$1\times2^3+0\times2^2+1\times2^1+0\times2^0+0\times2^{-1}+1\times2^{-2}$$

Solving this expression, we get its equivalent decimal number.

2.4.2 Decimal to Binary Conversion

When we enter the data through keyboard, the entered number is converted into binary end then stored in computer's memory. Let us see what the procedure is to accomplish this task.

2.4.2.1 Converting a Decimal Integer Number

Divisor	Quotient	Remainder
2	55	
2	27	1
2	13	1
2	6	1
2	3	0
2	1	1
	0	1

This shows that the decimal integer number 55 = binary integer number 110111

The decimal integer is successively divided by 2 till the quotient is reduced to zero. The remainder after each division, which is either 0 or 1, is noted down. The entire sequence of remainders taken in reverse order represents the equivalent binary number.

2.4.2.2 Converting a Decimal Fraction

The decimal fraction is successively multiplied by 2 till the new fraction is reduced to zero. The integer part, if any, after each multiplication, which is either 0 or 1, is noted down. The entire sequence of integer parts in order and prefixed with point, represents the equivalent binary fraction.

For example, consider a decimal fraction 125.

	.125
	× 2
0	.250
	× 2
0	.500
	× 2
1	.000

This shows that the decimal fraction .125 = binary fraction .001

Therefore, to convert a real number into binary, the integer part and the fractional part are converted separately, and then combined together to form the equivalent binary number. Therefore, the decimal real number 55.125 equals binary real number 110111.001.

2.4.3 Complements: *An Introduction*

In computers we need to represent and store positive as well as negative numbers. In order to store negative numbers, the complements are used. In binary number system, we can use 1's complement or 2's complement to represent a negative number.

2.4.3.1 1's Complement

The 1's complement of a binary number is a binary number obtained by inverting the bits i.e. 1 is changed to 0 and 0 to 1. Therefore, in order to find the representation of a negative integer number, first we obtain the binary representation of its magnitude, and then we take its 1's complement, which gives its final binary representation.

For example, to convert –44 to binary, first we convert 44 to binary, and then take its 1's complement that gives its final binary representation as illustrated next:

Decimal 44 = binary 101100

1's complement of 101100 = 010011

Therefore 010011 is the binary representation of –44 in 1's complement scheme.

2.4.3.2 2's Complement

The 2's complement of a binary number is taken by first taking 1's complement of it and then adding 1 to the least significant bit (LSB) position of the 1's complement. The 2's complement can also be obtained directly as stated — examine the bits from right to left (i.e. from least significant position towards most significant position), and as soon as the first non-zero bit is encountered, keep it unchanged and subsequent bits are inverted. For example, consider the representation of –44 in 2's complement scheme.

Decimal 44 = binary 101100

1's complement of 101100 = 010011

2's complement of 101100 = 010011 + 1 = 010100

Therefore 010100 is the binary representation of –44 in 2's complement scheme.

2.4.4 Storage of Integer Numbers

Most of the modern computers have a provision for representing small, large, and very large integer numbers. These integers are also referred to as 1-byte, 2-byte, and 4-byte integers. The following are the standard forms for representing the integer number.

- Signed representation
- 1's complement representation
- 2's complement representation

In each of these representations, the most significant bit is used to represent the sign of the integer number and the remaining bits are used to hold the magnitude of the number. In each of the above schemes if the sign bit is 0, then the integer number is positive, else it is negative.

2.4.4.1 Signed Representation

In signed representation, the bits reserved to hold the magnitude of the integer number holds the actual magnitude of the number. For example, the numbers 44 and –44 in 1-byte format will be represented as 00101100 and 10101100 respectively.

2.4.4.2 1's Complement Representation

In 1's complement representation, if the sign bit is 0 (i.e. integer number is positive), the bits reserved to hold the magnitude of the integer number holds the actual magnitude of the number. And, if the sign bit is 1 (i.e. integer number is negative), the bits reserved to hold the magnitude of the integer number holds the 1's complement of the magnitude of the number. For example, the numbers 44 and –44 in 1-byte format will be represented as 00101100 and 11010011 respectively.

2.4.4.3 2's Complement Representation

In 2's complement representation, if the sign bit is 0 (i.e. integer number is positive), the bits reserved to hold the magnitude of the integer number holds the actual magnitude of the number. And, if the sign bit is 1 (i.e. integer number is negative), the bits reserved to hold the magnitude of the integer number holds the 2's complement of the magnitude of the number.

For example, the numbers 44 and –44 in 1-byte format will be represented as 00101100 and 11010100 respectively.

Table 2.1: Type of Integers and range of values in 2's complement representation

Type	Size (in bytes)	Range
Small	1	–128 to +127
Large	2	–32,768 to +32,767
Very large	4	–4,294,967,296 to +4,294,967,295

In most of the computers, the 2's complement representation is used to represent the integers. This representation has advantages over the other representations as for as the implementation of the arithmetic operations are concerned. The technical details are beyond the scope of the book.

2.4.5 Storage of Real Numbers

As we know that real numbers can be written in two formats: *fixed-point* (without exponent) and *floating-point* (with exponent). The floating-point format is similar to scientific-form, where the real number consists of two parts — *mantissa* and an *exponent.* For example, the real number 1.257×10^5 in the scientific form will be written as 1.257E5 in the floating-point format.

Though the decimal point can be moved freely to the left or to the right, it is standard practice, also more convenient, to place the decimal point immediately before the first non-zero digit i.e. before the most significant digit.

The shifting of the decimal point to the left of the most significant digit is called *normalization* and the real numbers represented in this form are known as *normalized floating-point number.* The normalized floating-point numbers consists of two parts — mantissa that is a fractional part, and the exponent. Mantissas of the normalized floating-point numbers satisfy the following conditions:

for positive numbers : $0.1 \le \text{mantissa} < 1.0$
for negative numbers: $-1.0 < \text{mantissa} \le -0.1$

In other words, mantissa of a normalized floating-point number must satisfy

$$0.1 \le |\text{mantissa}| < 1.0$$

As an example, the floating-point number 0.00125e–2 is expressed as 0.125e–4 in the normalized floating-point representation.

The real numbers are represented in computer memory as normalized binary floating-point numbers. A floating-point number are represented in four consecutive bytes. Out of these four bytes, the mantissa is represented in lower order three bytes and the exponent in the most significant byte.

Figure 2.1: Sample floating point format (32-bit format) for storing real numbers in normalized floating-point format

The exponent of a floating-point number can be positive or negative. One way of taking care of the sign of the exponent is to reserve the most significant bit of the byte used to hold the exponent for sign. This scheme will be able to hold the exponent in the range –128 to +127. But this scheme is not used very often. Instead, an equivalent system is used where the actual exponent is increased i.e. *biased* by the addition of a constant. Then the modified (increased) exponent is stored. The constant added to the actual exponent is determined in such a way that the minimum possible integer in 8-bits is raised to zero. Thus in the above format, the constant is 128. With the addition of 128, the actual exponent, which may be from –128 to +127, is now modified and varies from 0 to 255. All these are positive values, and therefore, this approach eliminates the use of sign bit. The sign of the floating-point number depends only on the sign bit of the mantissa. If the sign bit is 0, then the floating-point number is positive else negative. The rest of 23 bits represent the bit pattern of the mantissa.

Example 2.1:

Let us see how the number 13.5 is represented in the above format.

Step 1. Decimal number 13.5 is converted into binary. Using the procedure described earlier, we obtain 1101.1 as binary equivalent of 13.5.

Step 2. Convert binary number 1101.1 into normalized floating-point number. We obtain 0.11011×2^4.

Step 3. The mantissa 11011, when extended to 23 bits, by adding 0's on the right, becomes 11011000000000000000000.

Step 4. Here actual exponent is 4. Adding constant 128 to it, the modified exponent becomes 132. The binary equivalent of 132 is 10000100.

Step 5. As the number is positive, the sign-bit is 0.

Step 6. Combining the result of all of the above steps, we obtain the final representation of 13.5 as 10000100010110000000000000000000.

In order to find the normalized floating-point representation of a negative real number, the procedure is to first find the binary equivalent of the absolute value of the number. This binary pattern is extended to 23-bit pattern by adding 0s on the right. Then 2's complement of this 23-bit pattern is taken. Since the number is negative, the sign-bit will be 1. Thus, these 24-bits will represent the mantissa of the normalized floating-point binary number. The procedure for the exponent is same as for positive number.

Example 2.2

Let us see how the number –13.5 is represented in the above format.

Step 1. Absolute value of number –13.5, which is 13.5 is converted into binary. Using the procedure described earlier, we obtain 1101.1 as binary equivalent of 13.5.

Step 2. Convert binary number 1101.1 into normalized floating-point number. We obtain 0.11011×2^4.

Step 3. The mantissa 11011, when extended to 23 bits, by adding 0's on the right, becomes 11011000000000000000000.

Step 4. Taking 2's complement of this 23-bit pattern, we obtain 00101000000000000000000.

Step 5. As the number is negative, the sign-bit is 1.

Step 6. Using the same steps for exponent, we obtain 10000100 as binary equivalent of the modified exponent.

Step 7. Combining the result of all of the above steps, we obtain the final representation of –13.5 as 10000100100101000000000000000000.

2.4.5.1 Possible Range of Values

In order to find the range of real numbers that can be represented in the above format, we will make some observations. The largest floating-point number that may be represented will have 0 as the sign-bit followed by all 1s for the mantissa, and 11111111 as the biased exponent i.e. 127 as the actual exponent. Here, instead of finding this number, we will attempt to find the order of magnitude expressed in decimal.

We proceed using the following equation

$$2^{127} = 10^x$$

Taking logarithm on both sides, we get

$$127 \times \log 2 = x \times \log 10$$

$$x = 127 \times 0.301029995664 = 38 \qquad \text{(nearest integer)}$$

Therefore, the maximum representable decimal floating-point number, positive as well as negative, is of the order of 10^{+38}.

The minimum positive number occurs when the mantissa is positive, and the biased exponent is 00000000, *i.e.*, –128 as the actual exponent. Proceeding on the similar lines, we find that the minimum positive number is of the order of 10^{-38}.

The range in the floating-point representation is defined as the range of values between the minimum positive floating-point number and the maximum positive floating-point number. In the above format, it is simply written as $10^{\pm 38}$. Thus, in the above format, the real numbers will be of the order of $10^{\pm 38}$.

2.4.5.2 Accuracy of Values

We may like to ask — given a decimal real number with certain number of digits, *how accurate will be its representation in binary*? Remember that 23 bits are used to represent the magnitude of the mantissa. To answer the above question, we use the following equation

$$2^{23} = 10^x$$

which on solving gives $x = 6.9$.

Thus, with 23 bits for the mantissa, we can be certain of maximum accuracy of about 6.9 (=7) digits in the binary representation of a real number. It is possible that some real numbers of six or even fewer digits, such as 0.4, may not be accurately represented because of the problem of conversion.

2.5 FLOATING-POINT ARITHMETIC

For the illustration of arithmetic operations on normalized floating-point numbers, we consider a hypothetical computer that can store normalized floating-point numbers with four digit mantissa and two-digit exponent. In this format, the mantissa can vary from –0.9999 to +0.9999 and the exponent can vary from –99 to +99.

For example, the number –0.2345E5 will be stored as

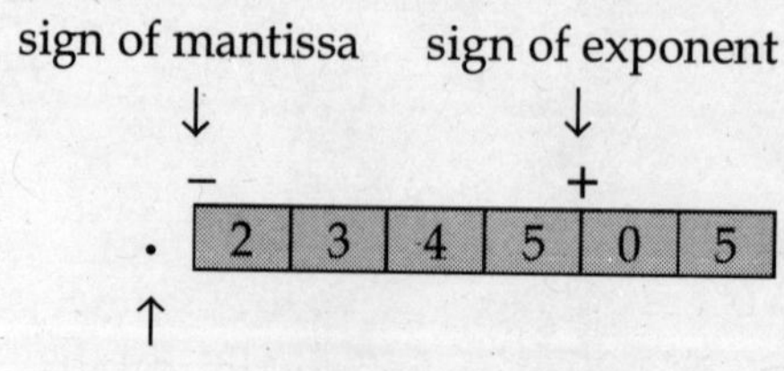

2.5.1 Addition Operation

Addition operation can be performed on normalized floating-point numbers if the exponents of these numbers are equal. If the exponents are not equal, then the exponent of the numbers with smaller exponent is made equal to the larger exponent and its mantissa is modified. Shifting the decimal point to the left does this modification by number of places equal to the positive difference between the two exponents. Now the mantissas of these numbers are added.

For example, if 0.7253E2 and 0.5467E5 are to be added, the decimal point of mantissa of 0.7253E2 is shifted by 3 (5–2) positions to the left and the exponent is incremented by 3. The number after this modification becomes 0.0007E5. These two numbers can now be added by adding their respective mantissas.

Note that the mantissa of the number representing sum of these two normalized floating-point numbers may not be normalized. Therefore, to complete the operation, the mantissa must be normalized. Further note that the mantissa of the sum (before normalization) can be maximum of +1.9999, there at most we need to shift the decimal point to left by one position in order to normalize it. As a result of this, the exponent of the sum is incremented by 1. Because of normalization process, the exponent of the sum may become greater than +99 that implies that the number is greater than the largest number that our hypothetical computer can store. This condition is called *overflow*, and the system will signal this error.

Example 2.3: Add 0.5467E5 to 0.7253E2

Solution: The decimal point of the mantissa of 0.7253E2 is shifted three positions to left. It becomes 0.0007, whereas the digits 3, 5, and 2 are chopped off i.e. truncated. The exponent is incremented by 3. Therefore, the number after normalization becomes 0.0007E5. These numbers can be added now as shown below:

Addend	0.0007E5
Augend	0.5467E5
Sum	0.5474E5

Example 2.4: Add 0.5467E2 to 0.7254E2

Solution: Since the exponents are already equal, they can be directly added as:

Addend	0.7254E2
Augend	0.5467E2
Sum	1.2721E2

In this case, the mantissa of the sum is greater than 1.0, so the decimal point is shifted to the left by one position and the exponent is increased by 1. The last digit of the mantissa of sum is chopped off giving the normalized sum as 0.1272E3.

Example 2.5: Add 0.5467E99 to 0.7254E99

Solution: Since the exponents are already equal, they can be directly added as:

Addend	0.7254E99
Augend	0.5467E99
Sum	1.2721E99

In this case, the mantissa of the sum is greater than 1.0, so the decimal point is shifted to the left by one position and the exponent is increased by 1. The last digit of the mantissa of sum is chopped off, giving the normalized sum as 0.1272E100. Since this number is greater than the largest number that our hypothetical computer can handle, it is a case of *overflow* and the system will indicate this condition.

Algorithm 2.1: To add two normalized floating-point numbers

Let us consider that x_1, x_2 be the mantissas and e_1, e_2 be the exponents of the two normalized floating-point numbers. Further consider that x and e represents the mantissa and exponent of the sum. The integer variable k is used as temporary variable.

```
Begin
    read: x1, e1, x2, e2
    set k  = | e1 - e2 |
    if (e1 > e2 ) then
        set x2 = x2 / 10^k
        set e = e1
    else
        set x1 = x1 / 10^k
        set e = e2
    endif
    set x = x1 + x2
    if ( x ≥ 1.0 ) then
        set x = x / 10
        set e = e + 1
    endif
    if ( e > 99 ) then
        write: "Overflow . . . "
    else
        write: x, e, "as mantissa and exponent of the sum"
    endif
End.
```

2.5.2 Subtraction Operation

Like addition operation, the subtraction operation can be performed on normalized floating-point numbers if the exponents of these numbers are equal. If the exponents are not equal, then the exponent of the numbers with smaller exponent is made equal to the larger exponent and its mantissa is modified. Shifting the decimal point to the left does this modification by number of places equal to the positive difference between the two exponents. Now the mantissas of these numbers are subtracted. Here also, the mantissa of the number representing difference of these two normalized floating-point numbers may not be normalized.

Therefore, to complete the operation, the mantissa must be normalized. Further note that the magnitude of mantissa of the difference (before normalization) will be in the range 0.0000 to 0.9999. These extreme values will occur in case the numbers are equal or the mantissa of the number to be subtracted is 0.0000 (zero) respectively. Therefore, the decimal may have to shift to right by more than one position. Correspondingly, the exponent is decremented by one for each shift of the mantissa. During this normalization process, the exponent of the difference may become less than –99 which implies that the number is smaller than the smallest number which our hypothetical computer can store. This condition is called *underflow*, and the system will signal this error.

Example 2.6: Subtract 0.7253E2 from 0.5467E5.

Solution: The number 0.7253E2 is normalized, thus resulting in number 0.0007E5. Now the number 0.0007E5 can be subtracted from 0.5467E5 as:

Minuend	0.5467E5
Subtrahend	0.0007E5
Difference	0.5460E5

In this case, the mantissa of the difference is already in the normalized form.

Example 2.7: Subtract 0.7254E5 from 0.7288E5.

Solution: Since the exponents are already equal, the number 0.7254E5 can be directly subtracted from 0.7288E5 as:

Minuend	0.7288E5
Subtrahend	0.7254E5
Difference	0.0034E5

In this case, the mantissa of the difference is less than 0.1, so the decimal point is shifted two positions to the right and the exponent is decreased by 2. The resultant difference becomes 0.3400E3.

Example 2.8: Subtract 0.7254E–99 from 0.7278E–99.

Solution: Since the exponents are already equal, the number 0.7254E5 can be directly subtracted from 0.7288E5 as:

Minuend	0.7278E–99
Subtrahend	0.7254E–99
Difference	0.0024E–99

In this case, the mantissa of the difference is less than 0.1, so the decimal point is shifted two positions to the right and the exponent is decreased by 2. The resultant difference becomes 0.3400E–101. Since this number is smaller than the smallest number that our hypothetical computer can handle, it is a case of *underflow* and the system will indicate this condition.

Algorithm 2.2: To compute difference of two normalized floating-point numbers

Let us consider that x_1, x_2 be the mantissas and e_1, e_2 be the exponents of the two normalized floating-point numbers. Further consider that x and e represents the mantissa and exponent of the difference. The integer variable k is used as temporary variable.

```
Begin
    read: x1, e1, x2, e2
    set k  = | e1 - e2 |
    if (e1 > e2 ) then
        set x2 = x2 / 10^k
        set e = e1
    else
        set x1 = x1 / 10^k
        set e = e2
    endif
    set x = x1 - x2
    while ( ( | x | < 0.1 ) and ( | x | > 0.0 ) ) do
        set x = x * 10
        set e = e - 1
    endwhile
    if ( e < -99 ) then
        write: "Underflow . . . "
    else
        write: x, e, "as mantissa and exponent of the difference"
    endif
End.
```

2.5.3 Multiplication Operation

In multiplication operation, the mantissas are multiplied and the exponents are added. The mantissa of the product will not be in normalized form. Further note than the magnitude of the product will be greater than 1.0 but less than 10.0. Therefore, at most, the decimal point of the mantissa of the product will shift one position to the left and the exponent is incremented by 1. As a result of this increment the exponent can become +99. Now if the mantissa is negative, this results in underflow else overflow.

Example 2.9: Multiply 0.6543E5 by 0.2255E3

Solution: 0.6543E5 × 0.2255E5 = (0.6543 × 0.2255)E(5+3) = 0.14754465E8

Thus, we obtain 0.1475E8 after truncating the mantissa of the product to four digits. This product is already in the normalized form.

Example 2.10: Multiply 0.1234E5 by 0.1111E13

Solution: 0.1234E5 × 0.1111E13 = (0.1234 × 0.1111)E(5+13) = 0.01370974E18

Thus, we obtain 0.0137E18 after truncating the mantissa of the product to four digits. And since this product is less than 0.1, the decimal point is shifted one position to the right and the exponent is decreased by 1. The resultant product becomes 0.1370E17.

Example 2.11: Multiply 0.1234E75 by 0.1111E37

Solution: 0.1234E75 × 0.1111E37 = (0.1234 × 0.1111)E(75+37) = 0.01370974E112

Thus, we obtain 0.0137E112 after truncating the mantissa of the product to four digits. And since this product is less than 0.1, the decimal point is shifted one position to the right and the exponent is decreased by 1. The resultant product becomes 0.1370E111. Since this number is greater than the largest number that our hypothetical computer can handle, it is a case of *overflow* and the system will indicate this condition.

Example 2.12: Multiply 0.1234E–75 by 0.1111E–37

Solution: 0.1234E–75 × 0.1111E–37 = (0.1234 × 0.1111)E[(–75)+(–37)]

= 0.01370974E–112

Thus, we obtain 0.0137E–112 after truncating the mantissa of the product to four digits. And since this product is less than 0.1, the decimal point is shifted one position to the right and the exponent is decreased by 1. The resultant product becomes 0.1370E–113. Since this number is smaller than the smallest number that our hypothetical computer can handle, it is a case of *underflow* and the system will indicate this condition.

Algorithm 2.3: To multiply two normalized floating-point numbers

Let us consider that x_1, x_2 be the mantissas and e_1, e_2 be the exponents of the two normalized floating-point numbers. Further consider that x and e represents the mantissa and exponent of the product.

```
Begin
    read: x1, e1, x2, e2
    set x = x1 × x2
    set e = e1 + e2
    if ( | x | ≥ 1.0 ) then
        set x = x / 10
        set e = e + 1
    endif
```

```
    if ( | x | < 0.1 ) then
        set x = x * 10
        set e = e - 1
    endif
    if ( e > 99 ) then
        write: "Overflow . . . "
    else if ( e < -99 ) then
        write: "Underflow . . . "
    else
        write: x, e, "as mantissa and exponent of the product"
    endif
End.
```

2.5.4 Division Operation

In division operation, the mantissa of the one number is divided by the mantissa of the second and the exponent of the second number is subtracted from the first. The mantissa of the quotient will not be in normalized form. Further note than the magnitude of the quotient may become greater than 1.0, but will always be less than 10.0. Therefore, at most, the decimal point of the mantissa of the quotient will shift one position to the left and the exponent is incremented by 1. As a result of this increment the exponent can become +99. Now if the mantissa is negative, this results in underflow else overflow.

Example 2.13: Divide 0.8888E5 by 0.2000E3

Solution: 0.8888E5 ÷ 0.2000E3 = (0.8888 ÷ 0.2000)E(5–3) = 4.4440E2

The mantissa of the quotient is greater than 1.0, therefore the decimal point is shifted one position to the left and the exponent is increased by 1. The resultant quotient becomes 0.4444E3.

Example 2.14: Divide 0.9998E5 by 0.1000E–99

Solution: 0.9998E5 ÷ 0.1000E–99 = (0.9998 ÷ 0.1000)E[5–(–99)] = 9.9980E104

The mantissa of the quotient is greater than 1.0, therefore the decimal point is shifted one position to the left and the exponent is increased by 1. The resultant quotient becomes 0.9998E105. This is a case of *overflow*.

Example 2.15: Divide 0.9998E–5 by 0.1000E99

Solution: 0.9998E-5 ÷ 0.1000E99 = (0.9998 ÷ 0.1000)E(–5–99) = 9.9980E–104

The mantissa of the quotient is greater than 1.0, therefore the decimal point is shifted one position to the left and the exponent is increased by 1. The resultant quotient becomes 0.9998E–103. This is a case of *underflow*.

Algorithm 2.4: To perform division of two normalized floating-point numbers

Let us consider that x_1, x_2 be the mantissas and e_1, e_2 be the exponents of the two normalized floating-point numbers. Further consider that x and e represents the mantissa and exponent of the quotient.

```
Begin
    read: x1, e1, x2, e2
    set x = x1 / x2
    set e = e1 - e2
    if ( | x | ≥ 1.0 ) then
        set x = x / 10
        set e = e + 1
    endif
    if ( | x | < 0.1 ) then
        set x = x * 10
        set e = e - 1
    endif
    if ( e > 99 ) then
        write: "Overflow . . . "
    else if ( e < -99 ) then
        write: "Underflow . . . "
    else
        write: x, e, "as mantissa and exponent of the product"
    endif
End.
```

2.6 PITFALLS OF FLOATING-POINT REPRESENTATION

We have seen in the preceding sections that the mantissas have to be truncated to four digits in order to fit into the normalized floating-point format of the hypothetical computer. This truncation leads to a number of surprising results.

1. It is well known fact that

   $$4x = x + x + x + x$$

   However, when arithmetic is performed using normalized floating-point representation, the above equation may not hold true.

   Example 2.16: Consider x = 0.6667E0, then $4x$ written as 0.4000E1 × 0.6667E0 using normalized floating-point representation gives a value of 0.2666E1. However, if we add 0.6667E0 four times as

   0.6667E0 + 0.6667E0 + 0.6667E0 + 0.6667E0

   answer will be 0.2665E1. This happens because of truncation in the intermediate and final results in order to fit into the normalized floating-point format.

2. The associative and distributive laws of arithmetic may not hold true. The proof of this fact is left as an exercise for the readers.

3. Another very important point to remember is that while performing arithmetic in normalized floating-point format, the equality of an expression to zero can never be assured.

Example 2.17: The quadratic equation

$$x^2 + 2x - 2 = 0$$

has roots $-1 \pm \sqrt{3}$. Expressing in normalized floating-point, the roots are 0.7320E0 and –0.2732E1.

If we substitute x = 0.732E0, and solve the expression $(x^2 + 2x) - 2$ as

$$(0.7320E0 \times 0.7320E0 + 0.2000E1 \times 0.7320E0) - 0.2000E1$$

$$= (0.5358E0 + 0.2000E1 \times 0.7320E0) - 0.2000E1$$

$$= (0.5358E0 + 0.1464E1) - 0.2000E1$$

$$= 0.1999E1 - 0.2000E1 = -0.1000E{-2}$$

Ideally, the expression should yield the value zero. The point, which we want to emphasize, is that in algorithms, it is not advisable to write a branching or looping instruction that compares the value of an expression with zero.

2.7 ERRORS

An *error* is defined as the difference between the actual value and the approximate value obtained from the experimental observation or from numerical computation. Consider that x represents some quantity and x_a is an approximation to x, then

$$\text{error} = \text{actual value} - \text{approximate value} = x - x_a$$

As mentioned earlier, the errors in computed results may be due to errors in input data and/or computational algorithm. It is important to note that it is impossible to remove the errors in the input data, however errors in calculations can be reduced to some extent.

The errors in the computed results can be classified in following categories:

1. Errors in input data
   - due to approximate measurements, known as *data errors.*
   - due to truncation of the digits, known as *truncation errors.*
   - due to rounding-off the digits, known as *round-off errors.*

2. Computational errors
   - due to pitfalls in computational algorithm.
   - due to truncation of digits during the arithmetic operation, because of limitation of storage.

2.7.1 Data Errors

These are errors that occur due to inaccurate measurements or observations that may be due to limitations of the measuring device. For example, vernier calliper, screw gauge, etc. can measure the quantity accurate to certain smallest value. The accuracy of the measurements also depends on the experience of the person.

2.7.2 Truncation Errors

Truncation errors occur when some digits from the number are discarded. There are mainly two situations when

i) During the representation of numbers in normalized floating-point form. Because only few digits in the mantissa can be accommodated, for example, only four digits in our hypothetical computer. For example, the number 0.0356879 takes the form 0.3568E–1 in our hypothetical computer in the normalized floating-point representation. The digits 7 and 9 have been discarded. This truncation introduces errors in the input data.

ii) During the conversion of a number from one system to another. For example, the decimal number 13.1 has equivalent binary representation 1101.0001100110011. . . It has a repeating fraction and therefore the conversion must be terminated after some digits. This introduces the truncation error.

2.7.3 Round-off Errors

Rounding-off the number also causes truncation but with some adjustment to the last digit retained depending on the values of the truncated digits.

The following numbers are rounded-off to two decimal places:

Given decimal number	0.667	4.3743	8.9954
Rounded-off number	0.67	4.37	9.00

Observe that one has been added to the second decimal place in the first and third number because the third decimal place is greater than five.

The procedure for rounding-off a number to given number of decimal places can therefore be stated as follows:

> Suppose a number is required to be rounded-off to n^{th} decimal place, then one is added to the nth decimal digit if the $(n+1)^{th}$ decimal digit is greater than or equal to 5, otherwise the nth decimal digit is kept unaltered.

The error introduced by rounding-off the numbers to given decimal places is known as round-off error.

2.7.4 Computational Errors

As mentioned earlier, the major sources of computational errors are:

- Normalized floating-point representation
- Truncation of infinite series expansion
- Inefficient algorithm

As we have seen in the preceding sections that while performing the basic arithmetic operations using the normalized floating-point representation, some digits in the mantissa have to be truncated in order to fit into the normalized floating-point form. Such truncations introduce varied amount of errors depending on the size of the numbers and the nature of the operations performed.

Similarly, errors are introduced because of finite representation of an inherently infinite series or process, such as sin(x), cos(x), log(x), e^x etc.. For example, the following infinite series

$$\cos(x) = 1-\frac{x^2}{2!}+\frac{x^4}{4!}-\frac{x^6}{6!}+...$$

is used to compute the value of cos(x). Since the series is infinite, it is not possible to use all the terms in the computations. In such cases, the process is terminated after a finite number of terms. Therefore, the terms omitted will introduce error in the computed result. Such errors are also called truncation errors, as they are introduced due to truncation of the infinite series. Errors due to truncation of an infinite can be estimated if we know the number of terms included in the computations.

And finally, the algorithms, if not properly selected or formulated, will also introduce errors in the various steps of computations. As an illustration, we consider an example of computing combinations.

Suppose there are n objects and r objects are to selected at a time, then possible number of combinations is given by

$$^{n}c_r = \frac{n!}{r!(n-r)!}$$

Simple and straight forward approach to compute the factorials is to compute $n!$, $r!$ and $(n-r)!$ separately and then substitute these values in the above formula to compute $^{n}c_r$.

The real problem with this approach occurs when the value of n becomes large. For large values of n, the $n!$ will become too large and may result in an overflow error.

However, this problem can be rectified we express the above formula recursively as

$$^{n}c_r = {}^{n}c_{r-1}\left[\frac{n-r+1}{r}\right]$$

where $^{n}c_r$ =1 when $r = 1$.

The above formula can also be expressed iteratively as

$$^{n}c_r = n\prod_{i=2}^{r}\left[\frac{n-i+1}{i}\right]$$

These formulations will compute $^{n}c_r$ without causing an overflow error provided the final answer is not too large.

2.8 MEASURES OF ACCURACY

It must be remembered that the round-off errors are most difficult to estimate and one has to follow some rules to reduce their effect on the final results. However, the truncation errors can be easily estimated and thus can be effectively reduced. In any case, one needs some measures of accuracy of the results. The two commonly used measures are described below.

2.8.1 Absolute Error

Absolute error is the positive difference between the actual value and the approximate value of a variable. If x is the actual value and x_a is the approximate value of a variable, then the absolute error is given by

$$\text{absolute error} = |\ x - x_a\ |$$

2.8.2 Relative Error

Relative error is the ratio of the error to the actual value of a variable. If x is the actual value and x_a is the approximate value of a variable, then the relative error is given by

$$\text{relative error} = \frac{x - x_a}{x}$$

2.9 SOME IMPORTANT RULES TO REMEMBER

Theoretical proof of these rules is beyond the scope of this book. However, these rules will be proved numerically in the examples to follow.

Consider x be any number expressed as

$$0.d_1d_2d_3 \ldots \times 10^n$$

where d_1, d_2, d_3, . . . are decimal digits.

Rule 1: *Absolute error because of truncation*

> If x_a is the approximate value of x after truncation to k digits, then
> $$| x - x_a | < 10^{n-k}$$

Rule 2: *Relative error because of truncation*

> If x_a is the approximate value of x after truncation to k digits, then
> $$\left| \frac{x - x_a}{x} \right| < 10^{-k+1}$$

Rule 3: *Absolute error because of rounding-off*

> If x_a is the approximate value of x after rounding-off to k digits, then
> $$| x - x_a | < 0.5 \times 10^{n-k}$$

Rule 4: *Relative error because of rounding-off*

> If x_a is the approximate value of x after rounding-off to k digits, then
> $$\left| \frac{x - x_a}{x} \right| < 0.5 \times 10^{-k+1}$$

This type of analysis is useful if we want to keep track of the errors that occur when the numbers in a calculation are truncated or rounded-off to a certain number of decimal digits.

Example 2.18: The solution of a problem is given as 3.436. It is known that the absolute error in the solution is less than 0.01. Find the interval within which the exact value must lie.

Solution: Given $x_a = 3.436$. If x is the actual value of the solution, then

$$| x - x_a | < 0.01 \quad \Rightarrow \quad x - 3.436 \; | < 0.01$$

From the definition of the absolute value

$$-0.01 < x - 3.436 < 0.01$$

Therefore

$$3.436 - 0.01 < x < 3.436 + 0.01$$

$$\Rightarrow \qquad 3.436 < x < 3.446$$

Hence, the exact value of the solution must lie the interval (3.426, 3.446).

Example 2.19: Let $x = 0.00458529$. Find the absolute error if x is truncated to three decimal digits.

Solution: Given $x = 0.00458529$

$= 0.458529 \times 10^{-2}$ [in the normalized floating-point form]

$x_a = 0.458 \times 10^{-2}$ [after truncating to three decimal places]

$\text{error} = x - x_a$

$= 0.458529 \times 10^{-2} - 0.458 \times 10^{-2}$

$= 0.000529 \times 10^{-2}$

Hence

$$| x - x_a | = 0.000529 \times 10^{-2} = 0.529 \times 10^{-5}$$

which is less than 10^{-2-3}, and hence proves rule 1.

Example 2.20: Let $x = 0.005998$. Find the relative error if x is truncated to three decimal digits.

Solution: Given $x = 0.005998$

$= 0.5998 \times 10^{-2}$ [in the normalized floating-point form]

$x_a = 0.599 \times 10^{-2}$ [after truncating to three decimal places]

$$\text{Relative error} = \left|\frac{x - x_a}{x}\right|$$

$$= \left|\frac{0.5998 \times 10^{-2} - 0.599 \times 10^{-2}}{0.5998 \times 10^{-2}}\right|$$

$$= 0.00133 = 0.133 \times 10^{-2}$$

which is less than 10^{-3+1}, and hence proves rule 2.

Example 2.21: Let x = 0.00458529. Find the absolute error if x is rounded-off to three decimal digits.

Solution: Given $x = 0.00458529$

$= 0.458529 \times 10^{-2}$ [in the normalized floating-point form]

$x_a = 0.459 \times 10^{-2}$ [after rounding-off to three decimal places]

error $= x - x_a$

$= 0.458529 \times 10^{-2} - 0.459 \times 10^{-2}$

$= -0.000471 \times 10^{-2}$

Hence

$$| x - x_a | = 0.000471 \times 10^{-2} = 0.471 \times 10^{-5}$$

which is less than $0.5 \times 10^{-2-3}$, and hence proves rule 3.

Example 2.22: Let x = 0.005998. Find the relative error if x is rounded-off to three decimal digits.

Solution: Given $x = 0.005998$

$= 0.5998 \times 10^{-2}$ [in the normalized floating-point form]

$x_a = 0.600 \times 10^{-2}$ [after rounding-off to three decimal places]

$$\text{Relative error} = \left|\frac{x - x_a}{x}\right| = \left|\frac{0.5998 \times 10^{-2} - 0.600 \times 10^{-2}}{0.5998 \times 10^{-2}}\right|$$

$$= 0.000333 = 0.333 \times 10^{-3}$$

which is less than $0.5 \times 10^{-3+1}$, and hence proves rule 4.

Example 2.23: Given the solution of a problem as $x_a = 35.25$ with relative error in the solution at most 2%. Find, to four decimal digits, the range of values within which the exact value of the solution must lie.

Solution: Given maximum relative error = 0.02

Therefore $-0.02 < \frac{x - x_a}{x} < 0.02$

If $\frac{x - x_a}{x} < 0.02$

then $x - x_a < 0.02x$ [since $x > 0$]

$\Rightarrow x(1 - 0.02) < x_a$

$$\Rightarrow x < \frac{x_a}{1 - 0.02} = \frac{35.25}{0.98} = 35.9693877551$$

However, if $-0.02 < \frac{x - x_a}{x}$ then $x - x_a > -0.02x$ [since $x > 0$]

$\Rightarrow x(1 + 0.02) > x_a$

$$\Rightarrow x > \frac{x_a}{1 + 0.02} = \frac{35.25}{1.02} = 34.3588235294$$

Hence $34.5588 < x < 35.9694$

2.10 RELATION BETWEEN RELATIVE ERROR AND SIGNIFICANT DIGITS

The number of significant digits required to express it, generally, determines the accuracy of a result. And the true index of accuracy of a computation is the relative error. We will see that what amount of relative error in the result is allowed to guarantee certain number of significant digits in the result. As an example, consider that $x = 0.1000$ is good to four significant digits.

Then $(0.1000 - 0.0005) \leq x \leq (0.1000 + 0.0005)$

$$0.09995 \leq x \leq 0.10005$$

And the relative error $= \frac{0.0001}{0.1000} = 0.001$

For the number 0.9999, which is also good to four significant digits, the relative error is

$$\frac{0.0001}{0.9999} = 0.000100010001 > 0.0001$$

Therefore, the relative error for a four digit normalized floating-point number lies between 0.0001 and 0.001. Thus, in order to guarantee four significant digit accuracy, we must ensure that the relative error is ≤ 0.0001 (i.e. 10^{-4}).

In general, if we demand a solution accurate to n significant digits, we must ensure that relative error is less than 10^{-n}.

In iterative methods of solving problems using normalized floating-point arithmetic, the iterative process stops when the required number of significant digits in the answer is assured.

2.11 ERROR PROPAGATION

Numerical solutions involve a series of computational steps. Therefore, it is necessary to understand the way the error propagates with progressive computations. Two demonstrate the error propagation in progressive computations; consider that x_a and y_a are approximate values of variable x and y respectively. Further, consider that

$$x - x_a = e_{ax} \quad \text{and} \quad y - y_a = e_{ay}$$

where e_{ax} denotes the error in taking x_a as the approximation to x and e_{ay} denotes the error in taking y_a as the approximation to y, and

$$\frac{x - x_a}{x} = e_{rx} \quad \text{and} \quad \frac{y - y_a}{y} = e_{ry}$$

where e_{rx} denotes the relative error in taking x_a as the approximation to x and e_{ay} denotes the relative error in taking y_a as the approximation to y.

2.11.1 Error Propagation in Addition Operation

$$x + y = (x_a + e_{ax}) + (y + e_{ay}) = x_a + y_a + e_{ax} + e_{ay}$$

Thus error in taking $x_a + y_a$ as approximation to $x + y$ is

$$(x + y) - (x_a + y_a) = e_{ax} + e_{ay}$$

The absolute error is

$$|(x + y) - (x_a + y_a)| = |e_{ax} + e_{ay}|$$

$$\leq |e_{ax}| + |e_{ay}| \qquad \text{[By the triangle inequality]}$$

This gives the important rule:

> The absolute error in using $x_a + y_a$ as approximation to $x + y$ is less than or equal to the sum of the absolute errors incurred by using x_a as an approximation to x, and y_a as an approximation to y.

2.11.2 Error Propagation in Subtraction Operation

$$x - y = (x_a + e_{ax}) - (y + e_{ay})$$
$$= x_a - y_a + e_{ax} - e_{ay}$$

Thus error in taking $x_a - y_a$ as approximation to $x - y$ is

$$(x - y) - (x_a - y_a) = e_{ax} - e_{ay}$$

The absolute error is

$$| (x - y) - (x_a - y_a) | = | e_{ax} - e_{ay} |$$
$$\leq | e_{ax} | + | e_{ay} | \qquad \text{[By the triangle inequality]}$$

This gives the important rule:

> The absolute error in using $x_a - y_a$ as approximation to $x - y$ is less than or equal to the sum of the absolute errors incurred by using x_a as an approximation to x, and y_a as an approximation to y.

While considering multiplication and division operations, it is convenient to work with the relative error.

$$\frac{x - x_a}{x} = e_{rx}$$

$$\Rightarrow \quad x - x_a = xe_{rx}$$

$$\Rightarrow \quad x_a = x - xe_{rx} = x(1 - e_{rx})$$

Since the relative error can be negative, therefore we can write

$$x_a = x(1+e_{rx})$$

Similarly

$$y_a = y(1+e_{ry})$$

With these notations, we will now see the error propagation in multiplication and division operations.

2.11.3 Error Propagation in Multiplication Operation

$$x_a \times y_a = [x(1+e_{rx})][\, y(1+e_{ry})]$$
$$= xy + (1+e_{rx})(1+e_{ry})$$
$$= xy + (1+ e_{rx} + e_{ry} + e_{rx}e_{ry})$$

If e_{rx} and e_{ry} are reasonably small then the term $e_{rx}e_{ry}$ can be ignored. Thus, we get

$$x_a \times y_a = xy + (1+ e_{rx} + e_{ry})$$

Hence, using approximation x_a and y_a to x and y respectively will cause xy to take the value $xy + (1+ e_r)$, where

$$e_r = e_{rx} + e_{ry}$$

The relative error e_r satisfies

$$|\, e_r \,| = |\, e_{rx} + e_{ry} \,| \;\leq\; |\, e_{rx} \,| + |\, e_{ry} \,|$$

This gives the important rule:

> The absolute error in using x_a y_a as approximation to xy is less than or equal to the sum of the magnitude of relative errors incurred by using x_a as an approximation to x, and y_a as an approximation to y.

2.11.4 Error Propagation in Division Operation

$$\frac{x_a}{y_a} = \frac{x(1+e_{rx})}{y(1+e_{ry})} = \frac{x}{y}(1+e_{rx})(1+e_{ry})^{-1}$$

$$= \frac{x}{y}(1+e_{rx})(1-e_{ry}+e_{ry}^2-...) = \frac{x}{y}(1+e_{rx}-e_{ry}-e_{rx}e_{ry}+e_{ry}^2-...)$$

If e_{rx} and e_{ry} are reasonably small then the term $e_{rx}e_{ry}$, e_{ry}^2, etc., can be ignored.

Thus, we get

$$\frac{x_a}{y_a} = \frac{x}{y}(1+e_{rx}-e_{ry})$$

Hence, using approximation x_a and y_a to x and y respectively will cause $\frac{x}{y}$ to take the value $\frac{x}{y}(1+e_r)$, where $e_r = e_{rx} - e_{ry}$.

Since e_{rx} and e_{ry} may be of opposite signs, the relative error e_r satisfies

$$| e_r | = | e_{rx} - e_{ry} | \leq | e_{rx} | + | e_{ry} |$$

This gives the important rule:

> The absolute error in using $\frac{x_a}{y_a}$ as approximation to $\frac{x}{y}$ is less than or equal to the sum of the magnitude of relative errors incurred by using x_a as an approximation to x, and y_a as an approximation to y.

EXERCISE . . .

1. What do you mean by *approximate numbers* and *significant digits*? Find the number of significant digits in each of the numbers given below:
   (*a*) 0.000234 (*b*) 1250000
   (*c*) 0.0002345×10^3 (*d*) 125.0000
   (*e*) 12500.00
2. Convert the following decimal numbers into binary numbers:
   (*a*) 125 (*b*) 250.125
   (*c*) 1000.25 (*d*) 106.50
   (*e*) 1111 (*f*) 111.0
3. Convert the following binary numbers into decimal numbers:
   (*a*) 110110 (*b*) 101010.01
   (*c*) 111111 (*d*) 10000.110
   (*e*) 111001.001 (*f*) 100001.010
4. How are numbers stored in computer? Give the range of integer and real values that can be stored in 32 bits.
5. Represent the following decimal integers in 1-byte format:
   (*a*) 125 (*b*) –75
   (*c*) 200 (*d*) –10
6. Represent the following decimal integers in 2-byte format:
   (*a*) 1001 (*b*) –5
   (*c*) –500 (*d*) 1250
7. What is the integer value being represented by the following 16-bit binary numbers?
   (*a*) 0000001010100011 (*b*) 1111111111010111
   (*c*) 0111111111111111 (*d*) 1000000000000000
8. Represent the following real numbers in 32-bit format using the most significant byte for representing the exponent.
   (*a*) 12.25 (*b*) –4.125
   (*c*) 100.125 (*d*) –100.125

9. What are the real values being represented by the following 32-bit binary numbers?
   (*a*) 10000110011001010000000000000000
   (*b*) 10000100100001110000000000000000
   (*c*) 10010000110001100000000000000000
   (*d*) 11000100100000010000000000000000
10. What is a normalized floating-point representation? Express the following numbers in the normalized floating-point representation.
   (*a*) 0.0002578 (*b*) −100.125
   (*c*) 101.005×10^3 (*d*) −1.125
11. Illustrate with suitable examples that with normalized floating-point representation, the following laws are not always valid.
   (*a*) Associative law [$(a + b) + c = a + (b + c)$]
   (*b*) Distributive law [$a(b + c) = ab + ac$]
12. Discuss the various types of errors that occur while performing numerical computations. What measures can be taken to improve the accuracy in the numerical computations?
13. Discuss the pitfalls in computing using normalized floating-point numbers.
14. What do you understand by underflow and overflow conditions? What happens when these conditions occur?
15. Differentiate between absolute error and relative error. Provide one example to illustrate when the absolute error would be a better measure and one example to illustrate when the relative error would be better measure.
16. Enumerate the possible sources of errors.
17. Use truncation to write the following decimal numbers in normalized floating-point from with three-digit mantissa and two-digit exponent.
   (*a*) 0.0045767 (*b*) 12.0234×10^0
   (*c*) 0.18181818×10^4 (*d*) 0.9999999×10^{-2}
18. Use rounding-off to write the following decimal numbers in normalized floating-point from with three-digit mantissa and two-digit exponent.
   (*a*) 0.00458767 (*b*) 12.07834×10^0
   (*c*) 0.18181818×10^4 (*d*) 0.9999999×10^{-2}
19. Given a number 0.05578, compute absolute error and relative error, if the number is
   (*a*) rounded-off to three decimal digits
   (*b*) truncated to three decimal digits
20. The value of x is given approximately as 16.58. The relative error in this approximation is less than 0.001. Find the interval within which the exact value of x must lie.
21. The solution of a problem is given as x = 3.345, and it is known that the relative error is not more than 1%. Find, to four decimal digits, the range of values within which the exact value of the solution must lie.

❑ ❑ ❑

3 Solution of Non-linear Equations

Learning Outcomes

After reading this chapter, students will have the ability to

- explain the types of non-linear equations
- differentiate between direct methods and iterative methods
- explain how to choose initial approximation for an iterative method
- explain the various termination criteria for iterative methods
- explain the concept of synthetic division
- explain the working of various iterative methods
- use iterative methods to find numerical solution manually
- write programs for various iterative methods
- find the rate of convergence for various iterative methods

3.1 INTRODUCTION

The determination of solution of non-linear equations is of interest not only to mathematicians but also to scientists and engineers. Though most of the non-linear equations can be solved algebraically, *i.e.*, analytically, using the techniques of algebra. There are many non-linear equations that cannot be solved algebraically.

As an example, consider the following equation

$$2^x - x - 3 = 0$$

which seems very simple but cannot be solved algebraically.

The solution obtained by algebraic manipulations is known as *algebraic solutions* or *analytical solutions*. There are many instances when the algebraic solution of the non-linear equation does exist but is extremely complicated and impractical for day-to-day purposes. In this chapter, we will consider some methods of obtaining solutions of non-linear equations. These solutions will be numerical not algebraic, and are called *numerical solutions*.

Before describing these methods, let us first review the various types of equations and other issues involved in finding their solutions.

In general, an equation in x is written as

$$f(x) = 0 \qquad \ldots (3.1)$$

The root(s) of this equation, also known as *zero(s) of equation*, is/are those values of x for which the equation is satisfied.

3.2 TYPES OF NON-LINEAR EQUATIONS

The non-linear equations fall in following categories:

- Polynomial
- Transcendental

3.2.1 Polynomial Equations

The polynomials are frequently occurring functions in science and engineering. A polynomial has the general form

$$a_n x^n + a_{n-1} x^{n-1} + a_{n-2} x^{n-2} + \ldots + a_2 x^2 + a_1 x + a_0 = 0 \qquad \text{where } a_n \neq 0 \qquad \ldots (3.2)$$

It is nth degree polynomial in x and has n roots. These roots may be

- Real and different
- Real and repeated
- Complex

The complex roots always appear in pairs and are of the form $a + ib$ and $a - ib$, where $i = \sqrt{-1}$, is an imaginary number, and a and b are real numbers and represent real and imaginary parts of the roots.

3.2.2 Transcendental Equations

A non-polynomial equation is called transcendental equations. Some examples of transcendental equations are

$$xe^x - x\sin x = 0 \qquad e^x\cos x - 3x = 0$$

$$e^{3x} - \frac{1}{2}x = 0 \qquad 2^x - x - 3 = 0$$

A transcendental equation may have finite/infinite number of real roots or may not have any real root at all.

3.3 METHODS OF FINDING SOLUTIONS OF NON-LINEAR EQUATIONS

In general, there are two kinds of methods to obtain solutions of non-linear equations. These are

- Direct methods
- Iterative methods

3.3.1 Direct Methods

Direct methods give the roots of non-linear equations in a finite number of steps. In addition, these methods are capable of giving all the roots at the same time.

For example, the roots of the quadratic equation

$$ax^2 + bx + c = 0 \qquad \text{where } a \neq 0$$

are given by

$$x_{1,2} = \frac{-b \pm \sqrt{b^2 - 4ac}}{2a}$$

3.3.2 Iterative Methods

Iterative methods, also known as *trial and error* methods, are based on the idea of successive approximations. They start with one or more initial approximations to the root and obtain a sequence of approximations by repeating a fixed sequence of steps till the solution with reasonable accuracy is obtained. Iterative methods, generally, give one root at a time.

Iterative methods are very cumbersome and time-consuming for solving non-linear equations manually. However, they are best suited for use on computers, due to following reasons:

1. Iterative methods can be concisely expressed as computational algorithms.
2. It is possible to formulate algorithms that can handle class of similar problems. For example, an algorithm can be developed to solve a polynomial equation of degree n.
3. Round-off errors are negligible in iterative methods as compared to direct methods.

3.4 CHOOSING INITIAL APPROXIMATION

The best way to choose an initial approximation to the root of equation

$$f(x) = 0$$

is either to plot the function $f(x)$ or to tabulate it. The roots of the equation are the points where the curve representing the function $f(x)$ intersects the x-axis. Therefore, any point in the interval where the function changes its sign can be taken as the initial approximation.

However, if the equation

$$f(x) = 0 \qquad \text{can be written as} \qquad f_1(x) = f_2(x)$$

then the points of intersection of the graphs of equations

$$y = f_1(x) \quad \text{and} \quad y = f_2(x)$$

give the root of the equation $f(x) = 0$.

Therefore, any value in the neighbourhood of this intersection point can be taken as the initial approximation.

Algorithm 3.1: To tabulate an arbitrary function

Let $f(x)$ is the given function, x_{min} and x_{max} are the end points of the interval over which the function $f(x)$ is to be tabulated, and dx is the increment for x.

```
Begin
   read: xmin, xmax, dx
   set x = xmin
   while ( x ≤ xmax ) do
      set y = f(x)
      write: x, y
      set x = x + dx
   endwhile
End.
```

For illustration, consider the equation

$$2^x - x - 3 = 0$$

which has two distinct roots. We are interested to find the intervals that contain the root. In order to do that, the values of

$$f(x) = 2^x - x - 3$$

are tabulated for various values of x. These values are listed in the Table 3.1.

Table 3.1: Tabulated data for function $f(x) = 2^x - x - 3$

x	–4	–3	–2	–1	0	1	2	3	4
$f(x)$	1.0625	0.125	–0.75	–1.5	–2	–2	–1	2	9

The root lies between those points where there is a sign change in $f(x)$. Hence, the root lies in the interval (–3, –2) and (2, 3).

We can state this result more concisely and in a form more suited for use on computers as

> If $f(x)$ is a continuous function and $f(a)\times f(b) < 0$, then the equation $f(x) = 0$ has a root (say μ) lying in the interval (a, b) or in the interval (b, a) if $b < a$.

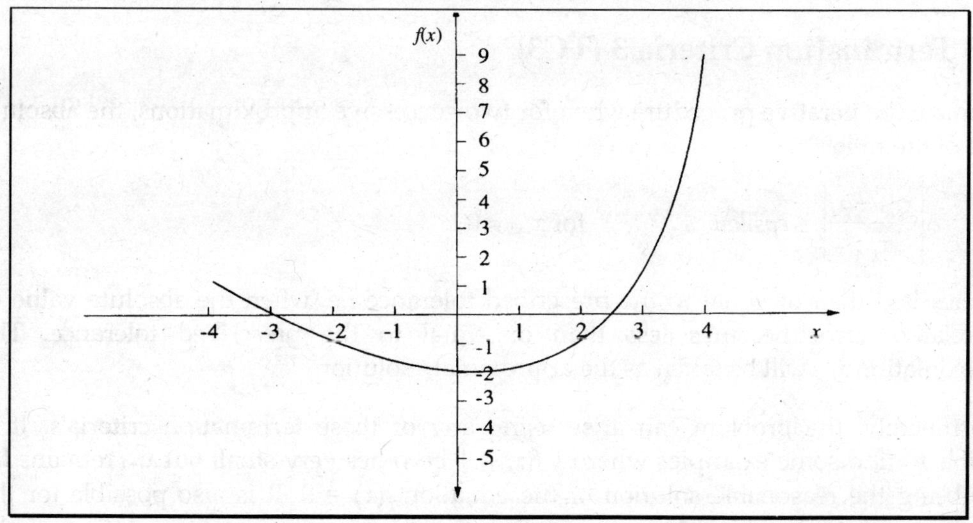

Figure 3.1: Plot of function $f(x) = 2^x - x - 3$

3.5 WHEN TO TERMINATE AN ITERATIVE PROCEDURE?

A very important question that must be addressed as regards to iterative methods is — *When to terminate the iterative procedure*? In general, we can say that iterative procedure is continued till the required degree of accuracy in the solution is achieved. *But how this degree of accuracy can be measured?* In order to obtain more practical answer, assume that the prescribed tolerance in the root is *epsilon*.

3.5.1 Termination Criteria 1 (TC1)

Suppose that starting with x_i as the current approximation, x_{i+1} is the next approximation, the iterative procedure would terminate when the inequality

$$| f(x_{i+1}) | \leq epsilon$$

is satisfied. The approximation x_{i+1} will be taken as the approximate solution.

3.5.2 Termination Criteria 2 (TC2)

Terminate the iterative procedure when two successive approximations differ by an amount less than or equal to the prescribe tolerance. If x_i and x_{i+1} are two successive approximation, the iterative procedure would terminate when the inequality

$$| x_{i+1} - x_i | \leq epsilon$$

is satisfied i.e. the absolute error is less than or equal to the prescribed tolerance. The approximation x_{i+1} will be taken as the approximate solution.

3.5.3 Termination Criteria 3 (TC3)

Terminate the iterative procedure when for two successive approximations, the absolute value of the ratio

$$\left| \frac{x_{i+1} - x_i}{x_{i+1}} \right| \leq epsilon \qquad \text{for } x_{i+1} \neq 0$$

becomes less than or equal to the prescribed tolerance i.e. when the absolute value of the relative error becomes less than or equal to the prescribed tolerance. The approximation x_{i+1} will be taken as the approximate solution.

Unfortunately, the problem can arise using any of these termination criteria's. It is possible to find some examples where $| f(x_{i+1}) |$ becomes very small but x_{i+1} remains far from being the reasonable solution of the equation $f(x) = 0$. It is also possible for the term $| x_{i+1} - x_i |$ to become very small without giving reasonable solution of the equation. In general, it is good to use *TC3*, and it has been proved to be the most reliable of the three termination criteria.

With this background, we will now discuss some popular iterative methods to find the solution of non-linear equations.

3.6 BISECTION METHOD REVISITED

Bisection method, also known as Bolzano method, is one of the simplest iterative methods.

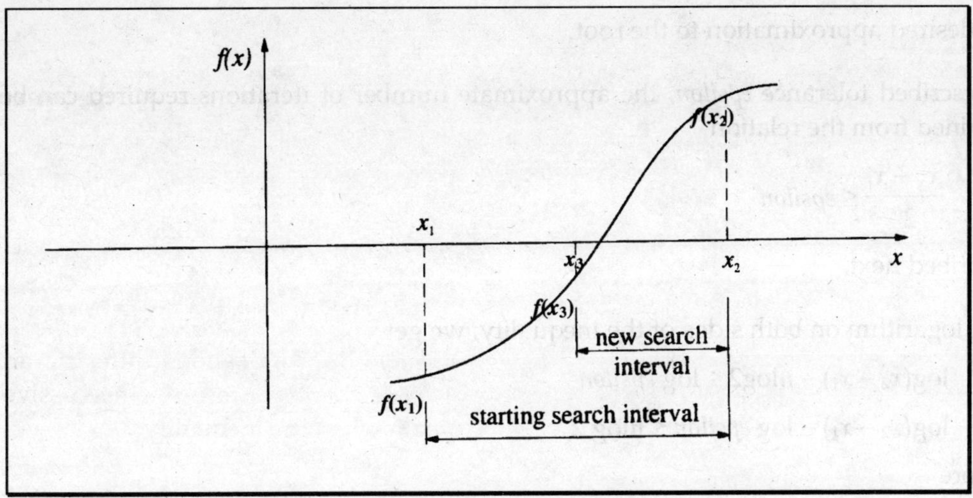

Figure 3.2: Root approximation by Bisection Method

To start with, two initial approximations, say x_1 and x_2 such that $f(x_1) \times f(x_2) < 0$ which ensures that root lies between x_1 and x_2, are taken. The next x-value, say x_3, as the mid point of the interval $[x_1, x_2]$ is computed.

There are three possibilities that can arise:

i) If $f(x_3) = 0$, then we have a root at x_3.

ii) If $f(x_1)$ and $f(x_3)$ are of opposite sign, then the root lies in the interval (x_1, x_3). Thus x_2 is replaced by x_3, and the new interval, which is half of the current interval, is again bisected.

iii) If $f(x_1)$ and $f(x_3)$ are of same sign, then the root lies in the interval (x_3, x_2). Thus x_1 is replaced by x_3, and the new interval is again bisected.

Therefore by repeating this interval bisection procedure, we keep enclosing the root in a new search interval, which is halved in each iteration. This iterative cycle is terminated when the search interval becomes smaller than the prescribed tolerance.

Hence, if *epsilon* is the prescribed tolerance in the required root, then the iterative cycle terminates when the absolute error becomes less than or equal to *epsilon* i.e.

$$|x_1 - x_2| \le epsilon$$

After repeating the bisection process *n* times, the search interval is reduced to length $\frac{x_2 - x_1}{2^n}$ that contains the required root. We take the mid point of the last search interval as the desired approximation to the root.

For prescribed tolerance *epsilon*, the approximate number of iterations required can be determined from the relation

$$\frac{x_2 - x_1}{2^n} \le epsilon$$

as described next.

Taking logarithm on both sides of the inequality, we get

$$\log(x_2 - x_1) - n\log 2 \le \log epsilon$$

$$\log(x_2 - x_1) - \log epsilon \le n\log 2$$

Therefore

$$n \ge \frac{\log(x_2 - x_1) - \log epsilon}{\log 2} \qquad \ldots (3.3)$$

As an example assume that the root lies in the interval (1, 2), then the minimum number of iterations required to find the root within the prescribed tolerance *epsilon* are listed in Table 3.2.

Table 3.2: Number of Iteration Required

epsilon	10^{-2}	10^{-3}	10^{-4}	10^{-5}	10^{-6}
n	7	10	14	17	20

This table shows that the Bisection method requires a large number of iterations to achieve a reasonable degree of accuracy for the desired root.

However, it requires only one function evaluation for each iteration. If the evaluation of function *f*(*x*) is rapid, then the use of Bisection method is strongly recommended.

Example 3.1: Given that one root of the non-linear equation

$$x^3 - 4x - 9 = 0$$

lies between 2.625 and 2.75. Find the root correct to four significant digits.

Solution: Since we want the solution correct to four significant digits, the iterative process will be terminated as soon as the successive iterations produces no change at first four significant positions or the function vanishes at new approximation.

Iteration 1: Starting with $x_1 = 2.625$ and $x_2 = 2.75$

$$f(x_1) = f(2.625) = 2.625^3 - 4 \times 2.625 - 9 = -1.4121$$

$$f(x_2) = f(2.75) = 2.75^3 - 4 \times 2.75 - 9 = 0.7969$$

$$x_3 = (2.625+2.75)/2 = 2.6875$$

$$f(x_3) = f(2.6875) = 2.6875^3 - 4 \times 2.6875 - 9 = -0.3391$$

Since $f(x_1) \times f(x_3) = -1.4121 \times (-0.3391) > 0$

Therefore replace x_1 by x_3. The new search interval becomes (2.6875, 2.75).

Iteration 2: Now we take $x_1 = 2.6875$ and $x_2 = 2.75$

$$f(x_1) = f(2.6875) = 2.6875^3 - 4 \times 2.6875 - 9 = -0.3391$$

$$f(x_2) = f(2.75) = 2.75^3 - 4 \times 2.75 - 9 = 0.7969$$

$$x_3 = (2.6875+2.75)/2 = 2.7186$$

$$f(x_3) = f(2.7186) = 2.7186^3 - 4 \times 2.7186 - 9 = 0.2182$$

Since $f(x_1) \times f(x_3) = -0.3391 \times 0.2181 < 0$

Therefore replace x_2 by x_3. The new search interval becomes (2.6875, 2.7186).

Iteration 3: Now we take $x_1 = 2.6875$ and $x_2 = 2.7186$

$$f(x_1) = f(2.6875) = 2.6875^3 - 4 \times 2.6875 - 9 = -0.3391$$

$$f(x_2) = f(2.7186) = 2.7186^3 - 4 \times 2.7186 - 9 = 0.2182$$

$$x_3 = (2.6875+2.7186)/2 = 2.7031$$

$$f(x_3) = f(2.7031) = 2.7031^3 - 4 \times 2.7031 - 9 = -0.0615$$

Since $f(x_1) \times f(x_3) = -0.3391 \times (-0.06215) > 0$

Therefore replace x_1 by x_3. The new search interval becomes (2.7031, 2.7186).

Iteration 4: Now we take $x_1 = 2.7031$ and $x_2 = 2.7186$

$$f(x_1) = f(2.7031) = 2.7031^3 - 4 \times 2.7031 - 9 = -0.0615$$
$$f(x_2) = f(2.7186) = 2.7186^3 - 4 \times 2.7186 - 9 = 0.2182$$
$$x_3 = (2.7031+2.7186)/2 = 2.7109$$
$$f(x_3) = f(2.7109) = 2.7109^3 - 4 \times 2.7109 - 9 = 0.0787$$

Since $f(x_1) \times f(x_3) = -0.06215 \times 0.0787 < 0$

Therefore replace x_2 by x_3. The new search interval becomes (2.7031, 2.7109).

Iteration 5: Now we take $x_1 = 2.7031$ and $x_2 = 2.7109$

$$f(x_1) = f(2.7031) = 2.7031^3 - 4 \times 2.7031 - 9 = -0.0615$$
$$f(x_2) = f(2.7109) = 2.7109^3 - 4 \times 2.7109 - 9 = 0.0787$$
$$x_3 = (2.7031+2.7109)/2 = 2.707$$
$$f(x_3) = f(2.707) = 2.707^3 - 4 \times 2.707 - 9 = 0.0085$$

Since $f(x_1) \times f(x_3) = -0.06215 \times 0.0085 < 0$

Therefore replace x_2 by x_3. The new search interval becomes (2.7031, 2.707).

Iteration 6: Now we take $x_1 = 2.7031$ and $x_2 = 2.707$

$$f(x_1) = f(2.7031) = 2.7031^3 - 4 \times 2.7031 - 9 = -0.0615$$
$$f(x_2) = f(2.707) = 2.707^3 - 4 \times 2.707 - 9 = 0.0085$$
$$x_3 = (2.7031+2.707)/2 = 2.7051$$
$$f(x_3) = f(2.7051) = 2.7051^3 - 4 \times 2.7051 - 9 = -0.0257$$

Since $f(x_1) \times f(x_3) = -0.06215 \times (-0.0257) > 0$

Therefore replace x_1 by x_3. The new search interval becomes (2.7051, 2.707).

Iteration 7: Now we take $x_1 = 2.7051$ and $x_2 = 2.707$

$$f(x_1) = f(2.7051) = 2.7051^3 - 4 \times 2.7051 - 9 = -0.0257$$
$$f(x_2) = f(2.707) = 2.707^3 - 4 \times 2.707 - 9 = 0.0085$$
$$x_3 = (2.7051+2.707)/2 = 2.7061$$

$f(x_3) = f(2.7061) = 2.7061^3 - 4 \times 2.7061 - 9 = -0.0077$

Since $f(x_1) \times f(x_3) = -0.0257 \times (-0.0077) > 0$

Therefore replace x_1 by x_3. The new search interval becomes (2.7061, 2.707).

Iteration 8: Now we take $x_1 = 2.7061$ and $x_2 = 2.707$

$f(x_1) = f(2.7061) = 2.7061^3 - 4 \times 2.7061 - 9 = -0.0077$

$f(x_2) = f(2.707) = 2.707^3 - 4 \times 2.707 - 9 = 0.0085$

$x_3 = (2.7061+2.707)/2 = 2.7066$

$f(x_3) = f(2.7066) = 2.7066^3 - 4 \times 2.7066 - 9 = 0.0013$

Since $f(x_1) \times f(x_3) = -0.0077 \times 0.0013 < 0$

Therefore replace x_2 by x_3. The new search interval becomes (2.7061, 2.7066).

Iteration 9: Now we take $x_1 = 2.7061$ and $x_2 = 2.7066$

$f(x_1) = f(2.7061) = 2.7061^3 - 4 \times 2.7061 - 9 = -0.0077$

$f(x_2) = f(2.7066) = 2.7066^3 - 4 \times 2.7066 - 9 = 0.0013$

$x_3 = (2.7061+2.7066)/2 = 2.7064$

$f(x_3) = f(2.7064) = 2.7064^3 - 4 \times 2.7064 - 9 = -0.0023$

Since $f(x_1) \times f(x_3) = -0.0077 \times (-0.0023) > 0$

Therefore replace x_1 by x_3. The new search interval becomes (2.7064, 2.7066).

Observe that iterations 7, 8 and 9 produce no change at the first four significant positions. Therefore, we take $x = 2.706$ as the desired solution which is correct to four significant digits.

Example 3.2: Find the root of equation

$$x^3 - x - 4 = 0$$

correct to four significant digits using Bisection method.

Solution: Since we want the solution correct to four significant digits, the iterative process will be terminated as soon as the successive iterations produces no change at first four significant positions or the function vanishes at new approximation.

Now $f(1.75) = 1.75^3 - 1.75 - 4 = -0.3906$

$f(1.8) = 1.8^3 - 1.8 - 4 = 0.0320$

Since $f(1.75)$ is –ve and $f(1.8)$ is +ve, therefore one root lies between 1.75 & 1.8.

Iteration 1: Starting with $x_1 = 1.75$ and $x_2 = 1.8$

$x_3 = (1.75+1.8)/2 = 1.775$

$f(x_3) = f(1.775) = 1.775^3 - 1.775 - 4 = -0.1826$

Thus, the first approximation to root is 1.775.

Since $f(x_1) \times f(x_3) = -0.3906 \times (-0.1826) > 0$

Therefore, replace x_1 by x_3. The new search interval becomes (1.775, 1.8).

Iteration 2: Now we take $x_1 = 1.775$ and $x_2 = 1.8$

$f(x_1) = f(1.775) = 1.775^3 - 1.775 - 4 = -0.1826$

$f(x_2) = f(1.8) = 1.8^3 - 1.8 - 4 = 0.0320$

$x_3 = (1.775+1.8)/2 = 1.7875$

$f(x_3) = f(1.7875) = 1.7875^3 - 1.7875 - 4 = -0.0762$

Thus, the second approximation to root is 1.7875.

Since $f(x_1) \times f(x_3) = -0.1826 \times -(0.0762) > 0$

Therefore, replace x_1 by x_3. The new search interval becomes (1.7875, 1.8).

Iteration 3: Now we take $x_1 = 1.7875$ and $x_2 = 1.8$

$f(x_1) = f(1.7875) = 1.7875^3 - 1.7875 - 4 = -0.0762$

$f(x_2) = f(1.8) = 1.8^3 - 1.8 - 4 = 0.0320$

$x_3 = (1.7875+1.8)/2 = 1.7937$

$f(x_3) = f(1.7937) = 1.7937^3 - 1.7937 - 4 = -0.0223$

Thus, the third approximation to root is 1.7937.

Since $f(x_1) \times f(x_3) = -0.0762 \times -(0.0223) > 0$

Therefore, replace x_1 by x_3. The new search interval becomes (1.7937, 1.8).

Iteration 4: Now we take $x_1 = 1.7937$ and $x_2 = 1.8$

$f(x_1) = f(1.7937) = 1.7937^3 - 1.7937 - 4 = -0.0223$

$f(x_2) = f(1.8) = 1.8^3 - 1.8 - 4 = 0.0320$

$x_3 = (1.7937+1.8)/2 = 1.7969$

$f(x_3) = f(1.7969) = 1.7969^3 - 1.7969 - 4 = 0.0048$

Thus, the fourth approximation to root is 1.7969.

Since $f(x_1) \times f(x_3) = -0.0223 \times 0.0048 < 0$

Therefore, replace x_2 by x_3. The new search interval becomes (1.7937, 1.7969).

Subsequent iterations will produce the following approximations to the root

1.7953, 1.7961, 1.7965, 1.7963, 1.7964

Observe the last four iterations, they produce no change at the first four significant positions. Therefore, we take $x = 1.796$ as the desired solution which is correct to four significant digits.

Example 3.3: Find the root of equation

$$x^3 - 5x + 3 = 0$$

correct to three decimal positions using Bisection method.

Solution: Since we want the solution correct to three decimal positions, the iterative process will be terminated as soon as the successive iterations produces no change at first three decimal positions or the function vanishes at new approximation.

Now $f(0.5) = 0.5^3 - 5\times0.5 + 3 = 0.625$

$f(1.0) = 1.0^3 - 5\times1.0 + 3 = -1.0$

Since $f(0.5)$ is +ve and $f(1.0)$ is –ve, therefore one root lies between 0.5 and 1.0.

Iteration 1: Starting with $x_1 = 0.5$ and $x_2 = 1.0$

$x_3 = (0.5+1.0)/2 = 0.75$

$f(0.75) = 0.75^3 - 5\times0.75 + 3 = -0.3281$

Thus, the first approximation to root is 0.75.

Since $\quad f(x_1) \times f(x_3) = 0.625 \times (-0.3281) < 0$

Therefore, replace x_2 by x_3. The new search interval becomes (0.5, 0.75).

Iteration 2: Now we take $x_1 = 0.5$ and $x_2 = 0.75$

$$f(x_1) = f(0.5) = 0.5^3 - 5\times0.5 + 3 = 0.625$$

$$f(x_2) = f(0.75) = 0.75^3 - 5\times0.75 + 3 = -0.3281$$

$$x_3 = (0.5+0.75)/2 = 0.625$$

$$f(0.625) = 0.625^3 - 5\times0.625 + 3 = 0.1191$$

Thus, the second approximation to root is 0.625.

Since $\quad f(x_1) \times f(x_3) = 0.625 \times 0.1191 > 0$

Therefore, replace x_1 by x_3. The new search interval becomes (0.625, 0.75).

Iteration 3: Now we take $x_1 = 0.625$ and $x_2 = 0.75$

$$f(x_1) = f(0.625) = 0.625^3 - 5\times0.625 + 3 = 0.1191$$

$$f(x_2) = f(0.75) = 0.75^3 - 5\times0.75 + 3 = -0.3281$$

$$x_3 = (0.625+0.75)/2 = 0.6875$$

$$f(x_3) = f(0.6875) = 0.6875^3 - 5\times0.6875 + 3 = -0.1125$$

Thus, the third approximation to root is 0.6875.

Since $\quad f(x_1) \times f(x_3) = 0.1191 \times (-0.1125) < 0$

Therefore, replace x_2 by x_3. The new search interval becomes (0.625, 0.6875).

Iteration 4: Now we take $x_1 = 0.625$ and $x_2 = 0.6875$

$$f(x_1) = f(0.625) = 0.625^3 - 5\times0.625 + 3 = 0.1191$$

$$f(x_2) = f(0.6875) = 0.6875^3 - 5\times0.6875 + 3 = -0.1125$$

$$x_3 = (0.625+0.6875)/2 = 0.6562$$

$$f(x_3) = f(0.6562) = 0.6562^3 - 5\times0.6562 + 3 = 0.0014$$

Thus, the fourth approximation to root is 0.6562.

Since $f(x_1) \times f(x_3) = 0.1191 \times 0.0014 > 0$

Therefore, replace x_1 by x_3. The new search interval becomes (0.6562, 0.6875).

Subsequent iterations will produce the following approximations to the root

0.6719, 0.6641, 0.6602, 0.6682, 0.6572, 0.6567, 0.6565. 0.6566

Observe the last three iterations, they produce no change at the first three decimal positions. Therefore, we take $x = 0.656$ as the desired solution which is correct to three decimal positions.

Algorithm 3.2: Bisection method

To find a root of $f(x) = 0$ within a prescribed tolerance say *epsilon*. Given values x_1 and x_2 such that $f(x_1) \times f(x_2) < 0$. The variable x_3 is used to store mid point of the interval.

```
Begin
  read: x1, x2                      // input value for x1and x2
  read: epsilon                     // input prescribed tolerance
  do
    set x3 = ( x1+x2)/2             // compute mid point
    if (f(x1)×f(x3)) < 0) then      // select appropriate subinterval
       set x2 = x3
    else
       set x1 = x3
    endif
  while ( | x1  - x2 | > epsilon )
  write: x3, "as the approximate root"  // output computed root
End.
```

3.7 FALSE POSITION METHOD

Though the bisection method guarantees that iterative process will converge, but its convergence is slow. The false position method, also known as *regula-falsi* or *method of linear interpolation,* is similar to the bisection method but faster than it. It also starts with two initial approximations to the root, say x_1 and x_2, for which $f(x)$ has opposite signs, and then by linear interpolation the next approximation is determined.

To describe its working assume that x_1 and x_2 are two initial approximations to root for which $f(x)$ has opposite signs. Join the points $(x_1, f(x_1))$ and $(x_2, f(x_2))$ by a straight line. The point where this line intersects the x-axis is the next approximation to the root. Let us suppose that the line intersects the x-axis at x_3. This is illustrated in Figure 3.3.

There are three possibilities:

i) If $f(x_3) = 0$, then we have a root at x_3.

ii) If $f(x_1)$ and $f(x_3)$ are of opposite sign, then the root lies in the interval (x_1, x_3). Thus x_2 is replaced by x_3 and the iterative procedure is repeated.

iii) If $f(x_1)$ and $f(x_3)$) are of same sign, then the root lies in the interval (x_3, x_2). Thus x_1 is replaced by x_3, and the iterative procedure is repeated.

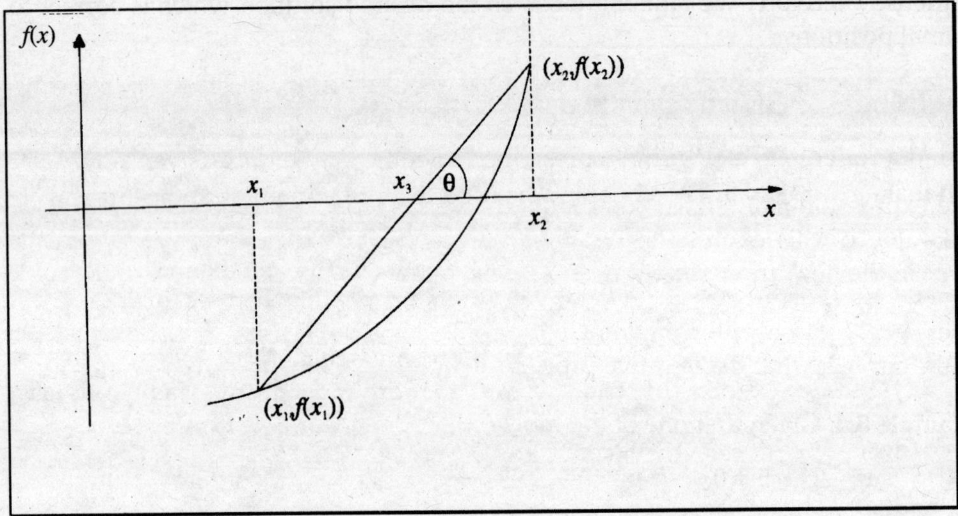

Figure 3.3: Root approximation by false position method

The iterative procedure terminates when the size of the search interval becomes less than the prescribed tolerance.

For derivation of the computational formula to find the intersection point, refer to Figure 3.3.

Slope of the line joining the points $(x_3, 0)$ and $(x_2, f(x_2))$ is given by

$$\tan\theta = \frac{f(x_2)}{x_2 - x_3} \qquad \ldots (3.4)$$

Slope of the line joining the points $(x_1, f(x_1))$ and $(x_2, f(x_2))$ is given by

$$\tan\theta = \frac{f(x_2) - f(x_1)}{x_2 - x_1} \qquad \ldots (3.5)$$

Since the line joining the points $(x_3, 0)$ and $(x_2, f(x_2))$ is a part of the line joining the points $(x_1, f(x_1))$ and $(x_2, f(x_2))$, therefore

$$\frac{f(x_2)}{x_2 - x_3} = \frac{f(x_2) - f(x_1)}{x_2 - x_1}$$

Solving for x_3, we get

$$x_3 = \frac{x_1 f(x_2) - x_2 f(x_1)}{f(x_2) - f(x_1)} \quad \ldots (3.6)$$

which gives the next approximation to the root.

In general, the $(i+1)^{th}$ approximation to the root is given by the formula

$$x_{i+1} = \frac{x_{i-1} f(x_i) - x_i f(x_{i-1})}{f(x_i) - f(x_{i-1})} \quad \ldots (3.7)$$

The expression $f(x_i) - f(x_{i-1})$ represents the slope of the curve in current interval. And if the value of this expression becomes very small, it will introduce a large amount of error in the new approximation.

Therefore, for computer implementation, we must put a limit on the permissible slope of the curve so that the iterative procedure does not fall in an endless loop.

Example 3.4: Given that one of the roots of the non-linear equation

$$x^3 - 2x - 5 = 0$$

lies in the interval (1.75, 2.5). Find the root correct to four significant digits.

Solution: Since we want the solution correct to four significant digits, the iterative process will be terminated as soon as the successive iterations produces no change at first four significant positions or the function vanishes at new approximation.

Iteration 1: Starting with $x_1 = 1.75$ and $x_2 = 2.5$

$$f(x_1) = f(1.75) = 1.75^3 - 2 \times 1.75 - 5 = -3.1406$$

$$f(x_2) = f(2.5) = 2.5^3 - 2 \times 2.5 - 5 = 5.625$$

$$x_3 = \frac{x_1 f(x_2) - x_2 f(x_1)}{f(x_2) - f(x_1)} = \frac{1.75 \times 5.625 - 2.5 \times (-3.1406)}{5.625 - (-3.1406)} = 2.0187$$

$$f(x_3) = f(2.0187) = 2.0187^3 - 2 \times 2.0187 - 5 = -0.8109$$

Since $\quad f(x_1) \times f(x_3) = -3.1406 \times (-0.8109) > 0$

Therefore replace x_1 by x_3. The new search interval becomes (2.0187, 2.5).

Iteration 2: Now we take $x_1 = 2.0187$ and $x_2 = 2.5$

$$f(x_1) = f(2.0187) = 2.0187^3 - 2 \times 2.0187 - 5 = -0.8109$$

$$f(x_2) = f(2.5) = 2.5^3 - 2 \times 2.5 - 5 = 5.625$$

$$x_3 = \frac{x_1 f(x_2) - x_2 f(x_1)}{f(x_2) - f(x_1)} = \frac{2.0187 \times 5.625 - 2.5 \times (-0.8109)}{5.625 - (-0.8109)} = 2.0793$$

$$f(x_3) = f(2.0793) = 2.0793^3 - 2 \times 2.0793 - 5 = -0.1688$$

Since $f(x_1) \times f(x_3) = -0.8109 \times (-0.1688) > 0$

Therefore replace x_1 by x_3. The new search interval becomes (2.0793, 2.5).

Iteration 3: Now we take $x_1 = 2.0793$ and $x_2 = 2.5$

$$f(x_1) = f(2.0793) = 2.0793^3 - 2 \times 2.0793 - 5 = -0.1688$$

$$f(x_2) = f(2.5) = 2.5^3 - 2 \times 2.5 - 5 = 5.625$$

$$x_3 = \frac{x_1 f(x_2) - x_2 f(x_1)}{f(x_2) - f(x_1)} = \frac{2.0793 \times 5.625 - 2.5 \times (-0.1688)}{5.625 - (-0.1688)} = 2.0916$$

$$f(x_3) = f(2.0916) = 2.0916^3 - 2 \times 2.0916 - 5 = -0.0329$$

Since $f(x_1) \times f(x_3) = -0.1688 \times (-0.0329) > 0$

Therefore replace x_1 by x_3. The new search interval becomes (2.0916, 2.5).

Iteration 4: Now we take $x_1 = 2.0916$ and $x_2 = 2.5$

$$f(x_1) = f(2.0916) = 2.0916^3 - 2 \times 2.0916 - 5 = -0.0329$$

$$f(x_2) = f(2.5) = 2.5^3 - 2 \times 2.5 - 5 = 5.625$$

$$x_3 = \frac{x_1 f(x_2) - x_2 f(x_1)}{f(x_2) - f(x_1)} = \frac{2.0916 \times 5.625 - 2.5 \times (-0.0329)}{5.625 - (-0.0329)} = 2.0940$$

$$f(x_3) = f(2.0940) = 2.0940^3 - 2 \times 2.0940 - 5 = -0.0062$$

Since $f(x_1) \times f(x_3) = -0.03299 \times (-0.0062) > 0$

Therefore replace x_1 by x_3. The new search interval becomes (2.0940, 2.5).

Iteration 5: Now we take $x_1 = 2.0940$ and $x_2 = 2.5$

$f(x_1) = f(2.0940) = 2.0940^3 - 2 \times 2.0940 - 5 = -0.0062$

$f(x_2) = f(2.5) = 2.5^3 - 2 \times 2.5 - 5 = 5.625$

$$x_3 = \frac{x_1 f(x_2) - x_2 f(x_1)}{f(x_2) - f(x_1)} = \frac{2.0940 \times 5.625 - 2.5 \times (-0.0062)}{5.625 - (-0.0062)} = 2.0944$$

$f(x_1) = f(2.0944) = 2.0944^3 - 2 \times 2.0944 - 5 = -0.0017$

Since $f(x_1) \times f(x_3) = -0.0062 \times (-0.0017) > 0$

Therefore replace x_1 by x_3. The new search interval becomes (2.0944, 2.5).

Iteration 6: Now we take $x_1 = 2.0944$ and $x_2 = 2.5$

$f(x_2) = f(2.0944) = 2.0944^3 - 2 \times 2.0944 - 5 = -0.0017$

$f(x_2) = f(2.5) = 2.5^3 - 2 \times 2.5 - 5 = 5.625$

$$x_3 = \frac{x_1 f(x_2) - x_2 f(x_1)}{f(x_2) - f(x_1)} = \frac{2.0944 \times 5.625 - 2.5 \times (-0.0017)}{5.625 - (-0.0017)} = 2.0945$$

$f(x_2) = f(2.0945) = 2.0945^3 - 2 \times 2.0945 - 5 = -0.0006$

Since $f(x_1) \times f(x_3) = -0.0017 \times (-0.0006) > 0$

Therefore replace x_1 by x_3. The new search interval becomes (2.0945, 2.5).

Observe that iterations 4, 5 and 6 produce no change at the first four significant positions in the successive approximations to the root. Therefore, we take $x = 2.094$ as the desired solution correct to four significant digits.

Example 3.5: Find the root of equation

$$\cos x - 3x + 1 = 0$$

correct to three decimal positions using False position method.

Solution: Since we want the solution correct to three decimal positions, the iterative process will be terminated as soon as the successive iterations produces no change at first three decimal positions or the function vanishes at new approximation.

Now $f(0) = \cos(0) - 3\times0 + 1 = 2.0$

$f(1) = \cos(1) - 3\times1 + 1 = -1.4597$

Since $f(0)$ is +ve and $f(1)$ is –ve, therefore one root lies between 0 and 1.

Iteration 1: Starting with $x_1 = 0$ and $x_2 = 1$

$$f(0) = \cos(0) - 3\times0 + 1 = 2.0$$

$$f(1) = \cos(1) - 3\times1 + 1 = -1.4597$$

$$x_3 = \frac{x_1 f(x_2) - x_2 f(x_1)}{f(x_2) - f(x_1)} = \frac{0\times(-1.4597) - 1\times2.0}{-1.4597 - 2.0} = 0.5781$$

$$f(x_3) = f(0.5781) = \cos(0.5781) - 3\times0.5781 + 1 = 0.1033$$

Thus, the first approximation to the root is 0.5781.

Since $f(x_1) \times f(x_3) = 2.0 \times 0.1033 > 0$

Therefore, replace x_1 by x_3. The new search interval becomes (0.5781, 1.0).

Iteration 2: Now we take $x_1 = 0.5781$ and $x_2 = 1.0$

$$f(x_1) = f(0.5781) = \cos(0.5781) - 3\times0.5781 + 1 = 0.1033$$

$$f(x_2) = f(1) = \cos(1) - 3\times1 + 1 = -1.4597$$

$$x_3 = \frac{x_1 f(x_2) - x_2 f(x_1)}{f(x_2) - f(x_1)} = \frac{0.5781\times(-1.4597) - 1\times0.1033}{-1.4597 - 0.1033} = 0.6060$$

$$f(x_3) = f(0.6060) = \cos(0.6060) - 3\times0.6060 + 1 = 0.0041$$

Thus, the second approximation to the root is 0.6060.

Since $f(x_1) \times f(x_3) = 0.1033 \times 0.0041 > 0$

Therefore, replace x_1 by x_3. The new search interval becomes (0.6060, 1.0).

Iteration 3: Now we take $x_1 = 0.6060$ and $x_2 = 1.0$

$$f(x_1) = f(0.6060) = \cos(0.6060) - 3\times0.6060 + 1 = 0.0041$$

$$f(x_2) = f(1) = \cos(1) - 3\times1 + 1 = -1.4597$$

$$x_3 = \frac{x_1 f(x_2) - x_2 f(x_1)}{f(x_2) - f(x_1)} = \frac{0.6060\times(-1.4597) - 1\times0.0041}{-1.4597 - 0.0041} = 0.6071$$

$$f(x_3) = f(0.6071) = \cos(0.6071) - 3\times0.6071 + 1 = 0.0002$$

Thus, the third approximation to the root is 0.6071.

Since $f(x_1) \times f(x_3) = 0.0041 \times 0.0002 > 0$

Therefore, replace x_1 by x_3. The new search interval becomes (0.6071, 1.0).

Iteration 4: Now we take $x_1 = 0.6071$ and $x_2 = 1.0$

$f(x_1) = f(0.6071) = \cos(0.6071) - 3{\times}0.6071 + 1 = 0.0002$

$f(x_2) = f(1) = \cos(1) - 3{\times}1 + 1 = -1.4597$

$$x_3 = \frac{x_1 f(x_2) - x_2 f(x_1)}{f(x_2) - f(x_1)} = \frac{0.6071 \times (-1.4597) - 1 \times 0.0002}{-1.4597 - 0.0002} = 0.6072$$

$f(x_3) = f(0.6071) = \cos(0.6071) - 3{\times}0.6071 + 1 = 0.0001$

Thus, the fourth approximation to the root is 0.6072.

Since last two iterations produce no change at the first three decimal positions, therefore, we take $x = 0.607$ as the desired solution correct to three decimal positions.

Example 3.6: Find the root of equation

$$\cos x - xe^x = 0$$

correct to three decimal places using False position method.

Solution: Since we want the solution correct to three decimal places, the iterative process will be terminated as soon as the successive iterations produces no change at three decimal places or the function vanishes at new approximation.

Now $f(0.5) = \cos(0.5) - 0.5 \times e^{0.5} = 0.0532$

$f(0.75) = \cos(0.75) - 0.75 \times e^{0.75} = -0.8561$

Since $f(0.5)$ is +ve and $f(0.75)$ is –ve, therefore one root lies between 0.5 & 0.75.

Iteration 1: Starting with $x_1 = 0.5$ and $x_2 = 0.75$

$f(x_1) = f(0.5) = \cos(0.5) - 0.5 \times e^{0.5} = 0.0532$

$f(x_2) = f(0.75) = \cos(0.75) - 0.75 \times e^{0.75} = -0.8561$

$$x_3 = \frac{x_1 f(x_2) - x_2 f(x_1)}{f(x_2) - f(x_1)} = \frac{0.5 \times (-0.8561) - 0.75 \times 0.0532}{-0.8561 - 0.0532} = 0.5146$$

$f(x_3) = f(0.5146) = \cos(0.5146) - 0.5146 \times e^{0.5146} = 0.0095$

Thus, the first approximation to the root is 0.5146.

Since $f(x_1) \times f(x_3) = 0.0532 \times 0.0095 > 0$

Therefore, replace x_1 by x_3. The new search interval becomes (0.5146, 0.75).

Iteration 2: Now we take $x_1 = 0.5146$ and $x_2 = 0.75$

$f(x_1) = f(0.5146) = \cos(0.5146) - 0.5146 \times e^{0.5146} = 0.0095$

$f(x_2) = f(0.75) = \cos(0.75) - 0.75 \times e^{0.75} = -0.8561$

$$x_3 = \frac{x_1 f(x_2) - x_2 f(x_1)}{f(x_2) - f(x_1)} = \frac{0.5146 \times (-0.8561) - 0.75 \times 0.0095}{-0.8561 - 0.0095} = 0.5172$$

$f(x_3) = f(0.5172) = \cos(0.5172) - 0.5172 \times e^{0.5172} = 0.0017$

Thus, the second approximation to the root is 0.5172.

Since $f(x_1) \times f(x_3) = 0.0532 \times 0.0017 > 0$

Therefore, replace x_1 by x_3. The new search interval becomes (0.5172, 0.75).

Iteration 3: Now we take $x_1 = 0.5172$ and $x_2 = 0.75$

$f(x_1) = f(0.5172) = \cos(0.5172) - 0.5172 \times e^{0.5172} = 0.0017$

$f(x_2) = f(0.75) = \cos(0.75) - 0.75 \times e^{0.75} = -0.8561$

$$x_3 = \frac{x_1 f(x_2) - x_2 f(x_1)}{f(x_2) - f(x_1)} = \frac{0.5172 \times (-0.8561) - 0.75 \times 0.0017}{-0.8561 - 0.0017} = 0.5177$$

$f(x_3) = f(0.5177) = \cos(0.5177) - 0.5177 \times e^{0.5177} = 0.0003$

Thus, the third approximation to the root is 0.5177.

Since $f(x_1) \times f(x_3) = 0.0532 \times 0.0003 > 0$

Therefore, replace x_1 by x_3. The new search interval becomes (0.5177, 0.75).

Iteration 4: Now we take $x_1 = 0.5177$ and $x_2 = 0.75$

$$f(x_1) = f(0.5177) = \cos(0.5177) - 0.5177 \times e^{0.5177} = 0.0003$$

$$f(x_2) = f(0.75) = \cos(0.75) - 0.75 \times e^{0.75} = -0.8561$$

$$x_3 = \frac{x_1 f(x_2) - x_2 f(x_1)}{f(x_2) - f(x_1)} = \frac{0.5177 \times (-0.8561) - 0.75 \times 0.0003}{-0.8561 - 0.0003} = 0.5178$$

$$f(x_3) = f(0.5178) = \cos(0.5178) - 0.5178 \times e^{0.5178} = 0.0000$$

Thus, the fourth approximation to the root is 0.5178.

Since, the last three iterations produce no change in first three decimal positions, and also the function $f(x)$ vanishes after fourth iteration. Therefore, we take $x = 0.517$ as solution correct to three decimal positions.

Algorithm 3.3: False Position Method

Here $f(x)$ is given function, x_1 and x_2 are two initial approximations, *epsilon* is the prescribed tolerance in the root, and *delta* is the prescribed lower bound for the slope of $f(x)$.

```
Begin
    read: x1, x2, epsilon, delta
    set f1 = f(x1)
    set f2 = f(x2)
    if ( f1 × f2 > 0 ) then
        write: "Initial approximation are unsuitable"
        exit
    endif
    do
        if ( | f2  - f1 | ≤ delta ) then
            write: "Slope of the function becomes too small"
            exit
        endif
        set x3 = (x1×f2  - x2×f1) / (f2  - f1)
        set f3 = f(x3)
        if ( f1 × f3 < 0 ) then
            set x2 = x3
            set f2 = f3
        else
            set x1 = x3
            set f1 = f3
        endif
    while ( | x1  - x2 | > epsilon ) )
    write: x3, "as the approximate root"
End.
```

3.8 SECANT METHOD

The secant method is similar to the false position method with only one difference. In the secant method, two most recent approximations to the root are used to find the new approximation instead of using only those two approximations that bound the interval containing the root. In addition, it is not necessary for the initial approximations to enclose the root. The secant method is much faster than the false position method, but the only difficulty with secant method is that convergence is not always assured.

To describe its working assume that x_1 and x_2 are two initial approximations to the root. Join the points $(x_1, f(x_1))$ and $(x_2, f(x_2))$ by a straight line. The point where this line intersects the x-axis gives the next approximation to the root. Let this intersection point be x_3. Now x_2 and x_3 are taken as the starting approximations for the next iteration. The line is drawn joining the points $(x_2, f(x_2))$ and $(x_3, f(x_3))$. Let this intersection point be x_4. The iterative procedure terminates when the relative error in two successive approximations is less than or equal to the prescribed tolerance.

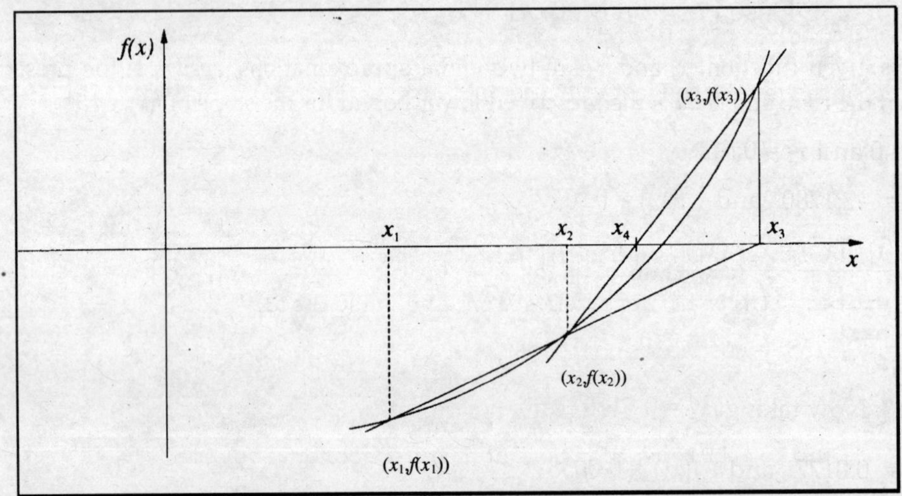

Figure 3.4: Root approximation by secant method

Proceeding on similar lines as in the case of false position method, the general formula for secant method can be derived as

$$x_{i+1} = \frac{x_{i-1} f(x_i) - x_i f(x_{i-1})}{f(x_i) - f(x_{i-1})} \quad \ldots (3.8)$$

Since the convergence is not always assured, therefore, we put a limit on the maximum number of iterations permitted.

Example 3.7: Given that one of the roots of the non-linear equation

$$\cos x - xe^x = 0$$

lies between 0.5 and 1.0. Find the root correct to three decimal places.

Solution: Since we want the solution correct to three decimal places, the iterative process will be terminated as soon as the successive iterations produces no change at three decimal places or the function vanishes at new approximation.

Iteration 1: Starting with $x_1 = 0.5$ and $x_2 = 1.0$

$$f(x_1) = f(0.5) = \cos(0.5) - 0.5 \times e^{0.5} = 0.0532$$

$$f(x_2) = f(1.0) = \cos(1.0) - 1.0 \times e^{1.0} = -2.1780$$

$$x_3 = \frac{x_1 f(x_2) - x_2 f(x_1)}{f(x_2) - f(x_1)} = \frac{0.5 \times (-2.1780) - 1.0 \times 0.0532}{-2.1780 - 0.0532} = 0.5119$$

$$f(x_3) = f(0.5119) = \cos(0.5119) - 0.5119 \times e^{0.5119} = 0.0177$$

Iteration 2: Since we take the last two approximations. Therefore, we have

$$x_1 = 1.0 \text{ and } x_2 = 0.5119$$

$$f(x_1) = -2.1780 \quad \text{and} \quad f(x_2) = 0.0177$$

$$x_3 = \frac{x_1 f(x_2) - x_2 f(x_1)}{f(x_2) - f(x_1)} = \frac{1.0 \times 0.0177 - 0.5119 \times (-2.1780)}{0.0177 - (-2.1780)} = 0.5158$$

$$f(x_3) = f(0.5158) = \cos(0.5158) - 0.5158 \times e^{0.5158} = 0.0058$$

Iteration 3: Now taking $x_1 = 0.5119$ and $x_2 = 0.5158$

$$f(x_1) = 0.0177 \quad \text{and} \quad f(x_2) = 0.0058$$

$$x_3 = \frac{x_1 f(x_2) - x_2 f(x_1)}{f(x_2) - f(x_1)} = \frac{0.5119 \times 0.0058 - 0.5158 \times 0.0177}{0.0058 - 0.0177} = 0.5178$$

$$f(x_3) = f(0.5178) = \cos(0.5178) - 0.5178 \times e^{0.5178} = 0.0000$$

Observe that after third iterations, the function $f(x)$ vanishes i.e. becomes zero.

Therefore, $x = 0.5178$ is taken as the approximate solution.

Example 3.8: Find the root of equation

$$x^3 - 5x + 3 = 0$$

correct to three decimal positions using Secant method.

Solution: Since we want the solution correct to three decimal positions, the iterative process will be terminated as soon as the successive iterations produces no change at first three decimal positions or the function vanishes at new approximation.

Now $f(0) = 0^3 - 5\times0 + 3 = 3.0$

$f(1) = 1^3 - 5\times1 + 3 = -1.0$

Since $f(0)$ is +ve and $f(1)$ is –ve, therefore one root lies between 0 and 1.

Iteration 1: Starting with $x_1 = 0$ and $x_2 = 1$

$f(0) = 0^3 - 5\times0 + 3 = 3.0$

$f(1) = 1^3 - 5\times1 + 3 = -1.0$

$$x_3 = \frac{x_1 f(x_2) - x_2 f(x_1)}{f(x_2) - f(x_1)} = \frac{0\times(-1.0) - 1\times3.0}{-1.0 - 3.0} = 0.75$$

$f(x_3) = f(0.75) = 0.75^3 - 5\times0.75 + 3 = -0.3281$

Thus, the first approximation to the root is 0.75.

Iteration 2: Since we take the last two approximations. Therefore, we have

$x_1 = 1.0$ and $x_2 = 0.75$

$f(x_1) = -1.0$ and $f(x_2) = -0.3281$

$$x_3 = \frac{x_1 f(x_2) - x_2 f(x_1)}{f(x_2) - f(x_1)} = \frac{1.0\times(-0.3281) - 0.75\times(-1.0)}{-0.3281 - (-1.0)} = 0.6279$$

$f(x_3) = f(0.6279) = 0.6279^3 - 5\times0.6279 + 3 = 0.1080$

Thus, the second approximation to the root is 0.6279.

Iteration 3: Since we take the last two approximations. Therefore, we have

$x_1 = 0.75$ and $x_2 = 0.6279$

$f(x_1) = -0.3281$ and $f(x_2) = 0.1080$

$$x_3 = \frac{x_1 f(x_2) - x_2 f(x_1)}{f(x_2) - f(x_1)} = \frac{0.75\times0.1080 - 0.6279\times(-0.3281)}{0.1080 - (-0.3281)} = 0.6582$$

$f(x_3) = f(0.6582) = 0.6582^3 - 5\times0.6582 + 3 = -0.0057$

Thus, the third approximation to the root is 0.6582.

Subsequent two iterations produce 0.6566 as the approximations to the root, where function also nearly vanishes.

Thus, we take $x = 0.656$ as the desired solution correct to three decimal positions.

Example 3.9: Using Secant method, find the approximate value $\sqrt{12}$ of correct to three decimal places.

Solution: Let $x = \sqrt{12} \quad \Rightarrow \quad x^2 = 12 \quad \Rightarrow \quad x^2 - 12 = 0$

Thus, the problem of approximating the value of $\sqrt{12}$ reduces to the problem of finding the root of equation $x^2 - 12 = 0$.

Since we want the solution correct to three decimal positions, the iterative process will be terminated as soon as the successive iterations produces no change at first three decimal positions or the function vanishes at new approximation.

Iteration 1: Starting with $x_1 = 3$ and $x_2 = 4$

$$f(3) = 3^2 - 12 = -3.0$$

$$f(4) = 4^2 - 12 = 4.0$$

$$x_3 = \frac{x_1 f(x_2) - x_2 f(x_1)}{f(x_2) - f(x_1)} = \frac{3 \times 4 - 4 \times (-3.0)}{4.0 - (-3.0)} = 3.4286$$

$$f(x_3) = f(3.4286) = 3.4286^2 - 12 = -0.2449$$

Thus, the first approximation to the root is 3.4286.

Iteration 2: Taking last two approximations, we write

$x_1 = 4.0$ and $x_2 = 3.4286$

$f(x_1) = 4.0$ and $f(x_2) = -0.2449$

$$x_3 = \frac{x_1 f(x_2) - x_2 f(x_1)}{f(x_2) - f(x_1)} = \frac{4.0 \times (-0.2449) - 3.4286 \times 4.0}{-0.2449 - 4.0} = 3.4615$$

$$f(x_3) = f(3.4615) = 3.4615^2 - 12 = -0.0178$$

Thus, the second approximation to the root is 3.4615.

Iteration 3: Taking last two approximations, we write

$x_1 = 4.0$ and $x_2 = 3.4286$

$f(x_1) = 4.0$ and $f(x_2) = -0.2449$

$$x_3 = \frac{x_1 f(x_2) - x_2 f(x_1)}{f(x_2) - f(x_1)} = \frac{4.0 \times (-0.2449) - 3.4286 \times 4.0}{-0.2449 - 4.0} = 3.4615$$

$f(x_3) = f(3.4615) = 3.4615^2 - 12 = -0.0178$

Thus, the second approximation to the root is 3.4615.

Iteration 4: Taking last two approximations, we write

$x_1 = 3.4286$ and $x_2 = 3.4615$

$f(x_1) = -0.2449$ and $f(x_2) = -0.0178$

$$x_3 = \frac{x_1 f(x_2) - x_2 f(x_1)}{f(x_2) - f(x_1)} = \frac{3.4286 \times (-0.0178) - 3.4615 \times (-0.2449)}{-0.0178 - 0.2449} = 3.4641$$

$f(x_3) = f(3.4641) = 3.4641^2 - 12 = 0.0001$

Thus, the second approximation to the root is 3.4641.

Iteration 5: Taking last two approximations, we write

$x_1 = 3.4615$ and $x_2 = 3.4641$

$f(x_1) = -0.0178$ and $f(x_2) = 0.0001$

$$x_3 = \frac{x_1 f(x_2) - x_2 f(x_1)}{f(x_2) - f(x_1)} = \frac{3.4615 \times 0.0001 - 3.4641 \times (-0.0178)}{0.0001 - 0.0178} = 3.4641$$

Since, this iteration also produces the same approximation as of previous iteration, therefore we terminate iteration here and take $x = 3.464$ as the desired solution correct to three decimal positions of equation $x^2 - 12 = 0$. Thus, the approximate value of $\sqrt{12}$ correct to three decimal positions is 3.464.

Algorithm 3.4: Secant Method

Here $f(x)$ is the given function, x_1 and x_2 are two initial approximations, *epsilon* is the prescribed tolerance, *delta* is the prescribed lower bound for the slope of $f(x)$, and n is the maximum number of iterations permitted.

```
Begin
   read: x1, x2, epsilon, delta, n
   set f1 = f(x1)
   set f2 = f(x2)
   for i = 1 to n by 1 do
      if ( | f2  - f1 | ≤ delta ) then
         write: "Slope of  the function becomes too small"
         exit
      endif
```

```
        set x3 = (x1×f2  - x2×f1) / (f2  - f1)
        set f3 = f(x3)
        if ( | ((x3  - x2)/x3) | ≤ epsilon ) then
            write: x3, "as the approximate root"
            exit
        endif
        set x1 = x2
        set f1 = f2
        set x2 = x3
        set f2 = f3
    endfor
    write: "Solution does not converge in", n, "iterations"
End.
```

3.9 NEWTON-RAPHSON METHOD

The Newton-Raphson method, also known as Newton's method of tangents, is one of the fastest iterative methods. This method begins with one initial approximation. Here one have to take due care while selecting the initial approximation, as it is very sensitive to the initial approximation. Once proper choice is made for the initial approximation, it converges faster than false position method and the secant method.

3.9.1 Analytical Derivation

Let $f(x) = 0$

be the given non-linear equation, and x_0 be the initial approximation, and h the correction to x_0 so that

$f(x_0+h) = 0$

Expanding $f(x_0+h)$ by Taylor's series, we get

$$f(x_0+h) = f(x_0) + hf^1(x_0) + \frac{h^2}{2!} f^2(x_0) + ... + \frac{h^n}{n!} f^n(x_0) \quad \text{. . . (3.9)}$$

where $f^1(x_0)$, $f^2(x_0)$, . . . are 1st order, 2nd order, ... derivatives of $f(x)$ at x_0.

If the values of h and $f^n(x_0)$ are such that

$$\lim_{n\to\infty} \frac{h^n}{n!} f^n(x_0) = 0 \quad \text{. . . (3.10)}$$

then neglecting the higher order terms in h, i.e., $O(h^2)$, we get

$$f(x_0) + hf^1(x_0) = 0 \ \Rightarrow \ h = -\frac{f(x_0)}{f^1(x_0)} \quad \text{. . . (3.11)}$$

Therefore if x_0 is the initial approximation, by adding to this the value of h, we get the next approximation to the root i.e.

$$x_1 = x_0 + h$$

$$\Rightarrow \quad x_1 = x_0 - \frac{f(x_0)}{f^1(x_0)} \qquad \text{... (3.12)}$$

Similarly, taking x_1 as the starting value, we get

$$x_2 = x_1 - \frac{f(x_1)}{f^1(x_1)}$$

In general, from the ith approximation, we obtain the $(i+1)^{th}$ approximation as

$$\boxed{x_{i+1} = x_i - \frac{f(x_i)}{f^1(x_i)}} \qquad \text{... (3.13)}$$

The iterative procedure terminates when the relative error for two successive approximations becomes less than or equal to the prescribed tolerance.

3.9.2 Geometric Derivation and Interpretation

Consider that the initial approximation is x_0. Now draw a tangent to the curve at the point $(x_0, f(x_0))$. The point, say $(x_1, 0)$, where this tangent intersects the x-axis gives the next approximation to the root. Starting with x_1 as the current approximation, we again draw a tangent to the curve at the point $(x_1, f(x_1))$. The intersection point of these tangent gives, says x_2, as the next approximation to the root. This iterative procedure is repeated till the relative error between two successive approximations is less than or equal to the prescribed tolerance.

If the proper choice of the initial approximation is not made, the method may diverge. This causes the iterative procedure to fall in endless loop. Therefore, to provide an exit in such circumstances, we put a limit on the maximum number of iterations permitted.

The slope of the curve at point $(x_0, f(x_0))$ is given by

$$\tan\theta = \frac{f(x_0)}{x_0 - x_1} = f^1(x_0) \qquad \text{... (3.14)}$$

Solving for x_1, we get

$$x_1 = x_0 - \frac{f(x_0)}{f^1(x_0)}$$

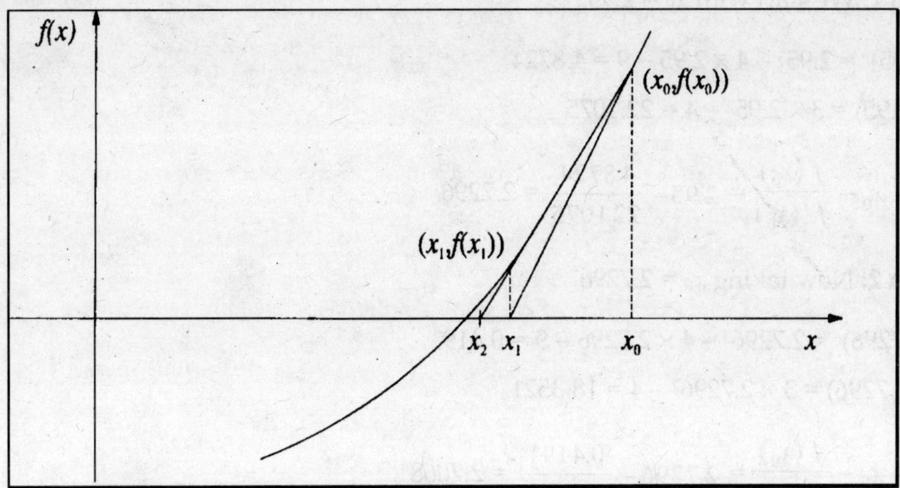

Figure 3.5: Root approximation by Newton-Raphson method

The next approximation would be

$$x_2 = x_1 - \frac{f(x_1)}{f^1(x_1)}$$

In general, the $(i+1)^{th}$ approximation is obtained from the ith approximation as

$$x_{i+1} = x_i - \frac{f(x_i)}{f^1(x_i)}$$

In the above formula, we assume that $f^1(x_i)$, the slope of the curve, is not equal to or nearly equal to zero. However, if the slope of the curve at x_i is too small, it may complicate the computations. Therefore, to avoid this we put a limit on the lower bound of the slope.

Example 3.10: Given that one of the roots of the non-linear equation

$$x^3 - 4x - 9 = 0$$

lies between 2.625 and 3.0. Find the root correct to four significant digits.

Solution: Since we want the solution correct to three significant digits, the iterative process will be terminated as soon as the successive iterations produces no change at first four significant positions or the function vanishes at new approximation.

Given $f(x) = x^3 - 4x - 9$

therefore

$$f^1(x) = 3x^2 - 4$$

Iteration 1: We start with $x_0 = 2.95$.

$$f(2.95) = 2.95^3 - 4 \times 2.95 - 9 = 4.8724$$

$$f^1(2.95) = 3 \times 2.95^2 - 4 = 22.1075$$

$$x_1 = x_0 - \frac{f(x_0)}{f^1(x_0)} = 2.95 - \frac{4.8724}{22.1075} = 2.7296$$

Iteration 2: Now taking $x_0 = 2.7296$

$$f(2.7296) = 2.7296^3 - 4 \times 2.7296 - 9 = 0.4191$$

$$f^1(2.7296) = 3 \times 2.7296^2 - 4 = 18.3521$$

$$x_1 = x_0 - \frac{f(x_0)}{f^1(x_0)} = 2.7296 - \frac{0.4191}{18.3521} = 2.7068$$

Iteration 3: Now taking $x_0 = 2.7068$

$$f(2.7068) = 2.7068^3 - 4 \times 2.7068 - 9 = 0.0049$$

$$f^1(2.7068) = 3 \times 2.7068^2 - 4 = 17.9803$$

$$x_1 = x_0 - \frac{f(x_0)}{f^1(x_0)} = 2.7068 - \frac{0.0049}{17.9803} = 2.7065$$

Iteration 4: Now taking $x_0 = 2.7065$

$$f(2.7065) = 2.7065^3 - 4 \times 2.7065 - 9 = -0.0005$$

$$f^1(2.7065) = 3 \times 2.7065^2 - 4 = 17.9754$$

$$x_1 = x_0 - \frac{f(x_0)}{f^1(x_0)} = 2.7065 - \frac{-0.0005}{17.9754} = 2.7065$$

Observe that iterations 2, 3, and 4 produce no change at first four significant digits in the successive approximations to the root. Therefore, we take $x = 2.706$ as the desired solution correct to four significant digits.

Example 3.11: Find the root of equation

$$x^3 - x - 4 = 0$$

correct to three decimal places using Newton-Raphson method.

Solution: Since we want the solution correct to three decimal places, the iterative process will be terminated as soon as the successive iterations produces no change at first three decimal places or the function vanishes at new approximation.

Given $f(x) = x^3 - x - 4$

therefore

$f^1(x) = 3x^2 - 1$

Iteration 1: We start with $x_0 = 2$

$f(2) = 2^3 - 2 - 4 = 2$

$f^1(2) = 3 \times 2^2 - 1 = 11$

$$x_1 = x_0 - \frac{f(x_0)}{f^1(x_0)} = 2 - \frac{2}{11} = 1.8182$$

Thus, the first approximation to the root is 1.8182.

Iteration 2: Now taking $x_0 = 1.8182$

$f(1.8182) = 1.8182^3 - 1.8182 - 4 = 0.1923$

$f^1(1.8182) = 3 \times 1.8182^2 - 1 = 8.9173$

$$x_1 = x_0 - \frac{f(x_0)}{f^1(x_0)} = 1.8183 - \frac{0.1923}{8.9173} = 1.7966$$

Thus, the second approximation to the root is 1.7966.

Iteration 3: Now taking $x_0 = 1.7966$

$f(1.7966) = 1.7976^3 - 1.7966 - 4 = 0.0025$

$f^1(1.7966) = 3 \times 1.7966^2 - 1 = 8.6834$

$$x_1 = x_0 - \frac{f(x_0)}{f^1(x_0)} = 1.7966 - \frac{0.0025}{8.6834} = 1.7963$$

Thus, the third approximation to the root is 1.7963.

Since for $x_0 = 1.7963$

$$f(1.7963) = 1.7973^3 - 1.7963 - 4 = 0.0000$$

We terminate the iterative procedure here and take $x = 1.796$ as the desired solution correct to three decimal places.

Example 3.12: Evaluate $\sqrt{12}$ correct to three decimal places using Newton-Raphson method.

Solution: Let $x = \sqrt{12} \Rightarrow x^2 = 12 \Rightarrow x^2 - 12 = 0$

Thus, the problem of evaluating $\sqrt{12}$ reduces to the problem of finding the root of equation $x^2 - 12 = 0$.

Since we want the solution correct to three decimal places, the iterative process will be terminated as soon as the successive iterations produces no change at first three decimal places or the function vanishes at new approximation.

Given $f(x) = x^2 - 12$ therefore $f^1(x) = 2x$

Iteration 1: We start with $x_0 = 3$

$$f(3) = 3^2 - 12 = -3$$

$$f^1(3) = 2 \times 3 = 6$$

$$x_1 = x_0 - \frac{f(x_0)}{f^1(x_0)} = 3 - \frac{-3}{6} = 3.5$$

Thus, the first approximation to the root is 3.5.

Iteration 2: Now taking $x_0 = 3.5$

$$f(3.5) = 3.5^2 - 12 = 0.25$$

$$f^1(3.5) = 2 \times 3.5 = 7$$

$$x_1 = x_0 - \frac{f(x_0)}{f^1(x_0)} = 3.5 - \frac{0.25}{7} = 3.4643$$

Thus, the second approximation to the root is 3.4643.

Iteration 3: Now taking $x_0 = 3.4643$

$f(3.4643) = 3.4643^2 - 12 = 0.0013$

$f^1(3.4643) = 2 \times 3.4643 = 6.9286$

$$x_1 = x_0 - \frac{f(x_0)}{f^1(x_0)} = 3.4643 - \frac{0.0013}{6.9286} = 3.4641$$

Thus, the third approximation to the root is 3.4641.

Since for $x_0 = 3.4641$

$f(3.4641) = 3.4641^2 - 12 = 0.0000$

We terminate the iterative procedure here as function vanishes at the new approximation and take $x = 3.464$ as the desired solution correct to three decimal places of equation $x^2 - 12 = 0$, hence the value of $\sqrt{12}$ correct to three decimal places is 3.464.

Algorithm 3.5: Newton-Raphson Method

Here $f(x)$ is the given function, $f^1(x)$ is first order derivative of function $f(x)$, x_0 is the initial approximations, *epsilon* is the prescribed tolerance, *delta* is the prescribed lower bound for the slope of $f(x)$, and n is the maximum number of iterations permitted.

```
Begin
   read: x0, epsilon, delta, n
   for i = 1 to n by 1 do
      if ( | f 1(x0) | ≤ delta ) then
         write: "Slope of f(x) becomes too small near x = ", x0
         exit
      endif

      set x1 = x0  - f(x0) / f 1(x0)
      set relative_error = | ((x1  - x0)/x1) |
      set x0 = x1

     if ( relative_error ≤ epsilon ) then
         write: x1, "as the approximate root."
         exit
      endif
   endfor
   write: "Solution does not converge in", n, "iterations"
End.
```

3.10 METHOD OF SUCCESSIVE APPROXIMATIONS

This method also known as the *direct substitution method* or *method of iterations* or *method of fixed iterations*, is applicable if the equation

$$f(x) = 0$$

can be expressed as

$$x = g(x)$$

If x_1 is the initial approximation to the root, then next approximation to the root is given by

$$x_2 = g(x_1)$$

and the next approximation will be

$$x_3 = g(x_2)$$

In general

$$x_{i+1} = g(x_i) \qquad \ldots (3.17)$$

The iterative cycle will terminate when the relative error in the new approximation is within the prescribed tolerance.

As we know that the solution of the equation

$$f(x) = 0$$

is given by the intersection point of the graph $f(x)$ and the x-axis.

We may regard such a solution as the point of intersection of two curves:

- One having the equation $y = 0$, which is the equation of line representing the x-axis.
- Other having the equation $y = f(x)$.

In the method of successive approximation, we may still regard the solution as being the intersection of two curves, but now the curves are:

- One having the equation $y = x$, which is the equation of line with slope +1.
- Other having the equation $y = g(x)$.

If the graphs of $y = x$ and $y = g(x)$ are as shown below, then at the point of intersection P, equations $y = x$ and $y = g(x)$ are satisfied simultaneously. At this point, we have $x = g(x)$, hence the x co-ordinate of this point is the required solution.

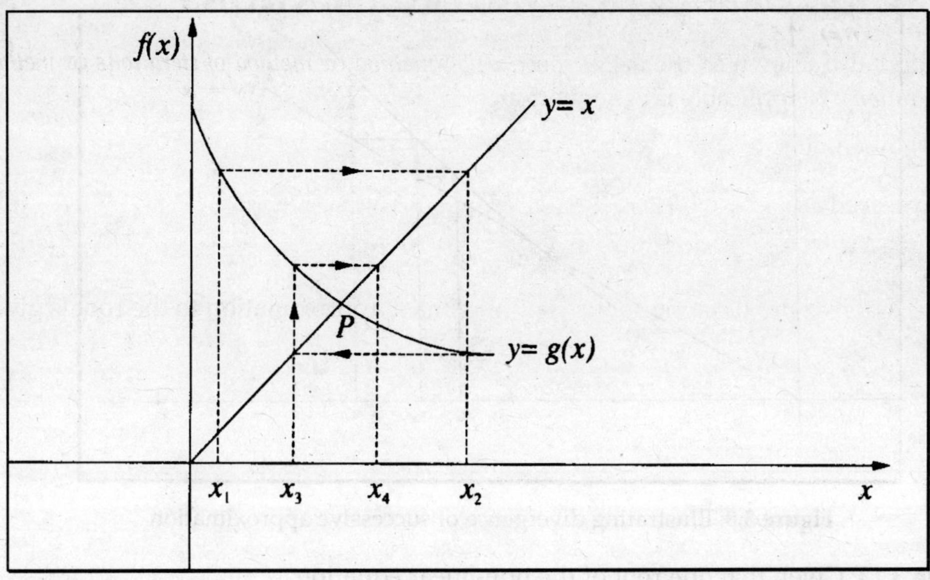

Figure 3.6: Root approximation by method of successive approximations

However, this procedure converge if and only if $| g^1(x_i) | < 1$.

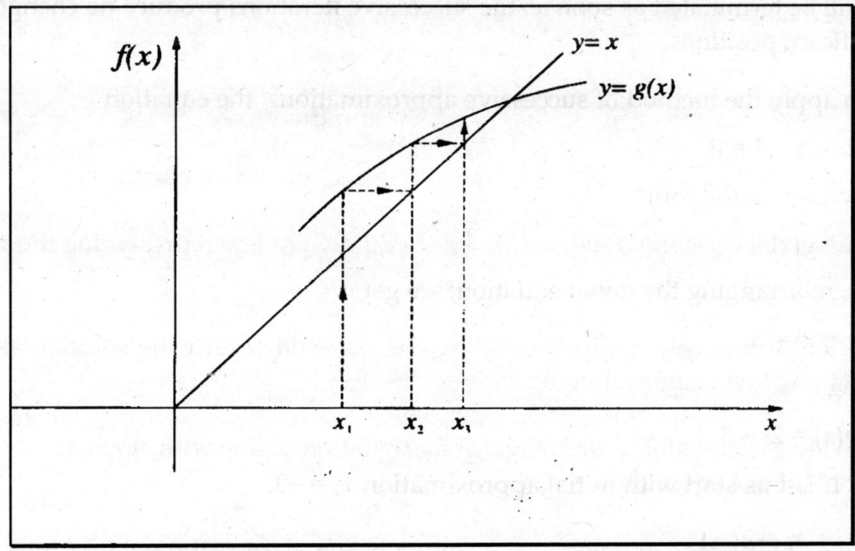

Figure 3.7: Illustrating convergence of successive approximation

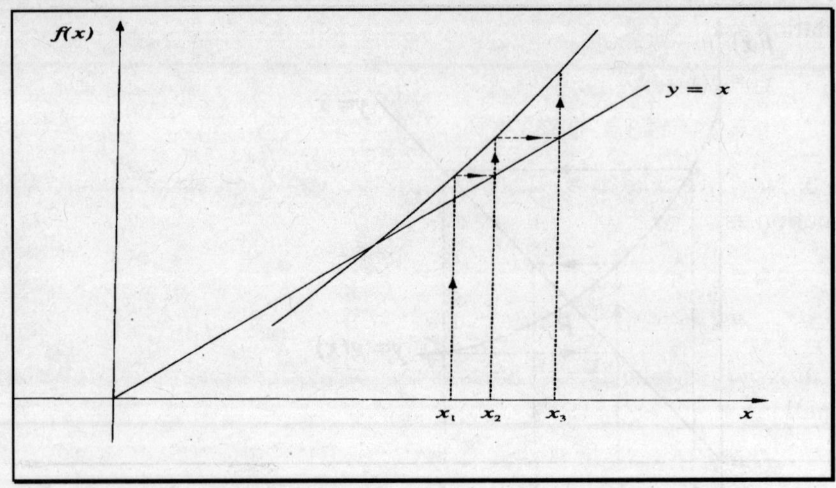

Figure 3.8: Illustrating divergence of successive approximation

Example 3.13: Given that one root of the non-linear equation

$$2^x - x - 3 = 0$$

lies in the interval (–3, –2). Find the root correct to three decimal places.

Solution: Since we want the solution correct to three significant digits, the iterative process will be terminated as soon as the successive iterations produce no change at first four significant positions.

Further to apply the method of successive approximations, the equation

$$2^x - x - 3 = 0$$

must be written in the form

$$x = g(x)$$

Therefore, re-arranging the given equation, we get

$$x = 2^x - 3$$

thus giving

$$g(x) = 2^x - 3$$

Iteration 1: Let us start with initial approximation $x_1 = -3$.

$$x_2 = g(x_1)$$

$$= 2^{-3} - 3 = -2.875$$

Iteration 2: Now we take $x_2 = -2.875$ as the current approximation to obtain the next approximation as

$$x_3 = g(x_2)$$
$$= 2^{-2.875} - 3 = -2.8637$$

Iteration 3: Now we take $x_3 = -2.8637$ as the current approximation to obtain the next approximation as

$$x_4 = g(x_3)$$
$$= 2^{-2.8637} - 3 = -2.8626$$

Iteration 4: Now we take $x_4 = -2.8626$ as the current approximation to obtain the next approximation as

$$x_5 = g(x_4)$$
$$= 2^{-2.8626} - 3 = -2.8625$$

Iteration 5: Now we take $x_5 = -2.8625$ as the current approximation to obtain the next approximation as

$$x_6 = g(x_5)$$
$$= 2^{-2.8625} - 3 = -2.8625$$

We see that after iteration number 3, there is no change at first four significant digits. Therefore, we take $x = -2.862$ as desired solution correct to four significant digits.

Example 3.14: Find the root of equation

$$2x - \cos x - 3 = 0$$

correct to three decimal places using Iteration method.

Solution: Let $f(x) = 2x - \cos x - 3$

$$f(0) = 0 - \cos 0 - 3 = -4$$
$$f(2) = 2 - \cos 2 - 3 = 1.42$$

Thus, a root lies between 0 and 2.

Further to apply Iteration method, the equation

$$2x - \cos x - 3 = 0$$

must be written in the form

$$x = g(x)$$

Therefore, re-arranging the given equation, we get

$$x = \frac{1}{2}(\cos x + 3)$$

thus giving

$$g(x) = \frac{1}{2}(\cos x + 3)$$

$$g'(x) = \frac{1}{2}(-\sin x)$$

$$\Rightarrow \quad |g'(x)| = \left|\frac{1}{2}(-\sin x)\right| < \frac{1}{2} < 1 \qquad [|\sin x| < 1]$$

Therefore, Iteration method can be applied.

Since we want the solution correct to three decimal places, the iterative process will be terminated as soon as the successive iterations produce no change at first three decimal places.

Iteration 1: Let us start with initial approximation $x_1 = 1.5$.

$$x_2 = g(x_1) = \frac{1}{2}(\cos 1.5 + 3) = 1.5354$$

Thus, the first approximation to the root is 1.5354.

Iteration 2: Now we take $x_2 = 1.5354$ as the current approximation to obtain the next approximation as

$$x_3 = g(x_2) = \frac{1}{2}(\cos 1.5354 + 3) = 1.5178$$

Thus, the second approximation to the root is 1.5178.

Iteration 3: Now we take $x_3 = 1.5178$ as the current approximation to obtain the next approximation as

$$x_4 = g(x_3) = \frac{1}{2}(\cos 1.5178 + 3) = 1.5665$$

Thus, the third approximation to the root is 1.5665.

The subsequent iterations produce the following approximations to the root

1.5665, 1.5221, 1.5243, 1.5232, 1.5238, 1.5235, 1.5236

Observe that the last four iterations produce no change at first three decimal places in the successive approximations to the root. Therefore, we take $x = 1.523$ as the desired solution correct to three decimal places.

Example 3.15: Find the root of equation

$$2x - \log_{10} x - 7 = 0$$

correct to three decimal places using Iteration method.

Solution: Let $f(x) = 2x - \log_{10} x - 7$

$$f(3) = 6 - \log_{10} 3 - 7 = 6 - 0.4771 - 7 = -1.4471$$

$$f(4) = 8 - \log_{10} 4 - 7 = 8 - 0.602 - 7 = 0.398$$

Thus, a root lies between 3 and 4.

Further to apply Iteration method, the equation

$$2x - \log_{10} x - 7 = 0$$

must be written in the form

$$x = g(x)$$

Therefore, re-arranging the given equation, we get

$$x = \frac{1}{2}(\log_{10} x + 7)$$

thus giving

$$g(x) = \frac{1}{2}(\log_{10} x + 7)$$

$$g'(x) = \frac{1}{2}\left(\frac{1}{x}\log_{10} e\right) \qquad [\log_{10} e = 0.4343]$$

$$\Rightarrow \quad |g'(x)| = \left|\frac{1}{2}\left(\frac{0.4343}{x}\right)\right| < 1 \qquad \text{when } 3 < x < 4$$

Therefore, Iteration method can be applied.

Since we want the solution correct to three decimal places, the iterative process will be terminated as soon as the successive iterations produce no change at first three decimal places.

Iteration 1: Let us start with initial approximation $x_1 = 3.5$.

$$x_2 = g(x_1) = \frac{1}{2}(\log_{10} 3.5 + 7) = 3.7720$$

Thus, the first approximation to the root is 3.7720.

Iteration 2: Now we take $x_2 = 3.7720$ as the current approximation to obtain the next approximation as

$$x_3 = g(x_2) = \frac{1}{2}(\log_{10} 3.7720 + 7) = 3.7883$$

Thus, the second approximation to the root is 3.7883.

Iteration 3: Now we take $x_3 = 3.7883$ as the current approximation to obtain the next approximation as

$$x_4 = g(x_3) = \frac{1}{2}(\log_{10} 3.7883 + 7) = 3.7892$$

Thus, the third approximation to the root is 3.7892.

Iteration 4: Now we take $x_4 = 3.7892$ as the current approximation to obtain the next approximation as

$$x_5 = g(x_4) = \frac{1}{2}(\log_{10} 3.7892 + 7) = 3.7893$$

Thus, the fourth approximation to the root is 3.7893.

Iteration 5: Now we take $x_5 = 3.7893$ as the current approximation to obtain the next approximation as

$$x_6 = g(x_5) = \frac{1}{2}(\log_{10} 3.7893 + 7) = 3.7893$$

Thus, the fifth approximation to the root is 3.7893.

Observe that the last two iterations produce the same approximations to the root. Therefore, we take $x = 3.789$ as the desired solution correct to three decimal places.

Algorithm 3.6: Method of Successive Approximations

Here $f(x)$ is given function and is expressed as $x = g(x)$, x_1 is the initial approximations, *epsilon* is the prescribed tolerance, and *n* is the maximum number of iterations permitted.

```
Begin
    read: x1, epsilon, n
    for i = 1 to n by 1 do
        set x2 = g(x1)
        set relative_error = | ((x2 - x1)/x2) |
        set x1 = x2
        if ( relative_error ≤ epsilon ) then
            write: "Root convergex in", i, "iterations"
            write: x2, "as the approximate root"
            exit
```

```
        endif
    endfor
    write: "Root does not converge in", n, "iterations"
End.
```

3.11 CONCEPT OF SYNTHETIC DIVISION

Division of a polynomial $f(x)$ by a factor $(x-r)$ can be carried rapidly by synthetic division. The remainder of the above division will give the value of the polynomial for $x = r$. However, if r happens to the root of the polynomial $f(x)$, then the remainder resulting from the division will be zero.

Similarly, the value of the derivative $f^1(x)$ of the polynomial $f(x)$ for $x = r$ is equal to the remainder obtained by second synthetic division of the result obtained from the first synthetic division, which will be a polynomial of degree one less than the original polynomial.

To understand the synthetic division, let us consider the following example.

Example 3.16: Divide the polynomial

$$x^5 - 3x^4 - 10x^3 + 5x^2 + 22x + 16$$

by $(x - 4)$.

Solution: The division by a factor $(x - 4)$ is carried out as shown below:

4	1	–3	–10	5	22	16
		4	4	–24	–76	–216
	1	1	–6	–19	–54	(–200)

The first row consists of the coefficients of the given polynomial. The second and third rows are generated from left to right using the following procedure:

The first coefficient is written as such in the third row. The first element in the third row is multiplied by 4. The product gives the first element of the second row and is written under the second element of the first row. Now the second element of first row and the first element of the second row are added. This sum gives the second element of the third row. Next the second element of the third row is multiplied by 4. This product gives the second element of the second row and is written under the third element of the first row. Then the third element of first row and the second element of the second row are added. This sum gives the third element of the third row. This procedure is repeated till the last element in the third row is obtained. This way, the second and third rows are generated from left to right.

The last element of the third row is the value of the polynomial at $x = 4$, i.e.

$$f(4) = -200$$

The reduced polynomial, a polynomial of degree one less than the given is obtained from the third row barring the last element as

$$x^4 + x^3 - 6x^2 - 19x - 54$$

Example 3.17: Obtain the derivative of the polynomial

$$x^5 - 3x^4 - 10x^3 + 5x^2 + 22x + 16$$

at $x = 4$.

Solution: As seen in Example 3.6, the first synthetic division gives polynomial

$$x^4 + x^3 - 6x^2 - 19x - 54$$

as quotient. The division of the reduced polynomial gives the value of the derivative of the original polynomial by a factor $(x-4)$ as shown below.

4	1	1	−6	−19	−54
		4	20	56	148
	1	5	14	37	(94)

The remainder 94 is the value of $f^1(4)$.

Similarly, we can obtain the values of higher derivatives as well.

Algorithm 3.7: Synthetic Division by Linear Factor $(x-r)$

Here array a represent the coefficients of the given polynomial of degree n of type

$$a_n x^n + a_{n-1}x^{n-1} + a_{n-2}x^{n-2} + \ldots + a_1 x + a_0$$

and array b represents the coefficients of the polynomial obtained as quotient after the division, and r is the point where value of the polynomial (obtained in b_0) is desired.

```
Begin
    read: n, r
    for i = n to 0 by -1 do
        read: a_i
    endfor
```

```
    set b_n = a_n
    for i = (n - 1) to 0 by -1 do
        set b_i = a_i + rxb_i+1
    endfor
    write: "Coefficients of the reduced polynomial"
    for i = n to 1 by -1 do
        write: b_i
    endfor
    write: b_0, "as the value of the polynomial"
End.
```

Now we consider the synthetic division by a quadratic factor.

Example 3.18: Suppose the polynomial is

$$x^5 - 3x^4 - 10x^3 + 10x^2 + 44x + 48$$

is to be divided by the quadratic factor $x^2 + 2x + 1$.

Solution: The division by a factor $x^2 + 2x + 1$ is carried out as shown below:

–2	1	–3	–10	10	44	48
–1		–2	10	2	–34	–22
			–1	5	1	–17
	1	–5	–1	17	11	9

The first row consists of the coefficients of the given polynomial. The second, third and fourth rows are generated from left to right using the following procedure:

The first coefficient is written as such in the fourth row. The first element in the fourth row is multiplied by –2. The product gives the first element of the second row and is written under the second element of the first row. Now the second element of first row and the first element of the second row are added. This sum gives the second element of the fourth row. Next the second element of the fourth row is multiplied by –2. This product gives the second element of the second row and is written under the third element of the first row. The first element of fourth row is multiplied by –1. This product gives the first element of third row and is written in the third column. Now the elements in the third column are added. This sum gives the third element of the fourth row. Next the third element of the fourth row is multiplied by –2.

This product gives the third element of the second row and is written under the fourth element, *i.e.*, fourth column, of the first row. The second element of fourth row is multiplied by –1. This product gives the second element of third row and is written in the fourth column. Now the elements in the fourth column are added. This sum gives the fourth element of the fourth row. This procedure is repeated till the last element in the fourth row is obtained. This way, the second, third and fourth rows are generated from left to right.

The reduced polynomial, a polynomial of degree less than by a factor of two is obtained from the fourth row barring the last two elements as

$$x^3 - 5x^2 - x + 17$$

The remainder as a result of above division is

$$(x+2)11 + 9 = \quad 11x + 31$$

Algorithm 3.8: Synthetic Division by Quadratic Factor (x^2+ux+v)

Here array a represent the coefficients of the given polynomial of degree n of type

$$a_nx^n + a_{n-1}x^{n-1} + a_{n-2}x^{n-2} + \ldots + a_1x + a_0$$

and array b represents the coefficients of the polynomial obtained as quotient after the division, u and v are the coefficients of the quadratic factor.

```
Begin
    read: n
    for i = n to 0 by -1 do
        read: a_i
    endfor
    read: u, v
    set b_n = a_n
    set b_n-1 = a_n-1 - u×b_n
    for i = (n - 2) to 0 by -1 do
        set b_i = a_i - u×b_i+1 - v×b_i+2
    endfor
    write: "Coefficients of the reduced polynomial"
    for i = n to 2 by -1 do
        write: b_i
    endfor
    write: (x + u)b_1 + b_0, "as the remainder"
End.
```

3.12 BAIRSTOW'S METHOD

Bairstow's method is capable of finding all the roots, real as well as complex, of polynomial equations only. It basically involves finding the quadratic factor of type

$$x^2 + ux + v$$

of the nth degree polynomial of type

$$x^n + a_1x^{n-1} + a_2x^{n-2} + a_3x^{n-3} \ldots + a_{n-1}x + a_n = 0 \qquad \ldots (3.18)$$

By taking arbitrary starting values for u and v, usually $u = v = 0$, the iterative process converges to correct values of u and v, with the result that the two roots can be determined from the quadratic factor as

$$r_{1,2} = \frac{-u \pm \sqrt{u^2 - 4v}}{2} \qquad \text{... (3.19)}$$

Thus, the first quadratic factor obtained from the iterative process gives two roots of the polynomial. The same iterative process can obtain the other roots after obtaining the quadratic factor of the polynomial of degree (n–2), resulting from the extraction of the quadratic factor for the first two roots. This new quadratic factor will yield another two roots. This procedure is continued to find all the roots. If the degree of original polynomial happens to be odd, then in the last iteration we will be left with a linear factor; otherwise the last factor will also be a quadratic factor, which can easily be solved.

We now develop the equations needed for the iterative process. The extraction of a quadratic factor

$$x^2 + ux + v$$

from the nth degree polynomial

$$x^n + a_1x^{n-1} + a_2x^{n-2} + a_3x^{n-3} ... + a_{n-1}x + a_n = 0$$

can be written in the form

$$(x^2 + ux + v)(x^{n-2} + b_1x^{n-3} + b_2x^{n-4} + \ldots + b_{n-3}x + b_{n-2}) + \text{remainder}$$
$$= x^n + a_1x^{n-1} + a_2x^{n-2} + a_3x^{n-3} ... + a_{n-1}x + a_n$$

By comparing coefficients on both sides, we get

$$\left.\begin{aligned} b_0 &= 1 \\ b_1 &= a_1 - u \\ b_2 &= a_2 - b_1u - v \\ b_3 &= a_3 - b_2u - b_1v \\ &\vdots \\ b_{n-1} &= a_{n-1} - b_{n-2}u - b_{n-3}v \\ b_n &= a_n - b_{n-1}u - b_{n-2}v \end{aligned}\right\} \qquad \text{... (3.20)}$$

and remainder = $(x + u)b_{n-1} + b_n$

If the values of u and v are such that the quadratic factor $x^2 + ux + v$ contains two roots of the nth degree polynomial, then the remainder will be zero. Therefore b_{n-1} and b_n must be zero.

The next step is to reduce the remainder term to nearly zero as required for the desired accuracy of the roots. This means that both b_{n-1} and b_n must be reduced to nearly zero. We see that b's are functions of both u and v. We can add du and dv as increments to u and v respectively to make the remainder term zero. Assuming that du and dv are too small, we can express b_{n-1} and b_n as Taylor's series expansion for a function of two variables as

$$b_{n-1}(u+du, v+dv) = 0 \approx b_{n-1} + \frac{\partial b_{n-1}}{\partial u} du + \frac{\partial b_{n-1}}{\partial v} dv \quad \ldots (3.21)$$

$$b_n(u+du, v+dv) = 0 \approx b_n + \frac{\partial b_n}{\partial u} du + \frac{\partial b_n}{\partial v} dv \quad \ldots (3.22)$$

ignoring higher order terms involving du and dv. The terms on the right hand side are evaluated for values of u and v and not for $u+du$ and $v+dv$.

The required partial derivatives can be obtained by differentiating the expression of Eqs. 3.20, which can be expressed as a set of c's, which in turn can be obtained from b's by synthetic division.

Taking partial derivatives of expressions of Eqs. 3.20 with respect to u and defining them in terms of c's, we get

$$\begin{aligned}
&\partial b_0/\partial u = 0 && \\
&\partial b_1/\partial u = -1 = -c_0 && \Rightarrow c_0 = 1 \\
&\partial b_2/\partial u = u - b_1 = -c_1 && \Rightarrow c_1 = b_1 - u \\
&\partial b_3/\partial u = -b_2 + c_1 u + v = -c_2 && \Rightarrow c_2 = b_2 - c_1 u - v \\
&\partial b_4/\partial u = -b_3 + c_2 u + c_1 v = -c_3 && \Rightarrow c_3 = b_3 - c_2 u - c_1 v \\
&\partial b_5/\partial u = -b_4 + c_3 u + c_2 v = -c_4 && \Rightarrow c_4 = b_4 - c_3 u - c_2 v \\
&\quad\vdots && \quad\vdots \\
&\partial b_n/\partial u = -b_{n-1} + c_{n-2} u + c_{n-3} v = -c_{n-1} && \Rightarrow c_{n-1} = b_{n-1} - c_{n-2} u - c_{n-3} v
\end{aligned}$$

Similarly, taking the partial derivatives with respect to v, we get

$$\begin{aligned}
&\partial b_0/\partial v = 0 && \\
&\partial b_1/\partial v = -1 = -c_0 && \Rightarrow c_0 = 1 \\
&\partial b_2/\partial v = u - b_1 = -c_1 && \Rightarrow c_1 = b_1 - u \\
&\partial b_3/\partial v = -b_2 + c_1 u + v = -c_2 && \Rightarrow c_2 = b_2 - c_1 u - v \\
&\partial b_4/\partial v = -b_3 + c_2 u + c_1 v = -c_3 && \Rightarrow c_3 = b_3 - c_2 u - c_1 v \\
&\partial b_5/\partial v = -b_4 + c_3 u + c_2 v = -c_4 && \Rightarrow c_4 = b_4 - c_3 u - c_2 v \\
&\quad\vdots && \quad\vdots \\
&\partial b_n/\partial v = -b_{n-1} + c_{n-2} u + c_{n-3} v = -c_{n-1} && \Rightarrow c_{n-1} = b_{n-1} - c_{n-2} u - c_{n-3} v
\end{aligned}$$

As you can observe, the c's can be expressed as

$$\left.\begin{aligned}
&c_0 = 1 \\
&c_1 = b_1 - u \\
&c_2 = b_2 - c_1 u - v \\
&c_3 = b_3 - c_2 u - c_1 v \\
&c_4 = b_4 - c_3 u - c_2 v \\
&\quad\vdots \\
&c_k = b_k - c_{k-1} u - c_{k-2} v \qquad \text{for } k = 2, 3, 4, \ldots, n-1
\end{aligned}\right\} \quad \ldots (3.23)$$

Therefore, Eqs. 3.21 and 3.22 can be expressed as

$$b_{n-1} = c_{n-2}du + c_{n-3}dv$$
$$b_n \ = c_{n-1}du + c_{n-2}dv$$

Solving these equations, wet get

$$du = \frac{\begin{vmatrix} b_n & c_{n-2} \\ b_{n-1} & c_{n-3} \end{vmatrix}}{\begin{vmatrix} c_{n-1} & c_{n-2} \\ c_{n-2} & c_{n-3} \end{vmatrix}} \quad \text{and} \quad dv = \frac{\begin{vmatrix} c_{n-1} & b_n \\ c_{n-2} & b_{n-1} \end{vmatrix}}{\begin{vmatrix} c_{n-1} & c_{n-2} \\ c_{n-2} & c_{n-3} \end{vmatrix}} \qquad \ldots (3.24)$$

So the values of u and v are updated as

$u = u + du$ and $v = v + dv$

and the process is repeated starting from the evaluation of b's and then c's, and finally the new values of du and dv are computed till these new values of du and dv become zero or approach to some pre-assigned value of *epsilon* such that

$(|du| + |dv|) \leq$ epsilon

Example 3.19: Find the roots of the polynomial

$$x^3 - 5x^2 - 2x + 24 = 0$$

correct to three significant digits i.e. prescribed tolerance is 0.001.

Solution: The degree of the given polynomial is $n = 3$. We start with $u = v = 0$

With these values of u and v, the synthetic division results in the following, for the first iteration.

$u = 0$	1	–5	–2	24	
$v = 0$	–	0	0	0	
		–	0	0	
	1	–5	–2	24	← b's
		0	0		
		–	0		
	1	–5	–2		← c's

$$du = \frac{\begin{vmatrix} 24 & -5 \\ -2 & 1 \end{vmatrix}}{\begin{vmatrix} -2 & -5 \\ -5 & 1 \end{vmatrix}} = -0.52 \quad \text{and} \quad dv = \frac{\begin{vmatrix} -2 & 24 \\ -5 & -2 \end{vmatrix}}{\begin{vmatrix} -2 & -5 \\ -5 & 1 \end{vmatrix}} = -4.59$$

After first iteration

$$u = u + du = 0.0 - 0.52 = -0.52$$

$$v = v + dv = 0.0 - 4.59 = -4.59$$

After two more iteration, the fourth iteration results in the following synthetic division

$$\begin{array}{l|rrrr}
u = 1.0 & 1 & -5 & -2 & 24 \\
v = 6.03 & - & 1 & -4 & 0.03 \\
 & & - & 6.03 & -24.12 \\
\hline
 & 1 & -4 & 0.03 & -0.09 \quad \longleftarrow b\text{'s} \\
 & & 1 & -3 & \\
 & & - & 6.03 & \\
\hline
 & 1 & -3 & 3.06 & \quad \longleftarrow c\text{'s}
\end{array}$$

$$du = \frac{\begin{vmatrix} -0.09 & -3 \\ 0.03 & 1 \end{vmatrix}}{\begin{vmatrix} 3.06 & -3 \\ -3 & 1 \end{vmatrix}} = 0.0 \quad \text{and} \quad dv = \frac{\begin{vmatrix} 3.06 & -0.09 \\ -3 & 0.03 \end{vmatrix}}{\begin{vmatrix} 3.06 & -3 \\ -3 & 1 \end{vmatrix}} = 0.0$$

After fourth iteration

$$u = u + du = -1.0 - 0.0 \quad = -1.0$$

$$v = v + dv = -6.03 + 0.03 = -6.0$$

The fifth iteration is described below:

$u = 1.0$	1	–5	–2	24	
$v = 6.0$	–	1	–4	0	
		–	6	–24	
	1	–4	0	0	← *b*'s
		1	–3		
		–	6		
	1	–3	–3		← *c*'s

$$du = \frac{\begin{vmatrix} 0 & -3 \\ 0 & 1 \end{vmatrix}}{\begin{vmatrix} 3 & -3 \\ -3 & 1 \end{vmatrix}} = 0.0 \quad \text{and} \quad dv = \frac{\begin{vmatrix} 3 & 0 \\ -3 & 0 \end{vmatrix}}{\begin{vmatrix} 3 & -3 \\ -3 & 1 \end{vmatrix}} = 0.0$$

After fifth iteration

$$u = u + du = -1.0 + 0.0 = -1.0$$

$$v = v + dv = -6.0 + 0.0 = -6.0$$

Since $(|du| + |dv|) = 0$

therefore the iteration process terminates here. Substituting these values of u and v into the expression

$$x^2 + ux + v$$

we get the quadratic factor as

$$x^2 - x - 6$$

which after solving gives two roots as

$$r_1 = 3 \qquad r_2 = -2$$

Note that in the last iteration

$$b_1 = 4$$

thus the factored form of the given polynomial is

$$(x^2 - x - 6)(x - 4) = 0$$

Thus, after extracting the quadratic factor, we are left with linear factor

$$x - 4$$

which gives

$$r_3 = 4$$

Thus, all roots of given polynomial $x^3 - 5x^2 - 2x + 24 = 0$ are $r_1 = 3, r_2 = -2$, and $r_3 = 4$.

Algorithm 3.9: Bairstow's Method

Here array *a* represent the coefficients of the given polynomial of degree *n* of type

$$x^n + a_1x^{n-1} + a_2x^{n-2} + a_3x^{n-3} ... + a_{n-1}x + a_n$$

and *u* and *v* are the coefficients of the quadratic factor.

```
Begin
    read: n, epsilon
    for i = 1 to n by 1 do
         read: ai
    endfor
    while ( n > 2 ) do
         set u  = 0, v = 0
         do
             set b1 = a1 - u
             set b2 = a2 - b1Xu  - v
             for k = 3 to n  by 1 do
                 set bk = ak - bk-1Xu  - bk-2Xv
             endfor
             set c1 = b1 - u
             set c2 = b2  - c1Xu  - v
             for k = 3 to (n  - 1) by 1 do
                  set ck = bk  - ck-1Xu  - ck-2Xv
             endfor
             set du = (bnX cn-2 - bn-1 X cn-2)/(cn-1 X cn-3 - cn-2 X cn-2)
             set dv = (bn-1X cn-1 - bn X cn-2)/(cn-1 X cn-3 - cn-2 X cn-2)
             set u = u + du
             set v = v + dv
         while ( ( |du| + |dv| ) ≥ epsilon )
         compute and print roots of quadratic factor x^2 + ux + v
         set n = n  - 2
         for i = 1 to n by 1 do
```

```
            set ai = bi
        endfor
    endwhile
    if ( n = 1 ) then
        write: b1, "as one of the roots"
    else
        compute and print roots of quadratic factor x2 + b1x + b2
    endif
End.
```

3.13 MULLER'S METHOD

Most of the methods that have discussed so far have approximated the function in the neighbourhood of the root by straight line. Obviously, this is not true in most of the cases. Muller's method is based on approximating the function in the neighbourhood of the root by a quadratic polynomial. This gives a much closer match to the actual function.

A second degree polynomial is made to fit three points viz. $[x_0, f(x_0)]$, $[x_1, f(x_1)]$, $[x_2, f(x_2)]$, near the root. The proper zero of this quadratic, using the quadratic formula is used as the improved estimate of the root. The process is then repeated using the set of three points nearest the root being evaluated.

The procedure for Muller's method is developed by writing a quadratic equation that fits through three points in the vicinity of a root, in the form av^2+bv+c. The development is simplified if we transform axes to pass through the middle point by letting $v = x - x_0$.

Let $h_1 = x_1 - x_0$ and $h_2 = x_0 - x_2$. We evaluate the coefficients by evaluating $av^2 + bv + c$ at the three points:

$$\left.\begin{aligned} v = 0: &\quad a(0)^2 + b(0) + c = f_0 \\ v = h_1: &\quad a(h_1)^2 + b(h_1) + c = f_1 \\ v = -h_2: &\quad a(h_2)^2 - b\,h_1 + c = f_2 \end{aligned}\right\} \quad \ldots (3.25)$$

From the first equation, we get $c = f_0$. If we take $h_2/h_1 = \gamma$, we can solve the other two equations for a and b.

$$\left.\begin{aligned} a &= \frac{\gamma f_1 - f_0(1+\gamma) + f_2}{\gamma h_1^2(1+\gamma)} \\ b &= \frac{f_1 - f_0 - ah_1^2}{h_1} \end{aligned}\right\} \quad \ldots (3.26)$$

After computing a, b, and c, we solve for the root of $av^2 + bv + c = 0$ by the quadratic formula, choosing the root nearest to the middle point x_0.

This value is

$$\text{root} = x_0 - \frac{2c}{b \pm \sqrt{b^2 - 4ac}} \qquad \ldots (3.27)$$

with the sign in the denominator taken to give the largest absolute value of denominator (that is, if $b > 0$, choose plus; if $b < 0$, choose minus; $b = 0$, choose either). The reason for using this somewhat unusual form of the quadratic formula is to make the next iteration closer to the root.

We take the root of the polynomial as one of a set of three points, for the next approximation. Taking these three points that are most closely spaced, *i.e.*, if the root is to the right of x_0, take x_0, x_1, and the root; if to the left, take x_0, x_2, and the root. We always reset the subscripts to make x_0 be the middle of the three values.

Example 3.20: One root of the equation

$$3x + \sin x - e^x = 0$$

lies between 0 and 1. Find that root accurate to four significant digits.

Solution: Let $x_0 = 0.5$ $\quad f(x_0) = 0.3307$ $\quad h_1 = 0.5$

$x_1 = 1.0$ $\quad f(x_1) = 1.1232$ $\quad h_2 = 0.5$

$x_2 = 0.0$ $\quad f(x_2) = -1$ $\quad \gamma = 1.0$

Then $$a = \frac{1.0 \times 1.1232 - 0.3307 \times 2.0 + (-1)}{1.0 \times (0.5)^2 \times 2.0} = -1.0764$$

$$b = \frac{1.1232 - 0.3307 - (-1.0764)(0.5)^2}{0.5} = 2.1232$$

$$c = 0.3307$$

$$root = 0.5 - \frac{2.0 \times 0.3307}{2.1232 + \sqrt{(2.1232)^2 - 4(-1.0764)(0.3307)}} = 0.3549$$

Therefore after first iteration, the approximation to root is 0.354914. Therefore, for next iteration, we have $x_0 = 0.354914$, $x_1 = 0.5$, $x_2 = 0.0$.

The following table shows the progress of Muller's method.

Table 3.3: Successive iterations of Muller's method

x_0	x_1	x_2	a	b	c	*root*
0.5	1.0	0.0	-1.0764	2.1232	0.3307	0.3549
0.3549	0.5	0.0	-0.8083	2.4918	-0.0138	0.3605
0.3605	0.5	0.3549	-0.9471	2.5014	0.0001	0.3604
0.3604	0.3605	0.3549	-0.8911	2.5018	0.0000	0.3604

Therefore, root correct to four significant digits is 0.3604.

Example 3.21: One root of the equation

$$\cos x - e^x = 0$$

lies between –1 and 1. Find that root accurate to three decimal places.

Solution: Let $x_0 = 0.0$ $\quad f(x_0) = 1.0$ $\quad h_1 = 1.0$

$x_1 = -1.0$ $\quad f(x_1) = -2.1780$ $\quad h_2 = 1.0$

$x_2 = 1.0$ $\quad f(x_2) = 0.9081$ $\quad \gamma = 1.0$

Then $$a = \frac{1.0 \times 1.0 - 1.0 \times 2.0 + 0.9081}{1.0 \times (1.0)^2 \times 2.0} = -1.6350$$

$$b = \frac{-2.1780 - 1.0 - (-1.6350)(1.0)^2}{0.5} = -1.5431$$

$$c = 1.0$$

$$root = 0 - \frac{2.0 \times 1.0}{-1.5431 + \sqrt{(-1.5431)^2 - 4(-1.6350)(1.0)}} = 0.4415$$

Therefore after first iteration, the approximation to root is 0.4415. Therefore, for next iteration, we have $x_0 = 0.4415$, $x_1 = 1.0$, $x_2 = 0.0$.

The following table shows the progress of Muller's method.

Table 3.4: Successive iterations of Muller's method

x_0	x_1	x_2	a	b	c	*root*
0.0	1.0	–1.0	–1.6350	–1.5431	1.0	0.4415
0.4415	1.0	0.0	–2.5170	–2.8836	0.2175	0.5125
0.5125	1.0	0.4415	–2.9729	–3.0513	0.0158	0.5177
0.5177	1.0	0.5125	–3.0515	–3.0422	0.0	0.5178

Since last two iterations produce no change at first three decimal places, therefore, we take root correct to three decimal places as 0.517.

Algorithm 3.10: Muller's Method

Here $f(x)$ is the given function, a_1 and b_1 are two initial approximations, *epsilon* is the prescribed tolerance.

```
Begin
   read: a1, b2, epsilon
   set x2 = a1
   set x0 = (a1+b1)/2
   set x1 = b1
   set xrp = 0
   do
      set h1 = x1 - x0, h2 = x0 - x2
      set γ = h2/h1
```

```
        set f0 = f(x0)
        set f1 = f(x1)
        set f2 = f(x2)
        set c = f0
        set a = (γf1-f0(1+γ)+f2)/(γh1^2(1+γ))
        set b = (f1-f0-ah1^2)/h1
        if ( b > 0 ) then
            set xrn = x0-2c/(b+√(b^2-4ac))
        else
            set xrn = x0-2c/(b-√(b^2-4ac))
        endif
        set relerror = |(xrp- xrn)/xrn|
        set xrp = xrn
        if ( xrn > x0 )then
            set x2 = x0, x0 = xrn
        else
            set x1 = x0, x0 = xrn
        endif
    while ( relerror > epsilon )
    print: "Approximate root = ", xrn
End.
```

3.14 QUOTIENT DIFFERENCE METHOD

Quotient difference method is another method for finding the roots of the polynomials and is very efficient one. To illustrate this, consider n^{th} degree polynomial

$$P_n(x) = a_nx^n + a_{n-1}x^{n-1} + a_{n-2}x^{n-2} + \ldots + a_1x + a_0 \qquad \ldots (3.28)$$

From this polynomial, we form array q and e terms, starting the tableau by calculating a first row of q's and second row of e's:

$$\left.\begin{aligned} q^{(0)} &= -\frac{a_{n-1}}{a_n} && \text{All other } q\text{'s are zero.} \\ e^{(i)} &= \frac{a_{n-i-1}}{a_{n-i}} && i = 1, 2, 3, \ldots, n-1. \\ e^{(0)} &= e^{(n)} = 0 \end{aligned}\right\} \qquad \ldots (3.29)$$

The start of the array is

$e^{(0)}$	$q^{(0)}$	$e^{(1)}$	$q^{(1)}$	$e^{(2)}$	$q^{(2)}$	...	$e^{(n-1)}$	$q^{(n-1)}$	$e^{(n)}$
	$\frac{-a_{n-1}}{a_n}$		0		0	...		0	
0		$\frac{a_{n-2}}{a_{n-1}}$		$\frac{a_{n-3}}{a_{n-2}}$			$\frac{a_0}{a_1}$		0

A new row of q's is computed by the equation

$$\text{New } q^{(i)} = e^{(i+1)} - e^{(i)} + q^{(i)} \qquad \ldots (3.30)$$

using terms from the e and q rows in the array. Note that this algorithm is "e to the right minus e to left plus q above".

A new row of e's is now computed by the equation

$$\text{New } e^{(i)} = \left(\frac{q^{(i)}}{q^{(i-1)}}\right) e^{(i)} \qquad \ldots (3.31)$$

"q to right over q to left times e above".

By continuing to compute rows of q's and then e's until all the e-values lies within the given tolerance, *i.e.*, prescribed precision. When this occurs, the q-values assume the values of the roots.

Example 3.22: Find all the real roots, correct to three decimal places, of the polynomial

$$128x^4 - 256x^3 + 160x^2 - 32x + 1 = 0$$

Solution: The initial values of e's and q's are

$e(0) = e(4) = 0$ $\quad$ $e(1) = -0.625$ $\quad$ $e(2) = -0.2$ $\quad$ $e(3) = -0.0312$

$q(0) = 2.0$ $\quad$ $q(1) = q(2) = q(3) = 0$

The results of different iterations are shown in Table 2.5.

Table 3.5: Successive iterations of Quotient-Difference method

$e^{(0)}$	$q^{(0)}$	$e^{(1)}$	$q^{(1)}$	$e^{(2)}$	$q^{(2)}$	$e^{(3)}$	$q^{(3)}$	$e^{(4)}$
0.0000	1.3750	-0.1932	0.4250	-0.0794	0.1688	-0.0058	0.0312	0.0000
0.0000	1.1818	-0.0881	0.5388	-0.0357	0.2424	-0.0009	0.0370	0.0000
0.0000	1.0937	-0.0476	0.5911	-0.0168	0.2772	0.0001	0.0379	0.0000
0.0000	1.0462	-0.0179	0.6220	-0.0037	0.2938	0.0000	0.0380	0.0000
0.0000	1.0179	-0.0117	0.6423	-0.0017	0.3017	0.0000	0.0381	0.0000
0.0000	1.0000	-0.0079	0.6565	-0.0008	0.3055	0.0000	0.0381	0.0000
0.0000	0.9883	-0.0054	0.6665	-0.0004	0.3072	0.0000	0.0381	0.0000
0.0000	0.9804	-0.0038	0.6736	-0.0002	0.3080	0.0000	0.0381	0.0000
0.0000	0.9749	-0.0027	0.6786	-0.0001	0.3084	0.0000	0.0381	0.0000
0.0000	0.9712	-0.0019	0.6823	0.0000	0.3085	0.0000	0.0381	0.0000
0.0000	0.9685	-0.0013	0.6848	0.0000	0.3086	0.0000	0.0381	0.0000
0.0000	0.9666	-0.0010	0.6867	0.0000	0.3086	0.0000	0.0381	0.0000
0.0000	0.9653	-0.0007	0.6880	0.0000	0.3086	0.0000	0.0381	0.0000
0.0000	0.9643	-0.0005	0.6889	0.0000	0.3087	0.0000	0.0381	0.0000

After 14th iteration, all values of e's are within the prescribed tolerance. Hence, the current value of q's gives the root of the given equation as 0.964, 0.689, 0.309, 0.038 (rounded to three significant digits).

Algorithm 3.11: Quotient-Difference Method

Here array *a* represent the coefficients of the given polynomial of degree *n* of type

$a_n x^n + a_{n-1} x^{n-1} + a_{n-2} x^{n-2} + \ldots + a_1 x + a_0$

and *epsilon* is the prescribed tolerance.

```
Begin
    read: n
    for i = n to 0 by -1 do
        read: a_i
    endfor
    read: epsilon
    set q_0 = -a_{n-1}/a_n
    for i = 1 to (n-1) by 1 do
        set q_i = 0
    endfor
    set e_0 = 0, e_n = 0
    for i = 1 to (n-1) by 1 do
        set e_i = a_{n-i-1}/a_{n-i}
    endfor
    do
        for i = 0 to (n-1) by 1 do
            set q_i = e_{i+1} - e_i + q_i
        endfor
        for i = 1 to (n-1) by 1 do
            set e_i = (q_i/q_{i-1}) e_i
        endfor
        set maxerror = |e_0|
        for i = 1 to (n-1) by 1 do
            if ( |e_i| > maxerror ) then
                set maxerror = |e_i|
            endif
        endfor
    while ( maxerror > epsilon ).
    print: "Roots of polynomial are"
    for i = 0 to n by 1 do
        print: q_i
    endfor
End.
```

3.15 CONVERGENCE OF ITERATIVE METHODS

The order or the rate at which the error between successive approximations to the root decreases judges convergence of an iterative method. The order of convergence of an iterative method is defined in terms of errors e_i and e_{i+1} in successive approximations.

An iterative method is said to be k^{th} order convergent if k is the largest integer such that

$$\lim_{i\to\infty} \frac{e_{i+1}}{e_i} \le M \qquad \ldots (3.32)$$

where M is a finite number.

In other words, the error in any step is proportional to the kth power of the error in the preceding step. Physically, the kth order convergence means that in each iteration, the number of significant digits in each approximation increases k times. For example, if the value of the root is good to n significant digits in the ith iteration ($| e_i | \le 10^{-n}$), then

$$| e_{i+1} | \le M \, | e_i^k | \qquad \ldots (3.33)$$

$$\le M \times 10^{-nk}$$

3.15.1 Convergence of Bisection Method

In each iteration, the original interval is divided into half, thus bracketing the root in new interval. If we take the midpoints of the successive intervals to be the approximations of the root, the one half of the current interval is the upper bound to the error. Thus if e_i is the error in i^{th} iteration, then the error e_{i+1} in the $(i+1)^{th}$ iteration is given as

$$e_{i+1} = 0.5e_i$$

Thus, the bisection method is linear order convergent.

3.15.2 Convergence of Secant Method

The general formula for the secant method is

$$x_{i+1} = \frac{x_{i-1} f(x_i) - x_i f(x_{i-1})}{f(x_i) - f(x_{i-1})}$$

For the purpose analysis, can be written as

$$x_{i+1} = x_i - \frac{(x_i - x_{i-1}) f(x_i)}{f(x_i) - f(x_{i-1})}$$

Suppose that y is the root of the equation $f(x) = 0$, $e_{i+1} = x_{i+1} - y$, $e_i = x_i - y$ and $e_{i-1} = x_{i-1} - y$, then the above formula can be written as

$$y + e_{i+1} = y + e_i - \frac{[(y + e_i) - (y - e_{i-1})] f(y + e_i)}{f(y + e_i) - f(y + e_{i-1})}$$

$$\Rightarrow \quad e_{i+1} = e_i - \frac{(e_i - e_{i-1}) f(y + e_i)}{f(y + e_i) - f(y + e_{i-1})} \qquad \ldots (3.34)$$

Expanding $f(y+e_i)$ and $f(y+e_{i-1})$ in Taylor's series around y up to third term, we get

$$f(y+e_i) = f(y) + e_i f^1(y) + \frac{e_i^2}{2} f^2(y)$$

$$f(y+e_{i-1}) = f(y) + e_{i-1} f^1(y) + \frac{e_{i-1}^2}{2} f^2(y)$$

The denominator is simplified as

$$f(y+e_i) - f(y+e_{i-1}) = \left[f(y) + e_i f^1(y) + \frac{e_i^2}{2} f^2(y)\right] - \left[f(y) + e_{i-1} f^1(y) + \frac{e_{i-1}^2}{2} f^2(y)\right]$$

$$= (e_i - e_{i-1}) f^1(y) + \frac{e_i^2 - e_{i-1}^2}{2} f^2(y)$$

Substituting the values of $f(y+e_i)$ and the denominator in Eq. (3.34), we get

$$e_{i+1} = e_i - \frac{(e_i - e_{i-1})\left[f(y) + e_i f^1(y) + \frac{e_i^2}{2} f^2(y)\right]}{(e_i - e_{i-1}) f^1(y) + \frac{e_i^2 - e_{i-1}^2}{2} f^2(y)}$$

Cancelling factor $(e_i - e_{i-1})$ from nominator and denominator, we get

$$e_{i+1} = e_i - \frac{f(y) + e_i f^1(y) + \frac{e_i^2}{2} f^2(y)}{f^1(y) + \frac{e_i + e_{i-1}}{2} f^2(y)}$$

As $f(y) = 0$, since y is a root of the equation $f(x) = 0$, therefore

$$e_{i+1} = e_i - \frac{e_i f^1(y) + \frac{e_i^2}{2} f^2(y)}{f^1(y) + \frac{e_i + e_{i-1}}{2} f^2(y)}$$

Dividing nominator and denominator by $f^1(y)$, we get

$$e_{i+1} = e_i - \frac{e_i + \frac{e_i^2}{2}\frac{f^2(y)}{f^1(y)}}{1 + \frac{e_i + e_{i-1}}{2}\frac{f^2(y)}{f^1(y)}} \quad \Rightarrow \quad e_{i+1} = e_i - \left[e_i + \frac{e_i^2}{2}\frac{f^2(y)}{f^1(y)}\right]\left[1 + \frac{e_i + e_{i-1}}{2}\frac{f^2(y)}{f^1(y)}\right]^{-1}$$

Using Binomial expansion, and retaining first two terms, we get

$$e_{i+1} = e_i - \left[e_i + \frac{e_i^2}{2}\frac{f^2(y)}{f^1(y)}\right]\left[1 - \frac{e_i + e_{i-1}}{2}\frac{f^2(y)}{f^1(y)}\right]$$

$$e_{i+1} = e_i - \left[e_i - \frac{e_i(e_i + e_{i-1})}{2} \frac{f^2(y)}{f^1(y)} + \frac{e_i^2}{2} \frac{f^2(y)}{f^1(y)} + \frac{e_i^2(e_i + e_{i-1})}{4} \left(\frac{f^2(y)}{f^1(y)} \right)^2 \right]$$

$$e_{i+1} = e_i e_{i-1} \frac{f^2(y)}{f^1(y)} + \text{terms containing } e_i e_{i-1},\ e_i^2,\ e_i^3, \text{ and } e_i^2 e_{i-1}$$

If e_i and e_{i-1} are very small, then ignoring the terms involving $e_i e_{i-1}$, e_i^2, e_i^3, and $e_i^2 e_{i-1}$, we get

$$e_{i+1} = e_i e_{i-1} \frac{f^2(y)}{f^1(y)}$$

which can be written as

$$e_{i+1} = e_i e_{i-1} K \qquad \text{... (3.35)}$$

where $K = \dfrac{f^2(y)}{f^1(y)}$, and is a constant.

In order to find the order of convergence, we now try to find some number m such that

$$e_{i+1} = A e_i^m \qquad \text{... (3.36)}$$

From Eq. (3.28) we can write

$$e_i = A e_{i-1}^m \quad \Rightarrow e_{i-1} = \left(\frac{e_i}{A} \right)^{\frac{1}{m}}$$

Substituting the values of e_{i+1} and e_{i-1} in Eq. (3.35), we get

$$A e_i^m = e_i^m \times \left(\frac{e_i}{A} \right)^{\frac{1}{m}} \times K \quad \Rightarrow e_i^m = K \times A^{-(1+\frac{1}{m})} \times e_i^{1+\frac{1}{m}} \qquad \text{... (3.37)}$$

Comparing the powers of e_i on both sides of Eq. (3.37), we get

$$m = 1 + \frac{1}{m} \quad \Rightarrow \quad m^2 - m - 1 = 0$$

which gives

$$m_{1,2} = \frac{(1 \pm \sqrt{5})}{2} \qquad \text{... (3.38)}$$

Taking the positive sign only, we have $m = 1.618$.

Hence, the rate of convergence of the secant method is 1.618.

3.15.3 Convergence of Newton-Raphson Method

The general formula for the *Newton-Raphson* method is

$$x_{i+1} = x_i - \frac{f(x_i)}{f^1(x_i)}$$

Suppose that y is the root of the equation $f(x) = 0$, e_i and e_{i+1} are the errors i^{th} and $(i+1)^{th}$ iteration then we can rewrite the above formula as

$$y + e_{i+1} = y + e_i - \frac{f(y+e_i)}{f^1(y+e_i)} \qquad \ldots (3.39)$$

Expanding $f(y+e_i)$ and $f^1(y+e_i)$ in Taylor's series around y up to third term, we get

$$y + e_{i+1} = y + e_i - \frac{f(y) + e_i f^1(y) + \frac{e_i^2}{2} f^2(y) + \frac{e_i^3}{6} f^3(y) + \ldots}{f^1(y) + e_i f^2(y) + \frac{e_i^2}{2} f^3(y) + \frac{e_i^3}{6} f^4(y) + \ldots}$$

As $f(y) = 0$, since y is a root of the equation $f(x) = 0$, therefore

$$e_{i+1} = e_i - \frac{e_i f^1(y) + \frac{e_i^2}{2} f^2(y) + \frac{e_i^3}{6} f^3(y) + \ldots}{f^1(y) + e_i f^2(y) + \frac{e_i^2}{2} f^3(y) + \frac{e_i^3}{6} f^4(y) + \ldots}$$

Dividing nominator and denominator by $f^1(y)$, we get

$$e_{i+1} = e_i - \frac{e_i + \frac{e_i^2}{2}\frac{f^2(y)}{f^1(y)} + \frac{e_i^3}{6}\frac{f^3(y)}{f^1(y)} + \ldots}{1 + e_i \frac{f^2(y)}{f^1(y)} + \frac{e_i^2}{2}\frac{f^3(y)}{f^1(y)} + \frac{e_i^3}{6}\frac{f^4(y)}{f^1(y)} + \ldots}$$

If $\frac{f^2(y)}{f^1(y)} > 0$ and e_i is very small, then the terms involving e_i^2, e_i^3,... can be ignored to obtain

$$e_{i+1} = e_i - \frac{e_i + \frac{e_i^2}{2}\frac{f^2(y)}{f^1(y)}}{1 + e_i \frac{f^2(y)}{f^1(y)}} \quad \Rightarrow \quad e_{i+1} = e_i - e_i\left[1 + \frac{e_i}{2}\frac{f^2(y)}{f^1(y)}\right]\left[1 + e_i \frac{f^2(y)}{f^1(y)}\right]^{-1}$$

Using Binomial expansion, and retaining first two terms, we get

$$e_{i+1} = e_i - e_i\left[1 + \frac{e_i}{2}\frac{f^2(y)}{f^1(y)}\right]\left[1 - e_i \frac{f^2(y)}{f^1(y)}\right]$$

$$e_{i+1} = e_i - e_i\left[1 - e_i\frac{f^2(y)}{f^1(y)} + \frac{e_i}{2}\frac{f^2(y)}{f^1(y)} - \frac{e_i^2}{2}\left(\frac{f^2(y)}{f^1(y)}\right)^2\right]$$

Again ignoring the terms involving e_i^2, we get

$$e_{i+1} = e_i - e_i\left[1 - \frac{e_i}{2}\frac{f^2(y)}{f^1(y)}\right] \Rightarrow e_{i+1} = \frac{e_i^2}{2}\frac{f^2(y)}{f^1(y)} \qquad \ldots (3.40)$$

$$\Rightarrow e_{i+1} = Ce_i^2 \quad \text{where } C = \frac{f^2(y)}{f^1(y)}$$

Hence, the rate of convergence of the Newton-Raphson method is 2.

3.15.4 Convergence of Method of Successive Approximations

The general formula for successive approximation is

$x_{i+1} = g(x_i)$

Suppose that y is the root of the equation $f(x) = 0$, e_i and e_{i+1} are the errors i^{th} and $(i+1)^{th}$ iteration then we can rewrite the above formula as

$y + e_{i+1} = g(y+e_i)$

Expanding $g(y+e_i)$ in Taylor's series around y, we get

$$y + e_{i+1} = g(y) + e_i g^1(y) + \frac{e_i^2}{2}g^2(y) + \frac{e_i^3}{6}g^3(y) + \ldots$$

If e_i is very small, then the terms involving e_i^2, e_i^3,... can be ignored to obtain

$$e_{i+1} = e_i g^1(y) - g(y) - y$$

Since y is a root of $f(x) = 0$, therefore $g(y) = y$, and the above equation reduces to

$$e_{i+1} = e_i g^1(y) \qquad \ldots (3.42)$$

Thus, the successive approximation method has first order convergence, and converges only if $|g^1(y)| < 1$.

3.15.5 Convergence of False Position Method

Like secant method, the general formula for the false position method is

$$x_{i+1} = \frac{x_{i-1}f(x_i) - x_i f(x_{i-1})}{f(x_i) - f(x_{i-1})}$$

which, for the purpose analysis, can be written as shown overleaf.

$$x_{i+1} = x_i - \frac{f(x_i)}{f(x_i) - f(x_{i-1})}(x_i - x_{i-1}) \qquad \ldots (3.43)$$

In each iteration, we replace one of the approximations with the new approximation such that the root is bracketed in the new interval. When this is done, one end of the interval does not change, therefore we can think of x_{i-1} (which of course can be x_i) as a constant in (i+1)th iteration. Thus Eq. (3.43) is of the form

$$x_{i+1} = g(x_i)$$

where

$$g(x_i) = x_i - \frac{f(x_i)}{f(x_i) - F}(x_i - X) \qquad \ldots (3.44)$$

and we have used X and F to indicate that these are fixed quantities. Differentiating it we get

$$g^1(x_i) = \frac{f^1(x_i) \times F \times (x_i - X) - F \times (f(x_i) - F)}{[f(x_i) - F]^2} \qquad \ldots (3.45)$$

We see that

$$|g^1(x_i)| \le 1 \qquad \text{for all } x_i$$

Therefore, proceeding in the same way as for method of successive approximations, we can prove that false position method is linearly convergent.

3.16 SYSTEM OF NON-LINEAR EQUATIONS

The real roots of system of non-linear equations in several unknowns can be found either by Newton-Raphson method or by method of successive approximations. The solution of the system of non-linear equations basically represents the intersections points of the curves represented by these equations. We will give the outline of each method for two unknowns and three unknowns. The reader will have no difficulty in extending these methods to any number of unknowns.

3.16.1 Newton-Raphson Method

Let us consider the case of two non-linear equations in two unknowns of type

$$f_1(x, y) = 0 \qquad \ldots (3.46)$$

$$f_2(x, y) = 0 \qquad \ldots (3.47)$$

Now if (x_0, y_0) be the approximate values of roots, and h, k be the corrections, such that

$$x = x_0 + h$$

$$y = y_0 + k$$

Then Eqs. (3.46) and (3.47) becomes

$$f_1(x_0+h, y_0+k)=0 \qquad \ldots (3.48)$$

$$f_2(x_0+h, y_0+k)=0 \qquad \ldots (3.49)$$

Expanding Eq. (3.48) and (3.49) by Taylor's series for a function of two variables, we get

$$f_1(x_0+h, y_0+k) = f_1(x_0, y_0) + h\left.\frac{\partial f_1}{\partial x}\right|_{(x_0, y_0)} + k\left.\frac{\partial f_1}{\partial y}\right|_{(x_0, y_0)} = 0$$

$$f_2(x_0+h, y_0+k) = f_2(x_0, y_0) + h\left.\frac{\partial f_2}{\partial x}\right|_{(x_0, y_0)} + k\left.\frac{\partial f_2}{\partial y}\right|_{(x_0, y_0)} = 0$$

The above equation can be written as

$$h\left.\frac{\partial f_1}{\partial x}\right|_{(x_0, y_0)} + k\left.\frac{\partial f_1}{\partial y}\right|_{(x_0, y_0)} = -f_1(x_0, y_0) \qquad \ldots (3.50)$$

$$h\left.\frac{\partial f_2}{\partial x}\right|_{(x_0, y_0)} + k\left.\frac{\partial f_2}{\partial y}\right|_{(x_0, y_0)} = -f_2(x_0, y_0) \qquad \ldots (3.51)$$

Solving these equations for h and k, we get

$$h = \frac{\begin{vmatrix} -f_1(x_0, y_0) & \left.\frac{\partial f_1}{\partial y}\right|_{(x_0, y_0)} \\ -f_2(x_0, y_0) & \left.\frac{\partial f_2}{\partial y}\right|_{(x_0, y_0)} \end{vmatrix}}{D} \qquad k = \frac{\begin{vmatrix} \left.\frac{\partial f_1}{\partial x}\right|_{(x_0, y_0)} & -f_1(x_0, y_0) \\ \left.\frac{\partial f_2}{\partial x}\right|_{(x_0, y_0)} & -f_2(x_0, y_0) \end{vmatrix}}{D}$$

where $D = \begin{vmatrix} \left.\frac{\partial f_1}{\partial x}\right|_{(x_0, y_0)} & \left.\frac{\partial f_1}{\partial y}\right|_{(x_0, y_0)} \\ \left.\frac{\partial f_2}{\partial x}\right|_{(x_0, y_0)} & \left.\frac{\partial f_2}{\partial y}\right|_{(x_0, y_0)} \end{vmatrix} = \left.\frac{\partial f_1}{\partial x}\right|_{(x_0, y_0)} \times \left.\frac{\partial f_2}{\partial y}\right|_{(x_0, y_0)} - \left.\frac{\partial f_2}{\partial x}\right|_{(x_0, y_0)} \times \left.\frac{\partial f_1}{\partial y}\right|_{(x_0, y_0)}$

Additional corrections can be found by repeated applications of these formulas with the improved values of x and y substituted at each step. The process will terminate when the new approximations to x, y and z are within the prescribed tolerance.

However, if three non-linear equations in three unknowns are given as

$$f_1(x, y, z) = 0 \qquad \ldots (3.52)$$

$$f_2(x, y, z) = 0 \qquad \ldots (3.53)$$

$$f_3(x, y, z) = 0 \qquad \ldots (3.54)$$

Now if (x_0, y_0, z_0) be the approximate values of the roots, and h, k, l be the corrections, such that

$$x = x_0 + h$$
$$y = y_0 + k$$
$$z = z_0 + l$$

Then Eqs. (3.52), (3.53) and (3.54) becomes

$$f_1(x_0 + h, y_0 + k, z_0 + l) = 0 \quad \ldots (3.55)$$
$$f_2(x_0 + h, y_0 + k, z_0 + l) = 0 \quad \ldots (3.56)$$
$$f_3(x_0 + h, y_0 + k, z_0 + l) = 0 \quad \ldots (3.57)$$

Expanding Eqs. (3.55), (3.56) and (3.57) by Taylor's series for a function of two variables, we get

$$f_1(x_0 + h, y_0 + k, z_0 + l) = f_1(x_0, y_0, z_0) + h\left.\frac{\partial f_1}{\partial x}\right|_{(x_0, y_0)} + k\left.\frac{\partial f_1}{\partial y}\right|_{(x_0, y_0)} + l\left.\frac{\partial f_1}{\partial z}\right|_{(x_0, y_0)} = 0$$

$$f_2(x_0 + h, y_0 + k, z_0 + l) = f_2(x_0, y_0, z_0) + h\left.\frac{\partial f_2}{\partial x}\right|_{(x_0, y_0)} + k\left.\frac{\partial f_2}{\partial y}\right|_{(x_0, y_0)} + l\left.\frac{\partial f_2}{\partial z}\right|_{(x_0, y_0)} = 0$$

$$f_3(x_0 + h, y_0 + k, z_0 + l) = f_3(x_0, y_0, z_0) + h\left.\frac{\partial f_3}{\partial x}\right|_{(x_0, y_0)} + k\left.\frac{\partial f_3}{\partial y}\right|_{(x_0, y_0)} + l\left.\frac{\partial f_3}{\partial z}\right|_{(x_0, y_0)} = 0$$

The above equation can be written as

$$h\left.\frac{\partial f_1}{\partial x}\right|_{(x_0, y_0)} + k\left.\frac{\partial f_1}{\partial y}\right|_{(x_0, y_0)} + l\left.\frac{\partial f_1}{\partial z}\right|_{(x_0, y_0)} = -f_1(x_0, y_0, z_0) \quad \ldots (3.58)$$

$$h\left.\frac{\partial f_2}{\partial x}\right|_{(x_0, y_0)} + k\left.\frac{\partial f_2}{\partial y}\right|_{(x_0, y_0)} + l\left.\frac{\partial f_2}{\partial z}\right|_{(x_0, y_0)} = -f_2(x_0, y_0, z_0) \quad \ldots (3.59)$$

$$h\left.\frac{\partial f_3}{\partial x}\right|_{(x_0, y_0)} + k\left.\frac{\partial f_3}{\partial y}\right|_{(x_0, y_0)} + l\left.\frac{\partial f_3}{\partial z}\right|_{(x_0, y_0)} = -f_3(x_0, y_0, z_0) \quad \ldots (3.60)$$

We can solve these equations for h, k and l. Then additional corrections can be found by repeated applications of these formulas with the improved values of x, y and z substituted at each step. The process will terminate when the new approximations to x, y and z are within the prescribed tolerance.

Example 3.23: Given two non-linear equations in two unknowns as

$$x^2 - y^2 + 7 = 0$$
$$x - xy + 9 = 0$$

with $(x_0, y_0) = (3.5, 4.5)$ as the initial approximation. Find the solution of the above system correct to four significant digits.

Solution: As we want the solution correct to four significant digits, the iterative process will be terminated as soon as we find that there is no change in the first four significant digits in the successive approximations.

Let $f_1(x, y) = x^2 - y^2 + 7$ $\quad f_2(x, y) = x - xy + 9$

then $\frac{\partial f_1}{\partial x} = 2x \quad \frac{\partial f_1}{\partial y} = -2y \quad \frac{\partial f_2}{\partial x} = 1 - y \quad \frac{\partial f_2}{\partial y} = -x$

Iteration 1: With $(x_0, y_0) = (3.5, 4.5)$, we have

$$f_1 = (3.5)^2 - (4.5)^2 + 7 = -1 \qquad f_2 = 3.5 - 3.5\times 4.5 + 9 = -3.25$$

$$\frac{\partial f_1}{\partial x} = 2\times 3.5 = 7 \quad \frac{\partial f_1}{\partial y} = -2\times 4.5 = -9 \quad \frac{\partial f_2}{\partial x} = 1 - 4.5 = -3.5 \quad \frac{\partial f_2}{\partial y} = -3.5$$

$$D = \frac{\partial f_1}{\partial x}\times\frac{\partial f_2}{\partial y} - \frac{\partial f_2}{\partial x}\times\frac{\partial f_1}{\partial y} = 7\times(-3.5) - (-3.5)\times(-9) = -56$$

$$h = \frac{\begin{vmatrix} 1 & -9 \\ 3.25 & -3.5 \end{vmatrix}}{-56} = -0.4598 \quad \text{and} \quad k = \frac{\begin{vmatrix} 7 & 1 \\ -3.5 & 3.25 \end{vmatrix}}{-56} = -0.4688$$

Hence

$$x_1 = x_0 + h = 3.5 - 0.4598 = 3.0402$$
$$y_1 = y_0 + k = 4.5 - 0.4688 = 4.0312$$

Thus we obtain $(x_1, y_1) = (3.0402, 4.0312)$ as the next approximation.

Iteration 2: Now starting with $(x_1, y_1) = (3.0402, 4.0312)$, we have

$$f_1 = (3.0402)^2 - (4.0312)^2 + 7 = -0.0078 \qquad f_2 = 3.0402 - 3.0402\times 4.0312 + 9 = -0.2155$$

$$\frac{\partial f_1}{\partial x} = 2\times 3.0402 = 6.0804 \qquad \frac{\partial f_1}{\partial y} = -2\times 4.0312 = -8.0624$$

$$\frac{\partial f_2}{\partial x} = 1 - 4.0312 = -3.0312 \qquad \frac{\partial f_2}{\partial y} = -3.0402$$

$$D = \frac{\partial f_1}{\partial x}\times\frac{\partial f_2}{\partial y} - \frac{\partial f_2}{\partial x}\times\frac{\partial f_1}{\partial y} = 6.0804\times(-3.0402) - (-3.0312)\times(-8.0624) = -42.9213$$

$$h = \frac{\begin{vmatrix} 0.0078 & -8.0624 \\ 0.2155 & -3.0402 \end{vmatrix}}{-42.9213} = -0.0399 \;\&\; k = \frac{\begin{vmatrix} 6.0804 & 0.0078 \\ -3.0312 & 0.2155 \end{vmatrix}}{-42.9213} = -0.0311$$

Hence

$x_2 = x_1 + h = 3.0402 - 0.0399 = 3.0003$
$y_2 = y_1 + h = 4.0312 - 0.0311 = 4.0001$

Thus we obtain $(x_2, y_2) = (3.0003, 4.0001)$ as the next approximation.

Iteration 3: Now starting with $(x_1, y_1) = (3.0402, 4.0312)$, we have

$f_1 = (3.0003)^2 - (4.0001)^2 + 7 = 0.001$ $\qquad f_2 = 3.0003 - 3.\ 0003 \times 4.0001 + 9 = -0.0012$

$\frac{\partial f_1}{\partial x} = 2 \times 3.0003 = 6.0006$ $\qquad \frac{\partial f_1}{\partial y} = -2 \times 4.0001 = -8.0002$

$\frac{\partial f_2}{\partial x} = 1 - 4.0001 = -3.0001$ $\qquad \frac{\partial f_2}{\partial y} = -3.0003$

$$D = \frac{\partial f_1}{\partial x} \times \frac{\partial f_2}{\partial y} - \frac{\partial f_2}{\partial x} \times \frac{\partial f_1}{\partial y} = 6.0006 \times (-3.0003) - (-3.0001) \times (-8.0002) = -42.005$$

$$h = \frac{\begin{vmatrix} -0.001 & -8.0002 \\ 0.0012 & -3.0003 \end{vmatrix}}{-42.005} = -0.0003 \;\&\; k = \frac{\begin{vmatrix} 6.0006 & -0.001 \\ -3.0001 & 0.0012 \end{vmatrix}}{-42.005} = -0.0001$$

Hence

$x_3 = x_2 + h = 3.0003 - 0.0003 = 3.0000$
$y_3 = y_2 + h = 4.0001 - 0.0001 = 4.0000$

Thus we obtain $(x_3, y_3) = (3.0000, 4.0000)$ as the next approximation.

If you compare the solution obtained in 3rd iteration with that of 2nd iteration, we observe that there is no difference at first four significant positions, thus we take $(x, y) = (3.0, 4.0)$ as the desired solution.

3.16.2 Method of Successive Approximations

Let us consider the case of two non-linear equations in two unknowns of type

$$f_1(x, y) = 0$$
$$f_2(x, y) = 0$$

This method can be applied if these equations can be written in the form

$$x = g_1(x, y) \qquad \text{... (3.61)}$$
$$y = g_2(x, y) \qquad \text{... (3.62)}$$

Now if (x_0, y_0) be the initial approximate values of the roots, then the next approximations are computed as

$$\left.\begin{aligned} x_1 &= g_1(x_0, y_0) \\ y_1 &= g_2(x_1, y_0) \end{aligned}\right\} \text{1st approximations}$$

$$\left.\begin{aligned} x_2 &= g_1(x_1, y_1) \\ y_2 &= g_2(x_2, y_1) \end{aligned}\right\} \text{2nd approximations}$$

$$\left.\begin{aligned} x_3 &= g_1(x_2, y_2) \\ y_3 &= g_2(x_3, y_2) \end{aligned}\right\} \text{3rd approximations}$$

$$\vdots$$

Continue iterating till the error in the new approximations is within the prescribed tolerance.

However, if three non-linear equations in three unknowns are given as

$$f_1(x, y, z) = 0$$
$$f_2(x, y, z) = 0$$
$$f_3(x, y, z) = 0$$

which can be written in the form

$$x = g_1(x, y, z)$$
$$y = g_2(x, y, z)$$
$$z = g_3(x, y, z)$$

Now if (x_0, y_0, z_0) be the initial approximate values of the roots, then the next approximations are computed as

$$\left.\begin{aligned} x_1 &= g_1(x_0, y_0, z_0) \\ y_1 &= g_2(x_1, y_0, z_0) \\ z_1 &= g_3(x_1, y_1, z_0) \end{aligned}\right\} \text{1st approximation}$$

$$\left.\begin{aligned} x_2 &= g_1(x_1, y_1, z_1) \\ y_2 &= g_2(x_2, y_1, z_1) \\ z_2 &= g_3(x_2, y_2, z_1) \end{aligned}\right\} \text{2nd approximation}$$

$$\left.\begin{aligned} x_3 &= g_1(x_2, y_2, z_2) \\ y_3 &= g_2(x_3, y_2, z_2) \\ z_3 &= g_3(x_3, y_3, z_2) \end{aligned}\right\} \text{3rd approximation}$$

$$\vdots$$

Continue iterating till the error in the new approximations is within the prescribed tolerance.

Example 3.24: Given two non-linear equations in two unknowns as

$$x + 3\log_{10} x - y^2 = 0$$
$$2x^2 - xy - 5x + 1 = 0$$

with $(x_0, y_0) = (3.4, 2.2)$ as the initial approximation. Find the solution of the above system correct to three significant digits.

Solution: As we want the solution correct to three significant digits, the iterative process will be terminated as soon as we find that there is no change in the first three significant digits in the successive approximations.

Let $f_1(x, y) = 2x^2 - xy - 5x + 1$ and $f_2(x, y) = x + 3\log_{10} x - y^2$

Rewrite these equations as

$$x = \sqrt{\frac{x(y+5)-1}{2}} = g_1(x, y) \quad \text{and} \quad y = \sqrt{x + 3\log_{10} x} = g_2(x, y)$$

Iteration 1: Taking $(x_0, y_0) = (3.4, 2.2)$ as the initial approximation, we proceed as

$$x_1 = g_1(x_0, y_0) = \sqrt{\frac{3.4 \times (2.2 + 5) - 1}{2}} = 3.4263$$

$$y_1 = g_2(x_1, y_0) = \sqrt{3.4263 + 3 \times \log_{10} 3.4263} = 2.2429$$

Thus we obtain $(x_1, y_1) = (3.4263, 2.2429)$ as the next approximation.

Iteration 2: Now starting with $(x_1, y_1) = (3.4263, 2.2429)$, we obtain

$$x_2 = g_1(x_1, y_1) = \sqrt{\frac{3.4263 \times (2.2429 + 5) - 1}{2}} = 3.4508$$

$$y_2 = g_2(x_2, y_1) = \sqrt{3.4508 + 3 \times \log_{10} 3.4508} = 2.2505$$

Thus we obtain $(x_2, y_2) = (3.4508, 2.2505)$ as the next approximation.

Iteration 3: Now starting with $(x_2, y_2) = (3.4508, 2.2505)$, we obtain

$$x_3 = g_1(x_2, y_2) = \sqrt{\frac{3.4508 \times (2.2505 + 5) - 1}{2}} = 3.4655$$

$$y_3 = g_2(x_3, y_1) = \sqrt{3.4655 + 3 \times \log_{10} 3.4655} = 2.2549$$

Thus we obtain $(x_3, y_3) = (3.4655, 2.2549)$ as the next approximation.

Iteration 4: Now starting $(x_3, y_3) = (3.4655, 2.2549)$, we obtain

$$x_4 = g_1(x_3, y_3) = \sqrt{\frac{3.4655 \times (2.2549 + 5) - 1}{2}} = 3.4743$$

$$y_4 = g_2(x_4, y_3) = \sqrt{3.4743 + 3 \times \log_{10} 3.4743} = 2.2576$$

Thus we obtain $(x_4, y_4) = (3.4743, 2.2576)$ as the next approximation.

Iteration 5: Now starting $(x_4, y_4) = (3.4743, 2.2576)$, we obtain

$$x_5 = g_1(x_4, y_4) = \sqrt{\frac{3.4743 \times (2.2576 + 5) - 1}{2}} = 3.4796$$

$$y_5 = g_2(x_5, y_4) = \sqrt{3.4796 + 3 \times \log_{10} 3.4796} = 2.2592$$

Thus we obtain $(x_5, y_5) = (3.4796, 2.2592)$ as the next approximation.

Iteration 6: Now starting $(x_5, y_5) = (3.4796, 2.2592)$, we obtain

$$x_6 = g_1(x_5, y_5) = \sqrt{\frac{3.4796 \times (2.2592 + 5) - 1}{2}} = 3.4826$$

$$y_6 = g_2(x_6, y_5) = \sqrt{3.4826 + 3 \times \log_{10} 3.4826} = 2.2602$$

Thus we obtain $(x_6, y_6) = (3.4826, 2.2602)$ as the next approximation.

Iteration 7: Now starting $(x_6, y_6) = (3.4826, 2.2602)$, we obtain

$$x_7 = g_1(x_6, y_6) = \sqrt{\frac{3.4826 \times (2.2602 + 5) - 1}{2}} = 3.4846$$

$$y_7 = g_2(x_7, y_6) = \sqrt{3.4846 + 3 \times \log_{10} 3.4846} = 2.2608$$

Thus we obtain $(x_7, y_7) = (3.4846, 2.2608)$ as the next approximation.

Iteration 8: Now starting $(x_7, y_7) = (3.4846, 2.2608)$, we obtain

$$x_8 = g_1(x_7, y_7) = \sqrt{\frac{3.4846 \times (2.2608 + 5) - 1}{2}} = 3.4858$$

$$y_8 = g_2(x_8, y_7) = \sqrt{3.4858 + 3 \times \log_{10} 3.4858} = 2.2611$$

Thus we obtain $(x_8, y_8) = (3.4858, 2.2611)$ as the next approximation.

Observe that there is no change in first three significant positions in iteration number 6, 7, and 8. Thus we take $(x, y) = (3.48, 2.26)$ as the desired solution.

EXERCISE . . .

1. What are the characteristics of iterative methods?
2. What are different parameters to be taken into account while selecting an iterative method?
3. Describe the *bisection* method. Suppose that one real root of an equation lies in the interval [1.5, 2.2] and the permitted tolerance in the solution is 0.0001, how much iteration will be performed.
4. Find the solution of $x^4 + 2x^2 - 16x + 5 = 0$ with tolerance 0.0001using *bisection* method.
5. One real root of the equation $e^{-x} - x = 0$ lies between 0 and 1. Find the root with tolerance 0.001 using bisection method. Can you guess the number of iterations required? 0.5674, 10
6. Describe the *Regula-falsi* method and derive its iterative formula. It is given that one real root of the equation $x^6 - x^4 - x^3 - 1 = 0$ lies between 1.4 and 1.5. Find that root correct to three significant digits.
7. The iterative formula for *Secant method* and *Regula-falsi* method is same, then why their order of convergence is different?
8. Describe the *Newton-Raphson* method and derive its iterative formula analytically. Is the convergence is always assured? One root of the equation $\sin x - x - 2 = 0$ lies near $x = -2.5$. Find that root with tolerance 0.001.
9. Describe the *successive approximation* method and derive its iterative formula. What is the condition for its convergence? One root of the equation $3x - 2e^x = 0$ lies between 0 and 1. Find that root with tolerance 0.001.
10. If $f(x) = x^6 - x^4 - x^3 - 1$. Find the value of $f^1(2)$ using synthetic division.
11. Describe the *Bairstow's* method. Can this method be used to find the solution of transcendental equation?
12. Solve the polynomial equation $x^4 - 8x^3 - 39x^2 - 62x + 50 = 0$ using *Bairstow's* method. The permitted tolerance is 0.001. (11.7404, 0.5754, -2.1580 +- 1.6566)
13. Prove that *Newton-Raphson* method is second order convergent.

14. Prove that the order of convergence of *secant* method is 1.618.
15. Describe the *Newton-Raphson* method to solve a system of non-linear equations.
16. Evaluate the value of $\sqrt{13}$ correct to three decimal places *using Newton-Raphson* method.
17. Describe the *successive approximation* method to solve a system of non-linear equations.
18. The function $f(x) = x^2 e^{-\frac{1}{2}x} - 1$ is known to have two real roots, one lie in the interval [0,2] and the second in the interval [2,10]. Use Muller's method to find these roots accurate to three decimal places.
19. Consider the function $f(x) = \frac{1}{5}(e^{-x} - x^3)$. Show that this function has at least one roots in the interval [–1,1].
20. Solve the polynomial equation $x^3 - x^2 - 2x + 1 = 0$ using *Quotient-Difference* method. The permitted tolerance is 0.001.
21. Find the roots of the polynomial

    $$x^3 - 5x^2 - 2x + 24 = 0$$

    correct to four significant digits by *Quotient-Difference* method.
22. Solve the following system of non-linear equation

    $$x^2 + y^2 + z^2 = 9$$
    $$x + y - z^2 = 0$$
    $$xyz = 3$$

    using Newton-Raphson method. Take $x_0 = 2.5$, and $y_0 = 0.2$, and $z_0 = 1.6$ as the initial approximation.
23. Solve the following system of non-linear equation

    $$x^2 + y^2 = 5$$
    $$xy = 3$$

    using *successive approximation* method. Take $x_0 = 2$ and $y_0 = 0$ as the initial approximation.
24. Give a comparative study of the following methods in the form of a table:
    (*a*) Bisection method
    (*b*) Method of false position
    (*c*) Newton-Raphson method
    (*d*) Secant method
    (*e*) Method of successive approximations

❑ ❑ ❑

$$x^3 - 3x^2 - 2x + 24 = 0$$

$$x^2 + y = 5$$

□ □ □

4 Matrices and Determinants

Learning Outcomes

After reading this chapter, students will have the ability to

- define a matrix
- explain various types of matrices
- explain various operations that can be performed on matrices
- define rank of a matrix
- explain the method to find the rank of a matrix
- explain the echelon form of a matrix
- define and find inverse of a matrix
- define eigen values for a matrix and how to compute them
- write programs for various operations on a given matrix
- define a determinant and explain its important properties
- define minors and co-factors of a determinant
- expand a given determinant
- explain pivoting and its usefulness

4.1 INTRODUCTION

The various kinds of problems occurring in the field of science, engineering, and applied mathematics can be solved efficiently by expressing them in the form of matrices and determinants. Therefore, the working knowledge of matrices and determinants is very essential. A brief description of matrices and determinants is presented in this chapter to enable the reader to appreciate the role which matrices and determinants play in numerical computing.

4.2 MATRICES

A *matrix* is a system in which $m.n$ elements are arranged in a rectangular formation of m rows and n columns bounded by brackets []. This formation is sometimes more explicitly known as m by n matrix, and written as $m \times n$ matrix. Each of the numbers of this formation is called an *element* of the matrix. A matrix is represented by a capital letter such as A, B, etc. For example

$$A = \begin{pmatrix} a_{11} & a_{12} & a_{13} & \cdot & \cdot & \cdot & a_{1n} \\ a_{21} & a_{22} & a_{23} & \cdot & \cdot & \cdot & a_{2n} \\ a_{31} & a_{32} & a_{33} & \cdot & \cdot & \cdot & a_{3n} \\ \cdot & \cdot & \cdot & \cdot & \cdot & \cdot & \cdot \\ \cdot & \cdot & \cdot & \cdot & \cdot & \cdot & \cdot \\ \cdot & \cdot & \cdot & \cdot & \cdot & \cdot & \cdot \\ a_{n1} & a_{n2} & a_{n3} & \cdot & \cdot & \cdot & a_{nn} \end{pmatrix}$$

is a matrix of order $m \times n$. Each element of the matrix is represented by the same lower-case letter with two suffixes. The first suffix denotes the row and the second suffix denotes the column.

Thus, a_{ij} denotes the element in the ith row and the jth column of the matrix A. In this notation, the matrix A is denoted by $[a_{ij}]$.

4.3 SPECIAL KIND OF MATRICES

4.3.1 Row matrix

Row matrix is a matrix having a single row. This type of matrix is also known as a *row vector*.

$A = [\, 1 \;\; 4 \;\; 3 \;\; 7 \;\; 9 \,]$ is an example of row matrix.

4.3.2 Column matrix

Column matrix is a matrix having a single column. This type of matrix is also known as a *column vector*.

$$A = \begin{pmatrix} 1 \\ 3 \\ 7 \\ 9 \end{pmatrix}$$ is an example of column matrix.

4.3.3 Square Matrix

A *square matrix* is a matrix with equal number of rows and columns. If square matrix have *n* rows and *n* columns, then it is called a square matrix of order *n*.

$$A = \begin{pmatrix} 9 & 4 & 3 & 7 \\ 2 & 1 & 5 & 0 \\ 6 & 7 & 3 & 1 \\ 9 & 2 & 7 & 5 \end{pmatrix}$$ is a square matrix of order 4.

The elements whose row and column number is same constitute the principal diagonal. In the above example, the elements 9, 1, 3, 5 constitute the principal diagonal.

4.3.4 Diagonal Matrix

Diagonal matrix is a square matrix whose all non-diagonal elements are zero.

$$A = \begin{pmatrix} 9 & 0 & 0 & 0 \\ 0 & 1 & 0 & 0 \\ 0 & 0 & 3 & 0 \\ 0 & 0 & 0 & 5 \end{pmatrix}$$ is a diagonal matrix of order 4.

4.3.5 Identity Matrix

Identity matrix is a square matrix whose elements on the principal diagonal elements are equal to 1 and rest all are zero. The identity matrix is also known as a *unit matrix*.

$$A = \begin{pmatrix} 1 & 0 & 0 & 0 \\ 0 & 1 & 0 & 0 \\ 0 & 0 & 1 & 0 \\ 0 & 0 & 0 & 1 \end{pmatrix}$$ is an identity matrix of order 4.

4.3.6 Symmetric Matrix

A square matrix is a *symmetric matrix* if

$a_{ij} = a_{ji}$ for all values of *i* and *j*.

$$A = \begin{pmatrix} 1 & 4 & 5 & 8 \\ 4 & 9 & 3 & 2 \\ 5 & 3 & 7 & 3 \\ 8 & 2 & 3 & 5 \end{pmatrix}$$ is a symmetric matrix of order 4.

4.3.7 Skew-symmetric Matrix

A square matrix is a *skew-symmetric matrix* if

$a_{ij} = -a_{ji}$

for all values of *i* and *j*, except for *i* = j. All the elements on the principal diagonal are zero.

$$A = \begin{pmatrix} 1 & -4 & 5 & -8 \\ 4 & 9 & -3 & 2 \\ -5 & 3 & 7 & 3 \\ 8 & -2 & -3 & 5 \end{pmatrix}$$

is a skew-symmetric matrix of order 4 and have only three non-zero elements.

4.3.8 Sparse Matrix

A *sparse matrix* is a matrix whose most of the elements are zero.

$$A = \begin{pmatrix} 0 & -4 & 0 & 0 \\ 0 & 0 & -3 & 0 \\ 0 & 0 & 0 & 0 \\ 0 & -2 & 0 & 0 \end{pmatrix}$$ is a sparse matrix of order 4×4 with.

4.3.9 Positive Definite Matrix

A square matrix is a *positive definite matrix* if

$$X^T AX \geq 0 \qquad \text{for all } X \neq 0$$

where X is column vector of order n.

However, if $X^T AX > 0$ for all $X \neq 0$

then the matrix is said to be *strictly positive definite.*

Alternatively to determine whether the given square matrix is positive definite or strictly positive definite we can proceed as follows:

Evaluate all the leading determinants of matrix A. If all the determinants are greater than or equal to zero, then the square matrix is positive definite. But if all the determinants are greater than zero, then the square matrix is strictly positive definite

$$A = \begin{pmatrix} 4 & 3 & 5 \\ 2 & 8 & 3 \\ 4 & 5 & 6 \end{pmatrix}$$

For this matrix, the determinant of order 1 is simply the element 4, which is positive.

The determinant of order 2 is

$$\begin{vmatrix} 4 & 3 \\ 2 & 8 \end{vmatrix} = 26 > 0$$

The determinant of order 3 is

$$\begin{vmatrix} 4 & 3 & 5 \\ 2 & 8 & 3 \\ 4 & 5 & 6 \end{vmatrix} = 22 > 0$$

Since all the determinants are greater than zero, the matrix A is strictly positive definite matrix. The reader is advised to check the values of these determinants.

4.3.10 Negative Definite Matrix

A square matrix is a *negative definite matrix* if

$$X^T AX \leq 0 \qquad \text{for all } X \neq 0$$

where X is column vector of order n.

However, if $X^TAX < 0$ for all $X \neq 0$

then the matrix is said to be *strictly negative definite.*

Alternatively to determine whether the given square matrix is positive definite or strictly positive definite we can proceed as follows:

Evaluate all the leading determinants of matrix *A*. If all the determinants are less than or equal to zero, then the square matrix is negative definite. But if all the determinants are less than zero, then the square matrix is strictly negative definite

$$A = \begin{pmatrix} -4 & 16 & 5 \\ -2 & 8 & 3 \\ -4 & 5 & 6 \end{pmatrix}$$

For this matrix, the determinant of order 1 is simply the element –4, which is negative.

The determinant of order 2 is

$$\begin{vmatrix} -4 & 16 \\ -2 & 8 \end{vmatrix} = 0$$

The determinant of order 3 is

$$\begin{vmatrix} -4 & 16 & 5 \\ -2 & 8 & 3 \\ -4 & 5 & 6 \end{vmatrix} = -22 < 0$$

Since all the determinants are either negative or zero, therefore the matrix *A* is negative definite matrix. The reader is advised to check the values of these determinants.

4.4 COMMON MATRIX OPERATIONS

This section describes the various kinds of operations performed on matrices such as addition, subtraction, multiplication, etc.

4.4.1 Addition of Matrices

If *A* and *B* are two matrices of the same order, then their sum *A* + *B*, is defined as the matrix whose each element is the sum of the corresponding elements of matrix *A* and matrix *B*, *i.e.,*

$$c_{ij} = a_{ij} + b_{ij} \qquad \text{for all values of } i \text{ and } j.$$

If $A = \begin{pmatrix} a_{11} & a_{12} & a_{13} \\ a_{21} & a_{22} & a_{23} \\ a_{31} & a_{32} & a_{33} \end{pmatrix}$ $\quad B = \begin{pmatrix} b_{11} & b_{12} & b_{13} \\ b_{21} & b_{22} & b_{23} \\ b_{31} & b_{32} & b_{33} \end{pmatrix}$

Then $A + B = \begin{pmatrix} a_{11}+b_{11} & a_{12}+b_{12} & a_{13}+b_{13} \\ a_{21}+b_{21} & a_{22}+b_{22} & a_{23}+b_{23} \\ a_{31}+b_{31} & a_{32}+b_{32} & a_{33}+b_{33} \end{pmatrix}$

Matrix addition is commutative as well as associative, *i.e.*, if *A*, *B* and *C* are three matrices of the same order, then

$$A + B = B + A$$

$$(A + B) + C = A + (B + C)$$

4.4.2 Subtraction of Matrices

If *A* and *B* are two matrices of the same order, then their difference *A* – *B*, is defined as the matrix whose each element is obtained by subtracting the corresponding elements of matrix *B* from matrix *A*, *i.e.*,

$c_{ij} = a_{ij} - b_{ij}$ for all values of *i* and *j*.

If $A = \begin{pmatrix} a_{11} & a_{12} & a_{13} \\ a_{21} & a_{22} & a_{23} \\ a_{31} & a_{32} & a_{33} \end{pmatrix}$ $\quad B = \begin{pmatrix} b_{11} & b_{12} & b_{13} \\ b_{21} & b_{22} & b_{23} \\ b_{31} & b_{32} & b_{33} \end{pmatrix}$

Then $A - B = \begin{pmatrix} a_{11}-b_{11} & a_{12}-b_{12} & a_{13}-b_{13} \\ a_{21}-b_{21} & a_{22}-b_{22} & a_{23}-b_{23} \\ a_{31}-b_{31} & a_{32}-b_{32} & a_{33}-b_{33} \end{pmatrix}$

4.4.3 Multiplying a Matrix by a Scalar

The multiplication of a matrix *A* by a scalar *k* gives a matrix whose each element is *a k time* than the corresponding element of matrix *A*.

If $A = \begin{pmatrix} a_{11} & a_{12} & a_{13} \\ a_{21} & a_{22} & a_{23} \\ a_{31} & a_{32} & a_{33} \end{pmatrix}$ then $kA = \begin{pmatrix} ka_{11} & ka_{12} & ka_{13} \\ ka_{21} & ka_{22} & ka_{23} \\ ka_{31} & ka_{32} & ka_{33} \end{pmatrix}$

4.4.4 Multiplication of Matrices

If A and B are two matrices, then the product $A.B$ is possible only if the number of columns of matrix A equals the number of rows of matrix B. Such matrices are said to be *conformable*.

If $$A = \begin{pmatrix} a_{11} & a_{12} & a_{13} \\ a_{21} & a_{22} & a_{23} \\ a_{31} & a_{32} & a_{33} \end{pmatrix} \qquad B = \begin{pmatrix} b_{11} & b_{12} \\ b_{21} & b_{22} \\ b_{31} & b_{32} \end{pmatrix}$$

Then $$A \times B = \begin{pmatrix} a_{11}b_{11} + a_{12}b_{21} + a_{13}b_{31} & a_{11}b_{12} + a_{12}b_{22} + a_{13}b_{32} \\ a_{21}b_{11} + a_{22}b_{21} + a_{23}b_{31} & a_{21}b_{12} + a_{22}b_{22} + a_{23}b_{32} \\ a_{31}b_{11} + a_{32}b_{21} + a_{33}b_{31} & a_{31}b_{11} + a_{32}b_{21} + a_{33}b_{31} \end{pmatrix}$$

In general, if

$$A = \begin{pmatrix} a_{11} & a_{12} & a_{13} & \cdot & \cdot & \cdot & a_{1n} \\ a_{21} & a_{22} & a_{23} & \cdot & \cdot & \cdot & a_{2n} \\ a_{31} & a_{32} & a_{33} & \cdot & \cdot & \cdot & a_{3n} \\ \cdot & \cdot & \cdot & \cdot & \cdot & \cdot & \cdot \\ \cdot & \cdot & \cdot & \cdot & \cdot & \cdot & \cdot \\ \cdot & \cdot & \cdot & \cdot & \cdot & \cdot & \cdot \\ a_{m1} & a_{m2} & a_{m3} & \cdot & \cdot & \cdot & a_{mn} \end{pmatrix} \qquad B = \begin{pmatrix} b_{11} & b_{12} & b_{13} & \cdot & \cdot & \cdot & b_{1p} \\ b_{21} & b_{22} & b_{23} & \cdot & \cdot & \cdot & b_{2p} \\ b_{31} & b_{32} & b_{33} & \cdot & \cdot & \cdot & b_{3p} \\ \cdot & \cdot & \cdot & \cdot & \cdot & \cdot & \cdot \\ \cdot & \cdot & \cdot & \cdot & \cdot & \cdot & \cdot \\ \cdot & \cdot & \cdot & \cdot & \cdot & \cdot & \cdot \\ b_{n1} & b_{n2} & b_{n3} & \cdot & \cdot & \cdot & b_{np} \end{pmatrix}$$

Then their product is defined as

$$C = \begin{pmatrix} c_{11} & c_{12} & c_{13} & \cdot & \cdot & \cdot & c_{1p} \\ c_{21} & c_{22} & c_{23} & \cdot & \cdot & \cdot & c_{2p} \\ c_{31} & c_{32} & c_{33} & \cdot & \cdot & \cdot & c_{3p} \\ \cdot & \cdot & \cdot & \cdot & \cdot & \cdot & \cdot \\ \cdot & \cdot & \cdot & \cdot & \cdot & \cdot & \cdot \\ \cdot & \cdot & \cdot & \cdot & \cdot & \cdot & \cdot \\ c_{m1} & c_{m2} & c_{m3} & \cdot & \cdot & \cdot & c_{mp} \end{pmatrix}$$

where

$$c_{ij} = a_{i1}b_{1j} + a_{i2}b_{2j} + a_{i3}b_{3j} + \ldots + a_{in}b_{nj}$$

is the sum of products of the elements of the ith row of matrix A and j^{th} column of matrix B. Using the summation notation, this product can be expressed as

$$c_{ij} = \sum_{k=1}^{n} a_{ik}b_{kj} \qquad \text{for } i = 1, 2, 3, \ldots, m$$

$$\text{for } j = 1, 2, 3, \ldots, p \text{ for each value of } i$$

Remember that matrix multiplication is not always commutative, i.e.

$$A \times B \neq B \times A$$

But matrix multiplication is associative, i.e.

$$(AB)C = A(BC)$$

4.4.5 Transpose of a Matrix

The *transpose* of a matrix A is a matrix obtained by interchanging its rows and columns. If the matrix A is of order $m \times n$, then its transpose, say matrix B, will be of order $n \times m$. The transpose of a matrix A is denoted by A' or A^T.

$$\text{If} \quad A = \begin{pmatrix} 9 & 4 & 3 & 7 \\ 2 & 1 & 5 & 0 \\ 6 & 7 & 3 & 1 \end{pmatrix} \quad \text{then} \quad B = A^T = \begin{pmatrix} 9 & 2 & 6 \\ 4 & 1 & 7 \\ 3 & 5 & 3 \\ 7 & 0 & 1 \end{pmatrix}$$

4.4.6 Trace of a Matrix

Trace of a square matrix A is defined as the sum of the elements in its principal diagonal. For example, if matrix A is of order $n \times n$, then

$$\text{Trace of } A, \text{ i.e., } tr(A) = \sum_{i=1}^{n} a_{ii}$$

4.5 DETERMINANT OF A MATRIX

With every matrix A, there is an associated quantity called *determinant* of A, and is denoted by $det(A)$ or $|A|$.

$$\text{If} \quad A = \begin{pmatrix} a_{11} & a_{12} & a_{13} \\ a_{21} & a_{22} & a_{23} \\ a_{31} & a_{32} & a_{33} \end{pmatrix} \quad \text{then} \quad |A| = \begin{vmatrix} a_{11} & a_{12} & a_{13} \\ a_{21} & a_{22} & a_{23} \\ a_{31} & a_{32} & a_{33} \end{vmatrix}$$

4.6 MINORS AND CO-FACTORS OF DETERMINANTS

The *minor* of any element of a given determinant is a determinant obtained by deleting the row and the column to which the element belong, from the original determinant.

If $|A| = \begin{vmatrix} a_{11} & a_{12} & a_{13} \\ a_{21} & a_{22} & a_{23} \\ a_{31} & a_{32} & a_{33} \end{vmatrix}$ then Minor of $a_{11} = \begin{vmatrix} a_{22} & a_{23} \\ a_{32} & a_{33} \end{vmatrix}$

The *co-factor* of any element of the determinant is equal to the corresponding minor with a proper sign attached to it. The general rule is

$$\text{Co-factor of } a_{ij} = (-1)^{i+j}|\Delta_{ij}|$$

where Δ_{ij} is the minor of a_{ij} and is denoted by A_{ij}.

$$\text{Co-factor of } a_{11} = A_{11} = (-1)^{1+1}\begin{vmatrix} a_{22} & a_{23} \\ a_{32} & a_{33} \end{vmatrix} = \begin{vmatrix} a_{22} & a_{23} \\ a_{32} & a_{33} \end{vmatrix}$$

4.7 SOME IMPORTANT PROPERTIES OF DETERMINANTS

We review here some of the important properties of determinants which will be useful in numerical computing.

- If the rows are changed to columns and columns into rows, the determinant remains unaltered.
- If any two rows of the determinant are interchanged, the determinant retains its value but the sign changes.
- A determinant vanishes if any two rows or columns are identical.
- If each element of a row is multiplied by the same factor, the whole determinant is multiplied by that factor.
- If an element consists of m terms, the determinant can be expressed as the sum of m determinants.
- If to each element of a row, an equal multiple of the corresponding element of one or more rows is added, the determinant remains unaltered.

4.8 ADJOINT OF A MATRIX

Let $A = [a_{ij}]$ be any given square matrix. The *adjoint* of a matrix A is the transpose of the matrix formed by replacing the element a_{ij} by its co-factor in the determinant $|A|$. It is also called adjugate matrix A. It is denoted by $adj(A)$.

If $|A| = \begin{vmatrix} a_{11} & a_{12} & a_{13} \\ a_{21} & a_{22} & a_{23} \\ a_{31} & a_{32} & a_{33} \end{vmatrix}$ then $adj(A) = \begin{vmatrix} A_{11} & A_{12} & A_{13} \\ A_{21} & A_{22} & A_{23} \\ A_{31} & A_{32} & A_{33} \end{vmatrix}$

where A_{ij} is the co-factor of the element a_{ij}.

4.9 RANK OF A MATRIX

Let $A = [a_{ij}]$ be any $m \times n$ matrix. It has square sub-matrices of different orders. The determinants of the sub-matrices are called minors of A. If all the minors of order $(r+1)$ are zero but there is at least one non-zero minor of order r, then r is called the rank of matrix A, and is denoted by $R(A)$. That is if we select any r rows and r columns from the matrix A, deleting all other rows and columns, then the determinant formed by these $r \times r$ elements is called the minor of A of order r.

Thus rank of a matrix A is defined as the highest order of any non-vanishing minor of A.

From the definition of the rank of a matrix A, the following conclusions can be drawn

- If A is null matrix, then $R(A) = 0$.
- If A is not a null matrix, then $R(A) \geq 0$.
- If A is a non-singular $n \times n$ matrix, then $R(A) = n$.
- If A is a $m \times n$ matrix, then $R(A) \leq$ minimum of m and n.
- If all minors of order r of matrix A are equal to zero, then $R(A) \leq r$.

4.9.1 Determination of Rank of a Matrix

The following methods can be used to find the rank of a matrix A of order $m \times n$:

- Start with the highest order minor of A. Let this order be r. If any one of them is non-zero, then $R(A) = r$. If all of them are zero, then start with minors of order $(r-1)$ and so on till you get a non-zero minor. The order of that minor is the rank of matrix A. This method usually involves a lot of computational work since we have to evaluate several determinants.
- If A is an $m \times n$ matrix and by series of elementary transformations on rows or columns or both, it can be put into one of the following forms, called *normal forms*:

  $\begin{pmatrix} I_r & O \\ O & O \end{pmatrix}$, $\begin{pmatrix} I_r \\ O \end{pmatrix}$, $(I_r \;\; O)$, (I_r), where I_r is the unit matrix of order r. Since the rank of a matrix is not changed as a result of elementary transformations, it follows that $R(A) = r$, because rth order minor $|I_r| = 1 \neq 0$.

Example 4.1: Find the rank of $A = \begin{pmatrix} 1 & 2 & 3 \\ -2 & 1 & 5 \end{pmatrix}$

Solution: Since A is a 2×3 matrix, therefore $R(A) \leq 2$. One of the second order minors is

$$\begin{vmatrix} 1 & 2 \\ -2 & 1 \end{vmatrix} = 4 \neq 0$$

Thus, $R(A) = 2$.

Example 4.2: Find the rank of $A = \begin{pmatrix} 1 & 2 & 3 \\ 1 & 4 & 2 \\ 2 & 6 & 5 \end{pmatrix}$

Solution: Applying row transformations $R_2 \to R_2 - R_1$ and $R_3 \to R_3 - 2R_1$, we get

$$A = \begin{pmatrix} 1 & 2 & 3 \\ 0 & 2 & -1 \\ 0 & 2 & -1 \end{pmatrix}$$

Now applying row transformation $R_3 \to R_3 - R_2$, we get

$$A = \begin{pmatrix} 1 & 2 & 3 \\ 0 & 2 & -1 \\ 0 & 0 & 0 \end{pmatrix}$$

Observe that there is only one-third order minor and that too is equal to zero, but the second order minor is

$$\begin{vmatrix} 1 & 2 \\ 0 & 2 \end{vmatrix} = 2 \neq 0$$

Thus, $R(A) = 2$.

4.10 ECHELON FORM OF A MATRIX

A matrix is said to be an echelon form if

- Number of zeros before the first non-zero element in a row is more than the number of zeros before the first non-zero element in the previous row, and
- Rows whose all the elements are zero occur after the non-zero rows.

Example 4.3: The matrices

$$A = \begin{pmatrix} 5 & 1 & 4 \\ 0 & 8 & 4 \\ 0 & 0 & 3 \end{pmatrix} \text{ and } B = \begin{pmatrix} -2 & 1 & 0 \\ 0 & 0 & 3 \\ 0 & 0 & 0 \end{pmatrix} \text{ are in echelon form.}$$

If a matrix A is in echelon form, then the rank of a matrix A is equal to the number of non-zero rows in the matrix. Therefore, to find the rank of a given matrix, reduce the matrix to echelon form by applying elementary transformations.

4.11 INVERSE OF A MATRIX

The inverse of a square matrix A, denoted by A^{-1}, exists only if

$$AA^{-1} = I$$

where I is the identity matrix. The Gauss-Jordon elimination technique is very effective in finding the inverse of a given matrix.

Suppose, we are interested to find the inverse of the following matrix

$$A = \begin{pmatrix} a_{11} & a_{12} & a_{13} & . & . & . & a_{1n} \\ a_{21} & a_{22} & a_{23} & . & . & . & a_{2n} \\ a_{31} & a_{32} & a_{33} & . & . & . & a_{3n} \\ . & . & . & . & . & . & . \\ . & . & . & . & . & . & . \\ . & . & . & . & . & . & . \\ a_{n1} & a_{n2} & a_{n3} & . & . & . & a_{nn} \end{pmatrix}$$

With the original matrix, we augment the identity matrix of same order as

$$\left(\begin{array}{ccccccc:ccccccc} a_{11} & a_{12} & a_{13} & . & . & . & a_{1n} & 1 & 0 & 0 & . & . & . & 0 \\ a_{21} & a_{22} & a_{23} & . & . & . & a_{2n} & 0 & 1 & 0 & . & . & . & 0 \\ a_{31} & a_{32} & a_{33} & . & . & . & a_{3n} & 0 & 0 & 1 & . & . & . & 0 \\ . & . & . & . & . & . & . & . & . & . & . & . & . & . \\ . & . & . & . & . & . & . & . & . & . & . & . & . & . \\ . & . & . & . & . & . & . & . & . & . & . & . & . & . \\ a_{n1} & a_{n2} & a_{n3} & . & . & . & a_{nn} & 0 & 0 & 0 & . & . & . & 1 \end{array}\right)$$

With the augmented matrix, which of order $n \times (n+n)$, the Gauss-Jordon elimination is carried out using first row as the pivot row.

The first step is to divide the first row by a_{11}, and then carry out the following transformations

$$R_i \to R_i - a_{i1}R_1 \qquad \text{for } i = 2, 3, 4, \ldots, n$$

we obtain the new matrix as

$$\left(\begin{array}{ccccccc|ccccc}
1 & a_{12} & a_{13} & . & . & . & a_{1n} & a_{1(n+1)} & 0 & 0 & . & . & . & 0 \\
0 & a_{22} & a_{23} & . & . & . & a_{2n} & a_{2(n+1)} & 1 & 0 & . & . & . & 0 \\
0 & a_{32} & a_{33} & . & . & . & a_{3n} & a_{3(n+1)} & 0 & 1 & . & . & . & 0 \\
. & . & . & . & . & . & . & . & . & . & . & . & . & . \\
. & . & . & . & . & . & . & . & . & . & . & . & . & . \\
. & . & . & . & . & . & . & . & . & . & . & . & . & . \\
0 & a_{n2} & a_{n3} & . & . & . & a_{nn} & a_{n(n+1)} & 0 & 0 & . & . & . & 1
\end{array}\right)$$

The next step is to use the second row as pivot row. Divide the second row by a_{22}, and then carry out the following transformations

$$R_i \to R_i - a_{i2}R_2 \qquad \text{for } i = 1, 3, 4, \ldots, n$$

we obtain the new matrix as

$$\left(\begin{array}{ccccccc|cccccc}
1 & 0 & a_{13} & . & . & . & a_{1n} & a_{1(n+1)} & a_{1(n+2)} & 0 & . & . & . & 0 \\
0 & 1 & a_{23} & . & . & . & a_{2n} & a_{2(n+1)} & a_{2(n+2)} & 0 & . & . & . & 0 \\
0 & 0 & a_{33} & . & . & . & a_{3n} & a_{3(n+1)} & a_{3(n+1)} & 1 & . & . & . & 0 \\
. & . & . & . & . & . & . & . & . & . & . & . & . & . \\
. & . & . & . & . & . & . & . & . & . & . & . & . & . \\
. & . & . & . & . & . & . & . & . & . & . & . & . & . \\
0 & 0 & a_{n3} & . & . & . & a_{nn} & a_{n(n+1)} & a_{n(n+2)} & 0 & . & . & . & 1
\end{array}\right)$$

The next step is to use the third row as pivot row. Divide the second row by a_{22}, and then carry out the following transformations

$$R_i \to R_i - a_{i3}R_3 \qquad \text{for } i = 1, 3, 4, \ldots, n$$

we obtain the new matrix as

$$\left(\begin{array}{ccccccc|cccccc}
1 & 0 & 0 & . & . & . & a_{1n} & a_{1(n+1)} & a_{1(n+2)} & a_{1(n+3)} & . & . & . & 0 \\
0 & 1 & 0 & . & . & . & a_{2n} & a_{2(n+1)} & a_{2(n+2)} & a_{2(n+3)} & . & . & . & 0 \\
0 & 0 & 1 & . & . & . & a_{3n} & a_{3(n+1)} & a_{3(n+1)} & a_{3(n+3)} & . & . & . & 0 \\
. & . & . & . & . & . & . & . & . & . & . & . & . & . \\
. & . & . & . & . & . & . & . & . & . & . & . & . & . \\
. & . & . & . & . & . & . & . & . & . & . & . & . & . \\
0 & 0 & 0 & . & . & . & a_{nn} & a_{n(n+1)} & a_{n(n+2)} & a_{n(n+3)} & . & . & . & 1
\end{array}\right)$$

This elimination process is carried out until the nth row is used as the pivot row.

Divide the nth row by a_{nn} and apply the following transformations.

$$R_i \rightarrow R_i - a_{in}R_n \qquad \text{for } i = 1, 2, 3, \ldots, (n-1)$$

This step yields the following matrix.

$$\left(\begin{array}{cccccc|cccccc}
1 & 0 & 0 & \cdot & \cdot & 0 & a_{1(n+1)} & a_{1(n+2)} & a_{1(n+3)} & \cdot & \cdot & a_{1(n+n)} \\
0 & 1 & 0 & \cdot & \cdot & 0 & a_{2(n+1)} & a_{2(n+2)} & a_{2(n+3)} & \cdot & \cdot & a_{2(n+n)} \\
0 & 0 & 1 & \cdot & \cdot & 0 & a_{3(n+1)} & a_{3(n+1)} & a_{3(n+3)} & \cdot & \cdot & a_{3(n+n)} \\
\cdot & \cdot & \cdot & \cdot & \cdot & \cdot & \cdot & \cdot & \cdot & \cdot & \cdot & \cdot \\
\cdot & \cdot & \cdot & \cdot & \cdot & \cdot & \cdot & \cdot & \cdot & \cdot & \cdot & \cdot \\
\cdot & \cdot & \cdot & \cdot & \cdot & \cdot & \cdot & \cdot & \cdot & \cdot & \cdot & \cdot \\
0 & 0 & 0 & \cdot & \cdot & 1 & a_{n(n+1)} & a_{n(n+2)} & a_{n(n+3)} & \cdot & \cdot & a_{n(n+n)}
\end{array}\right)$$

The matrix $[a_{ij}]$ for $i = 1$ to n and $j = 1$ to $(n+n)$ is the inverse of matrix A. The computational algorithm to find the inverse is given below.

Example 4.4: Find the inverse of the matrix $\begin{pmatrix} 1 & 1 & 0 \\ 1 & 0 & 1 \\ 1 & 2 & 2 \end{pmatrix}$

Solution: Write the augmented matrix as

$$\begin{pmatrix} 1 & 1 & 0 & 1 & 0 & 0 \\ 1 & 0 & 1 & 0 & 1 & 0 \\ 1 & 2 & 2 & 0 & 0 & 1 \end{pmatrix}$$

Starting with first row as the pivot row. Divide the first row with pivot element a_{11} (=1 in this case). This division causes no change in the elements of the first row.

Next apply the transformation

$$R_2 \rightarrow R_2 - a_{21}R_1 \qquad \text{for second row.}$$

Here $a_{21} = 1$. The elements of second row are computed as

$a_{21} = 1-1\times1 = 0$ $\qquad$ $a_{22} = 0-1\times1 = -1$

$a_{23} = 1-1\times0 = 1$ $\qquad$ $a_{24} = 0-1\times1 = -1$

$a_{25} = 1-1\times0 = 1$ $\qquad$ $a_{26} = 0-1\times0 = 0$

Now apply the transformation

$$R_3 \rightarrow R_3 - a_{31}R_1 \qquad \text{for third row.}$$

Here $a_{31} = 1$. The elements of third row are computed as

$a_{31} = 1 - 1\times1 = 0$ $\quad$ $a_{32} = 2 - 1\times1 = 1$

$a_{33} = 2 - 1\times0 = 2$ $\quad$ $a_{34} = 0 - 1\times1 = -1$

$a_{35} = 0 - 1\times0 = 0$ $\quad$ $a_{36} = 1 - 1\times0 = 1$

Hence after the first elimination step, the augmented matrix becomes

$$\begin{pmatrix} 1 & 1 & 0 & 1 & 0 & 0 \\ 0 & -1 & 1 & -1 & 1 & 0 \\ 0 & 1 & 2 & -1 & 0 & 1 \end{pmatrix}$$

Now taking second row as the pivot row. Divide the second row with pivot element a_{22} (=–1 in this case). The new elements of the second row becomes

$$0 \quad 1 \quad -1 \quad 1 \quad -1 \quad 0$$

Next apply the transformation

$R_1 \rightarrow R_1 - a_{12}R_2$ $\quad$ for first row.

Here $a_{12} = 1$. The elements of second row are computed as

$a_{11} = 1 - 1\times0 = 0$ $\quad$ $a_{12} = 1 - 1\times1 = 0$

$a_{13} = 0 - 1\times(-1) = 1$ $\quad$ $a_{14} = 1 - 1\times1 = 0$

$a_{15} = 0 - 1\times(-1) = 1$ $\quad$ $a_{16} = 0 - 1\times0 = 0$

Now apply the transformation

$R_3 \rightarrow R_3 - a_{32}R_2$ $\quad$ for third row.

Here $a_{32} = 1$. The new elements of third row are computed as

$a_{31} = 0 - 1\times0 = 0$ $\quad$ $a_{32} = 1 - 1\times1 = 0$

$a_{33} = 2 - 1\times(-1) = 3$ $\quad$ $a_{34} = -1 - 1\times1 = -2$

$a_{35} = 0 - 1\times(-1) = 1$ $\quad$ $a_{36} = 1 - 1\times0 = 1$

Hence after the second elimination step, the augmented matrix becomes

$$\begin{pmatrix} 1 & 0 & 1 & 0 & 1 & 0 \\ 0 & 1 & -1 & 1 & -1 & 0 \\ 0 & 0 & 3 & -2 & 1 & 1 \end{pmatrix}$$

Now taking third row as the pivot row. Divide the third row with pivot element a_{33} (= 3 in this case).

The new elements of the third row becomes

$$0 \quad 0 \quad 1 \quad -2/3 \quad 1/3 \quad 1/3$$

Next apply the transformation

$R_1 \rightarrow R_1 - a_{13}R_3$ for first row.

Here $a_{13} = 1$. The elements of second row are computed as

$a_{11} = 1{-}1{\times}0 = 0$ $\quad$ $a_{12} = 0{-}1{\times}0 = 0$

$a_{13} = 1{-}1{\times}1 = 0$ $\quad$ $a_{14} = 0{-}1{\times}(-2/3) = 2/3$

$a_{15} = 1{-}1{\times}1/3 = 2/3$ $\quad$ $a_{16} = 0{-}1{\times}1/3 = -1/3$

Now apply the transformation

$R_2 \rightarrow R_2 - a_{23}R_3$ for second row.

Here $a_{23} = -1$. The new elements of third row are computed as

$a_{21} = 0-(-1){\times}0 = 0$ $\quad$ $a_{22} = 1-(-1){\times}0 = 1$

$a_{23} = -1-(-1){\times}1 = 0$ $\quad$ $a_{24} = 1-(-1){\times}(-2/3) = 1/3$

$a_{25} = -1-(-1){\times}1/3 = -2/3$ $\quad$ $a_{26} = 0-(-1){\times}1/3 = 1/3$

Hence after the third elimination step, the augmented matrix becomes

$$\begin{pmatrix} 1 & 0 & 0 & 2/3 & 2/3 & -1/3 \\ 0 & 1 & 0 & 1/3 & -2/3 & 1/3 \\ 0 & 0 & 1 & -2/3 & 1/3 & 1/3 \end{pmatrix}$$

The matrix $[a_{ij}]$ for $i = 1$ to 3 and $j = 4$ to 6 is the inverse of the given matrix.

$$\text{Hence} \qquad A^{-1} = \begin{pmatrix} 2/3 & 2/3 & -1/3 \\ 1/3 & -2/3 & 1/3 \\ -2/3 & 1/3 & 1/3 \end{pmatrix}$$

Algorithm 4.1: Inverse of a matrix

To find the inverse of a square matrix A of order n. The augmented matrix is stored using a two-dimensional array named a of size $n{\times}2n$. The inverse of the matrix A is present in the augmented portion of the array a i.e. from columns $(n+1)$ to $2n$.

```
Begin
    read: n
    for i = 1 to n by 1 do
        for j = 1 to n by 1 do
```

```
            read: a_ij
        endfor
    endfor
    for i = 1 to n by 1 do                  // append identity matrix
        for j = (n+1) to (n+n) by 1 do
            if ((j-i) = n) then
                set a_ij = 1
            else
                set a_ij = 0
            endif
        endfor
    endfor
    for k = 1 to n by 1 do
        set temp = a_kk
        for j = k to (n+n) by 1 do
            set a_kj = a_kj / temp
        endfor
        for i = 1 to n by 1 do
            if ( i ≠ k ) then
                set comm = a_ik
                for j = k to (n+n) by 1 do
                    set a_ij = a_ij -temp × a_kj
                endfor
            endif
        endfor
    endfor
    write: "Elements of the inverse"
    for i = 1 to n by 1 do
        for j = (n+1) to (n+n) by 1 do
            write: a_ij
        endfor
    endfor
End.
```

4.12 INVERSE OF A MATRIX IN PLACE

If the matrices of the large size are to be inverted, then it becomes very important to save as much memory space as possible. A large amount of memory can be saved if the identity matrix is not stored at all. The matrix to be inverted is stored in memory and is replaced by its inverse. This technique is called *matrix inversion in place.*

For example, consider the following matrix $\begin{pmatrix} 1 & 1 & 0 \\ 1 & 0 & 1 \\ 1 & 2 & 2 \end{pmatrix}$

for which augmented matrix is $\begin{pmatrix} 1 & 1 & 0 & 1 & 0 & 0 \\ 1 & 0 & 1 & 0 & 1 & 0 \\ 1 & 2 & 2 & 0 & 0 & 1 \end{pmatrix}$

After the first reduction step, we obtain the following augmented matrix

$$\begin{pmatrix} 1 & 1 & 0 & 1 & 0 & 0 \\ 0 & -1 & 1 & -1 & 1 & 0 \\ 0 & 1 & 2 & -1 & 0 & 1 \end{pmatrix}$$

The first column of the reduced matrix plays no role in subsequent reduction steps and remains unaltered. Therefore, the fourth column can be stored in its place. Hence, only three columns (column number 2, 3 and 4) are the vital columns from this reduction step. Therefore, with matrix inversion in place, the matrix in memory at this stage will be

$$\begin{pmatrix} 1 & 1 & 0 \\ -1 & -1 & 1 \\ -1 & 1 & 2 \end{pmatrix}$$

The second reduction step produces the following augmented matrix

$$\begin{pmatrix} 1 & 0 & 1 & 0 & 0 & 0 \\ 0 & 1 & -1 & 1 & -1 & 0 \\ 0 & 0 & 3 & -1 & -2 & 1 \end{pmatrix}$$

Now, the fourth column can be stored at the first column place and the fifth column at the second column place. So with matrix inversion in place, the matrix in memory at this stage will be

$$\begin{pmatrix} 0 & 1 & 1 \\ 1 & -1 & -1 \\ -1 & -2 & 3 \end{pmatrix}$$

After the third reduction step, the final augmented matrix will be

$$\begin{pmatrix} 1 & 0 & 0 & 2/3 & 2/3 & -1/3 \\ 0 & 1 & 0 & 1/3 & -2/3 & 1/3 \\ 0 & 0 & 1 & -2/3 & 1/3 & 1/3 \end{pmatrix}$$

With matrix inverse in place, the final matrix in memory will be

$$\begin{pmatrix} 2/3 & 2/3 & -1/3 \\ 1/3 & -2/3 & 1/3 \\ -2/3 & 1/3 & 1/3 \end{pmatrix}$$

The above procedure can be accomplished by using following transformations in order.

1. $a_{kj} = a_{kj}/a_{kk}$ for $j = 1$ to n (except $j \neq k$)
2. $a_{kk} = 1/a_{kk}$
3. $a_{ij} = a_{ij} - a_{ik}a_{kj}$ for $i = 1$ to n (except $i \neq k$)
   for $j = 1$ to n for each value of i (except $j \neq k$)
4. $a_{ik} = - a_{ik}a_{kk}$ for $i = 1$ to n (except $i \neq k$)

Algorithm 4.2: Inverse of a matrix in place

To find the inverse of a square matrix A of order n. The matrix is stored using a two-dimensional array named a of size $n \times n$. The inverse of the matrix A is stored on the original matrix itself.

```
Begin
   read: n
   for i = 1 to n by 1 do
       for j = 1 to n by 1 do
           read: aij
       endfor
   endfor
   for k = 1 to n by 1 do
       for j = 1 to n by 1 do
           if ( j ≠ k ) then
               set akj = akj / akk
           endif
       endfor
       set akk = 1.0 / akk
       for i = 1 to n by 1 do
           if ( i ≠ k ) then
               for j = 1 to n by 1 do
                   if ( j ≠ k ) then
                       set aij = aij - aik × akj
                   endif
               endfor
           endif
       endfor
       for i = 1 to n by 1 do
           if ( i ≠ k ) then
               set aik = -aik × akk
```

```
            endif
        endfor
    endfor
    write: "Elements of the inverse"
    for i = 1 to n by 1 do
        for j = 1 to n by 1 do
            write: a_ij
        endfor
    endfor
End.
```

4.13 PIVOTING

The following important points we have not considered in above elimination process:

1. We are assuming that pivot element is non-zero. But if it happens to be equal to zero, then procedure must be modified so that the pivot row has non-zero pivot element and this can be accomplished by interchanging the pivot row with any other row.
2. If the pivot element has a non-zero value but is very small in magnitude, it will further increase the amount of error in the final result. Therefore, the magnitude of the pivot element has an effect on the accuracy of the result.

In order to get more accurate results, each elimination step should be made by using that row as the pivot row that has the largest pivot element in magnitude. This process of selecting the largest element in magnitude in the column under consideration as the pivot element and interchanging the rows is called *partial pivotal condensation* or simply *pivoting*. In contrast, selecting the largest element in magnitude between all row and columns and then interchanging the appropriate rows and columns is called *total pivoting* or *complete pivoting*. The total pivoting is bit complicated to implement, and it has been found that partial pivoting is capable of handling most of the cases quite reasonably.

These steps to perform pivoting are inserted in the beginning of the loop controlling the elimination process as shown in the following algorithm.

Algorithm 4.3: Partial Pivoting

```
Begin
    . . . . . . . . .
    for k = 1 to n by 1 do
        set max = | a_kk |
        set p = k
        if ( k ≠ n ) then
            for m = 1 to n by 1 do
                if (| a_mk | > max ) then
                    set max = | a_mk |
                    set p = m
```

```
            endif
        endfor
    endif
    if ( p ≠ k ) then
        for q = 1 to n by 1 do
            set temp = a_kq, a_kq = a_pq, a_pq = temp
        endfor
    endif
    . . . . . . . . .
End.
```

Remember that when the rows of matrix are interchanged, original matrix is disturbed. Therefore to retrieve original matrix, corresponding columns must be interchanged.

4.14 INVERSE OF A MATRIX IN PLACE WITH PIVOTING

Consider the following matrix

$$\begin{pmatrix} -3 & 2 & -1 \\ -4 & 1 & -1 \\ 2 & 0 & 1 \end{pmatrix}$$

Suppose we wish to interchange first and second row to utilise the largest value in magnitude in the pivot row as the pivot element. And further suppose that we are using the appended identity matrix. Now interchanging the first and second row will reshuffle the identity portion as

$$\begin{pmatrix} 0 & 1 & 0 \\ 1 & 0 & 0 \\ 0 & 0 & 1 \end{pmatrix}$$

Observe that we can obtain the same appended identity matrix by interchanging the first and second columns of this disturbed identity matrix. As with matrix inversion in place, the identity matrix is not in memory, we can not interchange its columns. The same effect can be obtained by interchanging the first and second column of the inverted matrix. Therefore, while inverting the matrix with pivoting, we keep track of row interchanges, and once the elimination process is complete, re-order the columns of the final matrix obtained at the end of the elimination process to get the final inverted matrix. The required steps are presented in the following algorithm.

Algorithm 4.4: Inverse of a matrix in place with pivoting

To find the inverse of a square matrix A of order n. The matrix is stored using a two-dimensional array named a of size $n \times n$. The inverse of the matrix A is stored on the original matrix itself. A two-dimensional array named *inter* of size $n \times 2$ is used to keep track of row interchanges.

```
Begin
   read: n
   for i = 1 to n by 1 do
      for j = 1 to n by 1 do
         read: a_ij
      endfor
   endfor

   for k = 1 to n by 1 do
      set max = | a_kk |
      set p = k

      if ( k ≠ n ) then
         for m = 1 to n by 1 do
            if (| a_mk | > max ) then
               set max =  | a_mk |
               set p = m
            endif
         endfor
      endif

      set inter_k1 = k
      set inter_k2 = p
      if ( p ≠ k ) then
         for q = 1 to n by 1 do
            set temp = a_kq,  a_kq = a_pq,  a_pq = temp
         endfor
      endif
      for j = 1 to n by 1 do
         if ( j ≠ k ) then
            set a_kj  = a_kj / a_kk
         endif
      endfor
      set a_kk = 1.0 / a_kk
      for i = 1 to n by 1 do
         if ( i ≠ k ) then
            for j = 1 to n by 1 do
               if ( j ≠ k ) then
                  set a_ij  = a_ij − a_ik × a_kj
               endif
            endfor
         endif
      endfor
      for i = 1 to n by 1 do
         if ( i ≠ k ) then
            set a_ik  = −a_ik × a_kk
         endif
      endfor
   endfor
```

```
    for q = n to 1 by -1 do
        set r = inter_q1
        set s = inter_q1
        if ( r ≠ s ) then
            for i = 1 to n by 1 do
                set temp = a_ir, a_ir = a_is, a_is = temp
            endfor
        endif
    endfor
    write: "Elements of the inverse"
    for i = 1 to n by 1 do
        for j = 1 to n by 1 do
            write: a_ij
        endfor
    endfor
End.
```

4.15 EXPANSION OF A DETERMINANT

$$|A| = \begin{vmatrix} a_{11} & a_{12} & a_{13} & \cdot & \cdot & \cdot & a_{1n} \\ a_{21} & a_{22} & a_{23} & \cdot & \cdot & \cdot & a_{2n} \\ a_{31} & a_{32} & a_{33} & \cdot & \cdot & \cdot & a_{3n} \\ \cdot & \cdot & \cdot & \cdot & \cdot & \cdot & \cdot \\ \cdot & \cdot & \cdot & \cdot & \cdot & \cdot & \cdot \\ \cdot & \cdot & \cdot & \cdot & \cdot & \cdot & \cdot \\ a_{n1} & a_{n2} & a_{n3} & \cdot & \cdot & \cdot & a_{nn} \end{vmatrix}$$

In many situations, it may be necessary to expand a given determinant, *i.e.*, find the value of the determinant. We know the method of expanding the determinant in terms of its co-factors from the college algebra. But this method is inefficient to handle large determinants. The expansion of the determinant using Gauss elimination technique is very efficient and can be generalised to expand a determinant of any size. The Gauss elimination technique is described below:

In the first step, apply following transformations

$$R_i - \frac{a_{i1}}{a_{11}} R_1 \qquad \text{for } i = 2, 3, 4, \ldots, n$$

taking the first row as the pivot row.

We obtain the determinant of type

$$|A| = \begin{vmatrix} a_{11} & a_{12} & a_{13} & . & . & . & a_{1n} \\ 0 & a_{22} & a_{23} & . & . & . & a_{2n} \\ 0 & a_{32} & a_{33} & . & . & . & a_{3n} \\ . & . & . & . & . & . & . \\ . & . & . & . & . & . & . \\ . & . & . & . & . & . & . \\ 0 & a_{n2} & a_{n3} & . & . & . & a_{nn} \end{vmatrix}$$

In the second step, apply following transformations

$$R_i \to R_i - \frac{a_{i2}}{a_{22}} R_2 \qquad \text{for } i = 3, 4, \ldots, n$$

taking the second row as the pivot row. We obtain the determinant of type

$$|A| = \begin{vmatrix} a_{11} & a_{12} & a_{13} & . & . & . & a_{1n} \\ 0 & a_{22} & a_{23} & . & . & . & a_{2n} \\ 0 & 0 & a_{33} & . & . & . & a_{3n} \\ . & . & . & . & . & . & . \\ . & . & . & . & . & . & . \\ . & . & . & . & . & . & . \\ 0 & 0 & a_{n3} & . & . & . & a_{nn} \end{vmatrix}$$

and so on.

Finally, apply the following transformation

$$R_i \to R_i - \frac{a_{i(n-1)}}{a_{(n-1)(n-1)}} R_{n-1} \qquad \text{for } i = n$$

taking the $(n-1)$th row as the pivot row.

We obtain the determinant of type

$$|A| = \begin{vmatrix} a_{11} & a_{12} & a_{13} & \cdots & a_{1n} \\ 0 & a_{22} & a_{23} & \cdots & a_{2n} \\ 0 & 0 & a_{33} & \cdots & a_{3n} \\ \cdot & \cdot & \cdot & \cdots & \cdot \\ \cdot & \cdot & \cdot & \cdots & \cdot \\ \cdot & \cdot & \cdot & \cdots & \cdot \\ 0 & 0 & 0 & \cdots & a_{nn} \end{vmatrix}$$

With this type of determinant, if we expand it on the first column, we will obtain

$$|A| = a_{11} \times a_{22} \times a_{33} \times \ldots \times a_{nn}$$

the product of the diagonal elements.

Example 4.5: Expand the determinant

$$|A| = \begin{vmatrix} 2 & 3 & -1 \\ 4 & 1 & 5 \\ 3 & -2 & 1 \end{vmatrix}$$

Solution: First apply transformation

$$R_2 \to R_2 - \frac{4}{2} R_1$$

The new elements of the second row are computed as

$$a_{21} = 4 - \frac{4}{2} \times 2 = 0 \qquad a_{22} = 1 - \frac{4}{2} \times 3 = -5 \qquad a_{23} = 5 - \frac{4}{2} \times (-1) = 7$$

Next apply transformation

$$R_3 \to R_3 - \frac{3}{2} R_1$$

The new elements of the third row are computed as

$$a_{31} = 3 - \frac{3}{2} \times 2 = 0 \qquad a_{32} = -2 - \frac{3}{2} \times 3 = -6.5 \qquad a_{33} = 1 - \frac{3}{2} \times (-1) = 2.5$$

Thus, after the first elimination step, the determinant becomes

$$|A| = \begin{vmatrix} 2 & 3 & -1 \\ 0 & -5 & 7 \\ 0 & -6.5 & 2.5 \end{vmatrix}$$

In the second and final elimination step, apply the transformation

$$R_3 \rightarrow R_3 - \frac{-6.5}{-5}R_2 \quad \Rightarrow \quad R_3 \rightarrow R_3 - \frac{6.5}{5}R_2$$

The element of the third row are computed as

$$a_{31} = 0 - \frac{6.5}{5} \times 0 = 0 \qquad a_{32} = -6.5 - \frac{6.5}{5} \times (-5) = 0 \qquad a_{33} = -2.5 - \frac{6.5}{5} \times 7 = -6.6$$

Therefore, the final determinant becomes

$$|A| = \begin{vmatrix} 2 & 3 & -1 \\ 0 & -5 & 7 \\ 0 & 0 & -6.6 \end{vmatrix}$$

Now multiplying the diagonal elements, we get

$$|A| = 2 \times (-5) \times (-6.6) = 66$$

Note that in the above elimination process, we have assumed that the pivot element to be non-zero. To avoid the zero division and to increase the accuracy in the final value of the determinant, pivoting will be done before the elimination process starts. But here the pivoting will be restricted from pivoting row to the last row. Since during pivoting, some rows will be interchanged, and as we know that with the row interchanges, the sign of the determinant changes. Therefore, in order to get its correct value, we count the number of row interchanges and change the sign of the determinant accordingly.

Incorporating partial pivotal, the algorithm for the expansion of a determinant by Gauss's elimination technique will look like as given below.

Algorithm 4.5: To expand a determinant

To expand the determinant of order *n*. The determinant is stored using a two-dimensional array named *a* of size *n*×*n*.

```
Begin
    read: n
    for i = 1 to n by 1 do
        for j = 1 to n by 1 do
```

```
            read: aij
        endfor
    endfor
    set count = 0
    for k = 1 to (n -1) by 1 do
        set max = | akk |
         set p = k
         for m = (k+1) to n by 1 do
             if (| amk | > max ) then
                 set max = | amk |
                 set p = m
             endif
         endfor
         if ( p ≠ k ) then
             set count = count +1
             for q = k to n by 1 do
                 set temp = akq
                 set akq = apq
                 set apq = temp
             endfor
         endif
         for i = (k+1) to n by 1 do
             set temp = aik/aii
             for j = k to n by 1 do
                 set aij = aij - temp × akj
             endfor
         endfor
    endfor
    set prod = 1.0
    for i = 1 to n by 1 do
        set prod = prod × aii
    endfor
    set det = (-1)^count × prod
    write: "Value of the determinant = ", det
End.
```

4.16 EIGEN VALUES OF A MATRIX

If A is a square matrix of order n, then we can form a matrix of type

$$A - \lambda I$$

where λ is a scalar and I is the identity matrix or order n. The determinant of this matrix equated to zero is called the *characteristic equation* of the matrix A, i.e.

$$|A - \lambda I| = 0$$

$$\text{If } A = \begin{pmatrix} a_{11} & a_{12} & a_{13} & . & . & . & a_{1n} \\ a_{21} & a_{22} & a_{23} & . & . & . & a_{2n} \\ a_{31} & a_{32} & a_{33} & . & . & . & a_{3n} \\ . & . & . & . & . & . & . \\ . & . & . & . & . & . & . \\ . & . & . & . & . & . & . \\ a_{n1} & a_{n2} & a_{n3} & . & . & . & a_{nn} \end{pmatrix}$$

$$\text{then } |A - \lambda I| = \begin{vmatrix} a_{11}-\lambda & a_{12} & a_{13} & . & . & . & a_{1n} \\ a_{21} & a_{22}-\lambda & a_{23} & . & . & . & a_{2n} \\ a_{31} & a_{32} & a_{33}-\lambda & . & . & . & a_{3n} \\ . & . & . & . & . & . & . \\ . & . & . & . & . & . & . \\ . & . & . & . & . & . & . \\ a_{n1} & a_{n2} & a_{n3} & . & . & . & a_{nn}-\lambda \end{vmatrix}$$

On expanding this determinant, the characteristic equation takes the form

$$\lambda^n + p_1\lambda^{n-1} + p_2\lambda^{n-2} + p_3\lambda^{n-3} + ... + p_n = 0$$

where p's can be expressed in terms of a_{ij}'s. The roots of this equation are called the *eigen values* of the matrix A.

4.16.1 Determination of Eigen Values

The determination of the eigen values involves two major steps:

1. To find the coefficients of the characteristic equation.
2. To solve the characteristic equation.

Numbers of methods are available to compute the coefficients of the characteristic equation. In the present text, we will discuss the *Faddeev-Leverrier* method, which is an efficient method and can be used for symmetric as well as asymmetric matrices.

This method computes the polynomial coefficients p_k's from the matrix A by forming a sequence of matrices $B_1, B_2, B_3, ..., B_n$ from which the p_k's are computed.

These values of p_k's are substituted in following basic form of characteristic equation

$$(-1)^n(\lambda^n - p_1\lambda^{n-1} - p_2\lambda^{n-2} - p_3\lambda^{n-3} - \dots - p_n) = 0$$

where $(-1)^n$ is used merely to give the terms of the characteristic equation the same sign as that they would have if the characteristic equation was generated by expanding the determinant. The p_k's values are determined as follows:

$B_1 = A$ and $p_1 = \text{tr}(B_1)$

$B_2 = A(B_1 - p_1 I)$ and $p_2 = \frac{1}{2}\text{tr}(B_2)$

$B_3 = A(B_2 - p_2 I)$ and $p_3 = \frac{1}{3}\text{tr}(B_3)$

$\vdots$ $\vdots$

$B_n = A(B_{n-1} - p_{n-1} I)$ and $p_n = \frac{1}{n}\text{tr}(B_n)$

Example 4.6: Find the eigen values of matrix $A = \begin{pmatrix} 3 & 2 & 4 \\ 2 & 0 & 2 \\ 4 & 2 & 3 \end{pmatrix}$

Solution: We know that $B_1 = A = \begin{pmatrix} 3 & 2 & 4 \\ 2 & 0 & 2 \\ 4 & 2 & 3 \end{pmatrix}$

$$p_1 = \text{tr}(B_1) = 3 + 0 + 3 = 6$$

Since $B_2 = A(B_1 - p_1 I)$

Substituting the values of A, B_1 and p_1 and simplifying, we get

$$B_2 = \begin{pmatrix} 11 & 2 & 4 \\ 2 & 8 & 2 \\ 4 & 2 & 11 \end{pmatrix}$$

$$p_2 = \tfrac{1}{2}\text{tr}(B_2) = \tfrac{1}{2}(11 + 8 + 11) = 15$$

Since $B_3 = A(B_2 - p_2 I)$

Substituting the values of A, B_2 and p_2 and simplifying, we get

$$B_2 = \begin{pmatrix} 8 & 0 & 0 \\ 0 & 8 & 0 \\ 0 & 0 & 8 \end{pmatrix}$$

$$p_3 = \tfrac{1}{3}\,\text{tr}(B_3) = \tfrac{1}{3}(8 + 8 + 8) = 8$$

Therefore, by substituting the values of p_1, p_2, and p_3 in characteristic equation, we get

$$(-1)^3(\lambda^3 - 6\lambda^2 - 15\lambda - 8) = 0$$

$$\Rightarrow \quad -\lambda^3 + 6\lambda^2 + 15\lambda + 8 = 0$$

This polynomial can be solved by any of the methods discussed in *Chapter 3*, for finding the roots of the above equation, hence the eigen values.

The above polynomial simply can be factorised as

$$(\lambda - 8)(\lambda + 1)(\lambda + 1) = 0$$

which gives

$$\lambda_1 = 8 \qquad \lambda_2 = -1 \qquad \lambda_2 = -1$$

as eigen values of the given matrix.

Algorithm 4.6: To determine the coefficients of the characteristic equation

The matrix A of order n is stored in a two-dimensional array named a of size $n{\times}n$. Another two-dimensional array b, also of size $n{\times}n$, is used to hold matrices $B_1, B_2, \ldots, B_n$. One-dimensional array p of size n is used to store the coefficients of the characteristic equation.

```
Begin
    read: n
    for i = 1 to n by 1 do
        for j = 1 to n by 1 do
            read: aij
        endfor
    endfor
    for i = 1 to n by 1 do
        for j = 1 to n by 1 do
            set bij = aij
        endfor
    endfor
    for k = 1 to (n -1) by 1 do
        set trace = 0
        for i = 1 to n by 1 do
            set trace = trace + bii
        endfor
        set pk = trace / k
        for i = 1 to n by 1 do
            set bii = aii - pk
        endfor
```

```
            for j = 1 to n by 1 do
                  for i = 1 to n by 1 do
                        set ci = bij
                  endfor
                  for i = 1 to n by 1 do
                        set sum = 0
                        for q = 1 to n by 1 do
                              set sum = sum + aiq × cq
                        endfor
                        set bij = sum
                  endfor
            endfor
      endfor
      set pn = b11
      write: "Coefficients of the characteristic equation"
      for i = 1 to n by 1 do
           write: pi
      endfor
End.
```

EXERCISE . . .

1. Define a matrix. Also give one example each of the following type of matrices:
   (*a*) Row matrix (*b*) Column matrix
   (*c*) Symmetric matrix (*d*) Positive definite matrix
   (*e*) Strictly negative definite matrix (*f*) Sparse matrix

2. If $A = \begin{pmatrix} 0 & 2 & 4 \\ 1 & 3 & 7 \\ 4 & 5 & 9 \end{pmatrix}$ and $B = \begin{pmatrix} 2 & 1 & 3 \\ 2 & 5 & 2 \\ 9 & 0 & 6 \end{pmatrix}$

   Compute
   (*a*) $A + B$ (*b*) $B + A$ (*c*) $A - B$ (*d*) $B - A$
   (*e*) AB (*f*) BA

3. If $A = \begin{pmatrix} 3 & 7 & -2 \\ 2 & 3 & 1 \\ 4 & 5 & 9 \end{pmatrix}$

   Find
   (*a*) Transpose of A (*b*) Trace of A (*c*) Adjoint of A (*d*) Rank of A

4. Expand the determinant

$$|A| = \begin{vmatrix} 2 & 5 & 4 \\ 2 & 5 & 7 \\ 5 & 1 & -6 \end{vmatrix}$$

   (*a*) by method of co-factors (*b*) by elimination technique

5. Does the inverse of the following matrix exist? If yes, find the same.

$$A = \begin{pmatrix} 3 & 7 & -2 \\ 2 & 3 & 1 \\ 4 & 5 & 9 \end{pmatrix}$$

6. Find the eigen values of the matrix

$$A = \begin{pmatrix} 1 & 2 & 3 \\ 2 & 3 & 1 \\ 5 & 1 & 6 \end{pmatrix}$$

7. Check whether the matrix A is positive definite, strictly positive definite, negative definite or strictly negative definite?

$$A = \begin{pmatrix} 4 & 2 & 3 \\ 2 & 0 & 1 \\ 5 & 2 & 6 \end{pmatrix}$$

8. What is pivoting? How is it useful?
9. Formulate an algorithm to check whether the given matrix is a symmetric matrix.
10. Formulate an algorithm to multiply two matrices.
11. Formulate an algorithm to check whether the given matrix is positive definite.
12. Formulate an algorithm to check whether the given matrix is invertible.
13. Formulate an algorithm to find the row having maximum number of negative elements in a given matrix.
14. Formulate an algorithm to find the arithmetic mean of elements of each row of a given matrix.
15. Formulate an algorithm to find the largest and smallest elements of a given matrix.
16. Formulate an algorithm that finds the sum of each row and each column of a given matrix and as its output print the row having largest sum of its elements and column having smallest sum of its elements.

❑ ❑ ❑

5. Does the inverse of the following matrix exist? If yes, find the same

[illegible]

6. Find the eigenvalues of the matrix

[illegible]

7. Check whether the matrix A is positive definite, strictly positive definite, negative definite or strictly negative definite?

[illegible]

8. What is pivoting? How is it useful?
9. Formulate an algorithm to check whether the given matrix is a symmetric matrix.
10. Formulate an algorithm to multiply two matrices.
11. Formulate an algorithm to check whether the given matrix is positive definite.
12. Formulate an algorithm to check whether the given matrix is invertible.
13. Formulate an algorithm to find the row having maximum number of negative elements in a given matrix.
14. Formulate an algorithm to find the arithmetic mean of elements of each row of a given matrix.
15. Formulate an algorithm to find the largest and smallest elements of a given matrix.
16. Formulate an algorithm that finds the sum of each row and each column of a given matrix and as its output print the row having largest sum of its elements and column having smallest sum of its elements.

Solution of System of Linear Equations

Learning Outcomes

After reading this chapter, students will have the ability to

- define a linear system of homogeneous and non-homogeneous equations
- explain the Crammer's method and its limitations
- explain various direct methods and iterative methods to find solution of non-homogeneous system of linear equations
- use various methods to solve non-homogeneous system of linear equations manually
- write programs for various methods to solve non-homogeneous system of linear equations manually
- define an eigen value problem
- explain the Power method to solve eigen value problem
- use Power method solve a to solve eigen value problem manually
- write a program to implement Power method

5.1 INTRODUCTION

In various fields of study, there are many problems involving the solution of system of algebraic equations. More often these equations are linear, therefore our discussion in this chapter is primarily concerned with the methods applicable for solving system of linear equations.

From study of algebra, the readers must be familiar with two common methods of solving system of linear equations:

- By the elimination of the variables by combining equations.
- By the use of determinants, method better known as *Cramer's rule.*

When smaller number of equations is to be solved, the Cramer's rule appears to have an edge over the elimination method. However, the Cramer's rule, method of determinants with expansion by minors, is completely impractical when large number of equations is to be solved. Forsythe has proved that in order to solve a system of n linear equations by Cramer's rule, evaluating determinants with expansion by minors, requires

$$[(n-1)(n+1)]!$$

multiplications. Thus, even the solution of 10 equations would require 35,92,51,200 multiplications. Obviously, a solution involving the use of determinants with expansion by minors is not practical. Even if determinants are evaluated by the use of Gauss elimination method, the use of Cramer's rule is very inefficient.

The system of linear equations of our concern will involve n equations in n unknowns having the general form

$$\begin{aligned}
a_{11}x_1 + a_{12}x_2 + a_{13}x_3 + \ldots + a_{1n}x_n &= b_1 \\
a_{21}x_1 + a_{22}x_2 + a_{23}x_3 + \ldots + a_{2n}x_n &= b_2 \\
a_{31}x_1 + a_{32}x_2 + a_{33}x_3 + \ldots + a_{3n}x_n &= b_3 \\
\vdots \qquad\qquad & \quad \vdots \\
a_{n1}x_1 + a_{n2}x_2 + a_{n3}x_3 + \ldots + a_{nn}x_n &= b_n
\end{aligned} \qquad \ldots (5.1)$$

If all the b's are not equal to zero, the system of linear equation is called *non-homogeneous.* A unique solution of non-homogeneous system of linear equations exists only if these equations are independent. If all the b's are zero, the system of linear equation is called *homogeneous.* A non-trivial solution (i.e. other than $x_1=x_2=x_3=\ldots=x_n=0$) exists only if all equations are not independent.

In this chapter, we will be concerned with the methods to obtain the solution of non-homogeneous system of linear equations.

5.2 SOLUTION OF NON-HOMOGENEOUS SYSTEM OF LINEAR EQUATIONS

The *coefficient matrix* of the system of linear equation is

$$A = \begin{pmatrix} a_{11} & a_{12} & a_{13} & \cdot & \cdot & \cdot & a_{1n} \\ a_{21} & a_{22} & a_{23} & \cdot & \cdot & \cdot & a_{2n} \\ a_{31} & a_{32} & a_{33} & & & & a_{3n} \\ \cdot & \cdot & \cdot & \cdot & \cdot & \cdot & \cdot \\ \cdot & \cdot & \cdot & \cdot & \cdot & \cdot & \cdot \\ \cdot & \cdot & \cdot & \cdot & \cdot & \cdot & \cdot \\ a_{n1} & a_{n2} & a_{n3} & \cdot & \cdot & \cdot & a_{nn} \end{pmatrix} \quad \ldots (5.2)$$

That is, the coefficient matrix is a matrix whose elements are the coefficients of the unknowns in the system of linear equations. If the constants of the Eqs. (5.1) i.e. b's, are appended to the coefficient matrix as a last column i.e.$(n+1)$th column, the resulting matrix is known as the *augmented matrix* as shown

$$[A\vdots B] = \begin{pmatrix} a_{11} & a_{12} & a_{13} & \cdot & \cdot & \cdot & a_{1n} & b_1 \\ a_{21} & a_{22} & a_{23} & \cdot & \cdot & \cdot & a_{2n} & b_2 \\ a_{31} & a_{32} & a_{33} & \cdot & \cdot & \cdot & a_{3n} & b_3 \\ \cdot & \cdot & \cdot & \cdot & \cdot & \cdot & \cdot & \cdot \\ \cdot & \cdot & \cdot & \cdot & \cdot & \cdot & \cdot & \cdot \\ \cdot & \cdot & \cdot & \cdot & \cdot & \cdot & \cdot & \cdot \\ a_{n1} & a_{n2} & a_{n3} & \cdot & \cdot & \cdot & a_{nn} & b_n \end{pmatrix} \quad \ldots (5.3)$$

In order to implement on computer, it is more convenient to express these b's as simply an additional column of the a_{ij}'s. In such a case, the matrix of Eqs. (5.3) might be expressed as

$$A = \begin{pmatrix} a_{11} & a_{12} & a_{13} & & & & a_{1n} & a_{1(n+1)} \\ a_{21} & a_{22} & a_{23} & \cdot & \cdot & \cdot & a_{2n} & a_{2(n+1)} \\ a_{31} & a_{32} & a_{33} & \cdot & \cdot & \cdot & a_{3n} & a_{3(n+1)} \\ \cdot & \cdot & \cdot & \cdot & \cdot & \cdot & \cdot & \cdot \\ \cdot & \cdot & \cdot & \cdot & \cdot & \cdot & \cdot & \cdot \\ \cdot & \cdot & \cdot & \cdot & \cdot & \cdot & \cdot & \cdot \\ a_{n1} & a_{n2} & a_{n3} & \cdot & \cdot & \cdot & a_{nn} & a_{n(n+1)} \end{pmatrix} \quad \ldots (5.4)$$

This form will be used in discussions in the sections to follow.

The general criteria for the existence of a solution of non-homogeneous linear equation in n unknowns is as follows:

- If rank of both coefficient and augmented matrices is equal to n, then the system of equations is consistent and has a *unique solution*.
- If rank of both coefficient and augmented matrices is less than n, then the system of equations is also consistent and has *infinite solutions*.
- If rank of coefficient matrix is less than rank of augmented matrix, then the system of equations is inconsistent and has *no solution*.

5.3 METHODS OF FINDING SOLUTION OF NON-HOMOGENEOUS SYSTEM OF LINEAR EQUATIONS

There are two types of methods to obtain solution of non-homogeneous system of linear equations. These are

- **Direct Methods**

Direct methods are also known as *reduction methods*. Under this category, we will discuss

- Gauss elimination method
- Gauss-Jordon method
- Matrix Inversion Method
- Crout's method (method of *LU* decomposition)

- **Iterative Methods**

Under this category, we will discuss

- Jacobi's method
- Gauss-Seidel method

Since a system of non-homogeneous linear equations usually consists of linearly independent equations, we will assume that a unique solution does exist. Remember that a unique solution for such as system of equations exists only if the coefficient matrix is *non-singular*, i.e. if the determinant of the coefficient matrix is non-zero.

5.4 DIRECT METHODS

The direct methods give the solution of system of non-homogeneous linear equation in a fixed number of steps. In this section, the various direct methods are described.

5.4.1 Gauss Elimination Method

This is one of the most widely used methods. This method is a systematic process of eliminating unknowns from the linear equations. This method is divided into two parts:

- Triangularization
- Back substitution

The system of n equations in n unknowns is reduced to an equivalent triangular system (an equivalent system is a system having identical solution) of equation of type

$$
\begin{aligned}
a_{11}x_1 + a_{12}x_2 + a_{13}x_3 + \ldots + a_{1n}x_n &= b_1 \\
a_{22}x_2 + a_{23}x_3 + \ldots + a_{2n}x_n &= b_2 \\
a_{33}x_3 + \ldots + a_{3n}x_n &= b_3 \\
&\vdots \\
a_{nn}x_n &= b_n
\end{aligned} \qquad \ldots (5.5)
$$

Using the back substitution procedure can easily solve this new equivalent system of equations.

The details of the Gauss elimination method are given below.

Step 1. Eliminate x_1 from 2nd equation onwards. This is done as follows:

- Subtract from the second equation $\frac{a_{21}}{a_{11}}$ times the first equation. This results in

$$\left[a_{21} - \frac{a_{21}}{a_{11}}a_{11}\right]x_1 + \left[a_{22} - \frac{a_{21}}{a_{11}}a_{12}\right]x_2 + \ldots + \left[a_{2n} - \frac{a_{21}}{a_{11}}a_{1n}\right]x_n = \left[b_2 - \frac{a_{21}}{a_{11}}b_1\right]$$

  which on simplifying gives equation of type

$$a_{22}x_2 + a_{23}x_3 + \ldots + a_{2n}x_n = b_2$$

- Similarly, subtract from the third equation $\frac{a_{21}}{a_{11}}$ times the first equation. This will give equation of type

$$a_{32}x_2 + a_{33}x_3 + \ldots + a_{3n}x_n = b_3$$

- If we repeat this process till the nth equation is operated, we get the new system of equation of type, as shown overleaf.

$$
\begin{aligned}
a_{11}x_1 + a_{12}x_2 + a_{13}x_3 + \ldots + a_{1n}x_n &= b_1 \\
a_{22}x_2 + a_{23}x_3 + \ldots + a_{2n}x_n &= b_2 \\
a_{32}x_2 + a_{33}x_3 + \ldots + a_{3n}x_n &= b_3 \\
&\vdots \\
a_{n2}x_2 + a_{n3}x_3 + \ldots + a_{nn}x_n &= b_n
\end{aligned}
\qquad \ldots (5.6)
$$

The solution of these equations is same as that of the original equations.

Step 2. Eliminate x_2 from 3rd equation onwards from the system of Eqs. (5.6). This is done as follows:

- Subtract from the third equation $\frac{a_{32}}{a_{22}}$ times the second equation.
- Subtract from the fourth equation $\frac{a_{42}}{a_{22}}$ times the second equation.
- and so on.

We will thus get the new system of equations of type

$$
\begin{aligned}
a_{11}x_1 + a_{12}x_2 + a_{13}x_3 + \ldots + a_{1n}x_n &= b_1 \\
a_{22}x_2 + a_{23}x_3 + \ldots + a_{2n}x_n &= b_2 \\
a_{33}x_3 + \ldots + a_{3n}x_n &= b_3 \\
&\vdots \\
a_{n3}x_3 + \ldots + a_{nn}x_n &= b_n
\end{aligned}
\qquad \ldots (5.7)
$$

This process will continue till the last equation contains only one unknown, namely x_n. The final form of equation will look like

$$
\begin{aligned}
a_{11}x_1 + a_{12}x_2 + a_{13}x_3 + \ldots + a_{1n}x_n &= b_1 \\
a_{22}x_2 + a_{23}x_3 + \ldots + a_{2n}x_n &= b_2 \\
a_{33}x_3 + \ldots + a_{3n}x_n &= b_3 \\
&\vdots \\
a_{nn}x_n &= b_n
\end{aligned}
\qquad \ldots (5.8)
$$

This process is called *triangularization*. In order to implement this method on computer, we consider the augmented matrix of Eqs. (5.4), which after elimination takes the following form as shown overleaf:

$$A = \begin{pmatrix} a_{11} & a_{12} & a_{13} & \cdots & a_{1n} & a_{1(n+1)} \\ 0 & a_{22} & a_{23} & \cdots & a_{2n} & a_{2(n+1)} \\ 0 & 0 & a_{33} & \cdots & a_{3n} & a_{3(n+1)} \\ \cdot & \cdot & \cdot & \cdots & \cdot & \cdot \\ \cdot & \cdot & \cdot & \cdots & \cdot & \cdot \\ \cdot & \cdot & \cdot & \cdots & \cdot & \cdot \\ 0 & 0 & 0 & \cdots & a_{nn} & a_{n(n+1)} \end{pmatrix} \quad \ldots (5.9)$$

Algorithm 5.1: Triangularizing n equations in n unknowns

The given augmented matrix A of order $n\times(n+1)$ is stored using a two-dimensional array named a of size $n\times(n+1)$.

```
Begin
    read: n
    for i = 1 to n by 1 do
        for j = 1 to (n +1) by 1 do
            read: aij
        endfor
    endfor
    for k = 1 to (n −1) by 1 do
        for i = (k+1) to n by 1 do
            set temp = aik/akk
            for j = k to (n +1) by 1 do
                set aij = aij − temp × akj
            endfor
        endfor
    endfor
End.
```

From the triangular system of linear equations, find the values of x_n from the nth equation as

$$x_n = \frac{a_{n(n+1)}}{a_{nn}} \quad \ldots (5.10)$$

Then the value of x_n is substituted back in the $(n-1)$th equation to obtain the value of x_{n-1}. This back substitution is continued till we get the solution for x_1. The general formula for back substitution is

$$x_i = \frac{\left[a_{i(n+1)} - \sum_{j=i+1}^{n} a_{ij}x_j\right]}{a_{ii}} \qquad \text{for } i = n-1, n-2, \ldots, 3, 2, 1 \quad \ldots (5.11)$$

This process is called *back substitution*.

Algorithm 5.2: Back Substitution

The reduced augmented matrix A of order $n\times(n+1)$ is stored using a two-dimensional array named a of size $n\times(n+1)$. The solution of the system of linear equations is stored in one-dimensional array named x of size n.

```
Begin
    set x_n = a_n(n+1) / a_nn
    for i = (n -1) to 1 by -1 do
        set sum = 0
        for j = (i+1) to n by 1 do
            set sum = sum + a_ij x_j
        endfor
        set x_i = (a_i(n+1) - sum) / a_ii
    endfor
End.
```

If we combine the steps for triangularization and back substitution, the complete algorithm for the Gauss elimination method may look like as given below:

Algorithm 5.3: Gauss elimination method

The augmented matrix A of order $n\times(n+1)$ is stored using a two-dimensional array named a of size $n\times(n+1)$. The solution of the system of linear equations is stored in one-dimensional array named x of size n.

```
Begin
    read: n
    for i = 1 to n by 1 do
        for j = 1 to (n +1) by 1 do
            read: a_ij
        endfor
    endfor
    for k = 1 to (n -1) by 1 do
        for i = (k+1) to n by 1 do
            set temp = a_ik / a_kk
            for j = k to (n +1) by 1 do
                set a_ij = a_ij - temp × a_kj
            endfor
        endfor
    endfor
    set x_n = a_n(n+1) / a_nn
    for i = (n -1) to 1 by -1 do
        set sum = 0
        for j = (i+1) to n by 1 do
            set sum = sum + a_ij x_j
        endfor
        set x_i = (a_i(n+1) - sum) / a_ii
```

```
        endfor
        write: "Solution of equations"
        for i = 1 to n by 1 do
            write: xi
        endfor
End.
```

Example 5.1: Solve the following system of linear equations

$$2x_1 + 8x_2 + 2x_3 = 14$$
$$x_1 + 6x_2 - x_3 = 13$$
$$2x_1 - x_2 + 2x_3 = 5$$

Solution: In order to eliminate x_1 from the second and third equation, first apply transformation

$$R_2 - \frac{a_{21}}{a_{11}} R_1 \quad \Rightarrow \quad R_2 - \frac{1}{2} R_1$$

The coefficients of the second equation are computed as

$$a_{21} = a_{21} - \frac{1}{2} a_{11} = 1 - \frac{1}{2} \times 2 = 0 \qquad a_{22} = a_{22} - \frac{1}{2} a_{12} = 6 - \frac{1}{2} \times 8 = 2$$

$$a_{23} = a_{23} - \frac{1}{2} a_{13} = -1 - \frac{1}{2} \times 2 = -2 \qquad b_2 = b_2 - \frac{1}{2} b_1 = 13 - \frac{1}{2} \times 14 = 6$$

Now apply transformation

$$R_3 - \frac{a_{31}}{a_{11}} R_1 \quad \Rightarrow \quad R_3 - \frac{2}{2} R_1 \quad \Rightarrow \quad R_3 - R_1$$

The elements of the third equation are computed as

$$a_{31} = a_{31} - a_{11} = 2 - 2 = 0 \qquad a_{32} = a_{32} - a_{12} = -1 - 8 = -9$$

$$a_{33} = a_{33} - a_{13} = 2 - 2 = 0 \qquad b_3 = b_3 - b_1 = 5 - 14 = -9$$

Thus eliminating x_1 from second and third equation, we obtain the new system of linear equations as

$$2x_1 + 8x_2 + 2x_3 = 14$$
$$2x_2 - 2x_3 = 6$$
$$-9x_2 + 0x_3 = -9$$

Finally, in order to eliminate x_2 from the third equation, apply transformation

$$R_3 - \frac{a_{32}}{a_{22}} R_2 \quad \Rightarrow \quad R_3 - \frac{-9}{2} R_2 \quad \Rightarrow \quad R_3 + \frac{9}{2} R_2$$

The elements of the third equation are computed as

$$a_{32} = a_{32} + \frac{9}{2} a_{22} = -9 + \frac{9}{2} \times 2 = 0$$

$$a_{33} = a_{33} + \frac{9}{2} a_{23} = 0 + \frac{9}{2} \times (-2) = -9$$

$$b_3 = b_3 + \frac{9}{2} b_2 = -9 + \frac{9}{2} \times 6 = 18$$

The final system will look as

$$\begin{aligned} 2x_1 + 8x_2 + 2x_3 &= 14 \\ 2x_2 - 2x_3 &= 6 \\ -9x_3 &= 18 \end{aligned}$$

Through back substitution, the following solution values are obtained:

$$x_3 = \frac{18}{-9} = -2 \qquad x_2 = \frac{(6 + 2x_3)}{2} = 1 \qquad x_1 = \frac{(14 - 2x_3 - 8x_2)}{2} = 5$$

Incorporating pivoting, the algorithm for the implementation of the modified Gauss elimination method may look like

Algorithm 5.3: Gauss elimination method with Pivoting

The augmented matrix *A* of order $n \times (n+1)$ is stored using a two-dimensional array named *a* of size $n \times (n+1)$. The solution of the system of linear equations is stored in one-dimensional array named *x* of size *n*.

```
Begin
    read: n
    for i = 1 to n by 1 do
        for j = 1 to (n +1) by 1 do
            read: a_ij
        endfor
    endfor
    for k = 1 to (n -1) by 1 do
        set max = | a_kk |
        set p = k
        for m = (k+1) to n by 1 do
            if (| a_mk | > max ) then
                set max = | a_mk |
                set p = m
            endif
        endfor
```

```
        if ( p ≠ k ) then
            for q = k to (n+1)by 1 do
                set temp = a_kq
                set a_kq = a_pq
                set a_pq = temp
            endfor
        endif
        for i = (k+1) to n by 1 do
            set temp = a_ik/a_kk
            for j = k to (n +1) by 1 do
                set a_ij  = a_ij  - temp × a_kj
            endfor
        endfor
    endfor
    set x_n = a_n(n+1)/a_nn
    for i = (n -1) to 1 by -1 do
        set sum = 0
        for j = (i+1) to n by 1 do
            set sum = sum + a_ij x_j
        endfor
        set x_i = (a_i(n+1) - sum)/a_ii
    endfor
    write: "Solution of equations"
    for i = 1 to n by 1 do
        write: x_i
    endfor
End.
```

5.4.2 Gauss-Jordon Method

The difference between the Gauss-Jordon and the Gauss elimination is that when an unknown is eliminated, it is eliminated from all other equations not only from equations to follow. Therefore, in each equation we are left with only one unknown, and this eliminates the use of back substitution process. Jordon did this modification to the Gauss elimination method, from where it derived its name Gauss-Jordon method.

Consider again the Eqs.(5.1). This system of equations is reduced to the following system

$$\begin{aligned} x_1 \qquad\qquad\qquad\qquad &= b_1 \\ x_2 \qquad\qquad\qquad &= b_2 \\ x_3 \qquad\qquad &= b_3 \\ &\vdots \\ x_n &= b_n \end{aligned} \qquad \ldots (5.12)$$

whose solution is, trivially, given by the right hand side.

The details of the Gauss-Jordon method are given below.

Step 1. Eliminate x_1 from all equations except first equation. This is done as follows:

- Divide the first equation by a_{11}. Subtract from the second equation a_{21} times the first equation, subtract from the third equation a_{31} times the first equation, and so on. Finally subtract from the nth equation a_{n1} times the first equation.

We get the new system of equations as

$$\begin{aligned} x_1 + a_{12}x_2 + a_{13}x_3 + \ldots + a_{1n}x_n &= b_1 \\ 0x_1 + a_{22}x_2 + a_{23}x_3 + \ldots + a_{2n}x_n &= b_2 \\ 0x_1 + a_{32}x_2 + a_{33}x_3 + \ldots + a_{3n}x_n &= b_3 \\ &\vdots \\ 0x_1 + a_{n2}x_2 + a_{n3}x_3 + \ldots + a_{nn}x_n &= b_n \end{aligned} \qquad \ldots (5.13)$$

Step 2. Eliminate x_2 from all equations except second equation. This is done as follows:

- Divide the second equation by a_{22}. Subtract from the first equation a_{12} times the second equation, subtract from the third equation a_{32} times the second equation, and so on. Finally subtract from the nth equation a_{n2} times the second equation.

We get the new system of equations as

$$\begin{aligned} x_1 + 0x_2 + a_{13}x_3 + \ldots + a_{1n}x_n &= b_1 \\ 0x_1 + x_2 + a_{23}x_3 + \ldots + a_{2n}x_n &= b_2 \\ 0x_1 + 0x_2 + a_{33}x_3 + \ldots + a_{3n}x_n &= b_3 \\ &\vdots \\ 0x_1 + 0x_2 + a_{n3}x_3 + \ldots + a_{nn}x_n &= b_n \end{aligned} \qquad \ldots (5.14)$$

⋮

This process will continue till x_n is eliminated from the first to $(n-1)$th equation. The final form of system of equations looks like

$$\begin{aligned} x_1 + 0x_2 + 0x_3 + \ldots + 0x_n &= b_1 \\ 0x_1 + x_2 + 0x_3 + \ldots + 0x_n &= b_2 \\ 0x_1 + 0x_2 + x_3 + \ldots + 0x_n &= b_3 \\ &\vdots \\ 0x_1 + 0x_2 + 0x_3 + \ldots + x_n &= b_n \end{aligned} \qquad \ldots (5.15)$$

It is clear that the values of the unknowns are given by the coefficients on the right hand side of the equations, i.e.,

$$x_1 = b_1 \qquad x_2 = b_2 \qquad x_3 = b_3 \qquad \ldots \qquad x_n = b_n$$

Example 5.2: Solve the following system of equations

$$2x_1 - 2x_2 + 5x_3 = 13$$
$$2x_1 + 3x_2 + 4x_3 = 20$$
$$3x_1 - x_2 + 3x_3 = 10$$

Solution: Dividing first equation by 2, we get

$$x_1 - x_2 + 2.5x_3 = 6.5$$

Now multiplying this equation by 2 and subtracting from the second equation, we get

$$0x_1 + 5.0x_2 - x_3 = 7$$

Similarly, multiply the first equation by 3 and subtract from the third equation, we get

$$0x_1 + 2x_2 - 4.5x_3 = -9.5$$

So after eliminating x_1 from second and third equation, new system of equation becomes

$$x_1 - x_2 + 2.5x_3 = 6.5$$
$$0x_1 + 5.0x_2 - x_3 = 7$$
$$0x_1 + 2x_2 - 4.5x_3 = -9.5$$

Now dividing the second row by 5.0, we get the new second row as

$$0x_1 + x_2 - 0.2x_3 = 1.4$$

In order to eliminate x_2 from first and the third equation, proceed as follows:

Multiplying the new second equation by −1 and subtracting from the first equation, we get the new first equation as

$$x_1 + 0x_2 + 2.3x_3 = 7.9$$

Now multiplying the new second equation by 2 and subtracting from the third equation, we get the new third equation as

$$0x_1 + 0x_2 - 4.1x_3 = -12.3$$

So after eliminating x_2 from first and third equation, the new system of equation becomes

$$x_1 + 0x_2 + 2.3x_3 = 7.9$$
$$0x_1 + x_2 - 0.2x_3 = 1.4$$
$$0x_1 + 0x_2 - 4.1x_3 = -12.3$$

Now dividing the third row by −4.1, we get the new third row as

$$0x_1 + 0x_2 + x_3 = 3.0$$

In order to eliminate x_3 from first and second equation, proceed as follows:

Multiplying the new third equation by 2.3 and subtracting from the first equation, we get the new first equation as

$$x_1 + 0x_2 + 0x_3 = 1.0$$

Now multiplying the new third equation by –0.2 and subtracting from the second equation, we get the new second equation as

$$0x_1 + x_2 + 0x_3 = 2.0$$

So after eliminating x_3 from first and second equation, final system of equation becomes

$$x_1 + 0x_2 + 0x_3 = 1.0$$
$$0x_1 + x_2 + 0x_3 = 2.0$$
$$0x_1 + 0x_2 + x_3 = 3.0$$

Thus, the required solution is $x_1 = 1.0$, $x_2 = 2.0$, $x_3 = 3.0$.

Algorithm 5.5: Gauss-Jordon Method

The augmented matrix A of order $n \times (n+1)$ is stored using a two-dimensional array named a of size $n \times (n+1)$. The solution of the system of linear equations is stored in one-dimensional array named x of size n.

```
Begin
   read: n
   for i = 1 to n by 1 do
      for j = 1 to (n+1) by 1 do
         read: aij
      endfor
   endfor
   for k = 1 to n by 1 do
      set temp = akk
      for j = k to (n +1) by 1 do
         set akj  = akj/temp
      endfor
      for i = 1 to n by 1 do
         if ( i ≠ k ) then
            set comm = aik
            for j = k to (n+1) by 1 do
               set aij = aij  - comm × akj
            endfor
         endif
      endfor
   endfor
   write: "Solution of equations"
   for i = 1 to n by 1 do
      write: xi
   endfor
End.
```

5.4.3 Matrix Inversion Method

Consider again the system of linear equation given by Eqs. (5.1). This system of linear equations can be written in the matrix form as

$$AX = B \qquad \text{... (5.16)}$$

where

$$A = \begin{pmatrix} a_{11} & a_{12} & a_{13} & \cdot & \cdot & \cdot & a_{1n} \\ a_{21} & a_{22} & a_{23} & \cdot & \cdot & \cdot & a_{2n} \\ a_{31} & a_{32} & a_{33} & \cdot & \cdot & \cdot & a_{3n} \\ \cdot & \cdot & \cdot & \cdot & \cdot & \cdot & \cdot \\ \cdot & \cdot & \cdot & \cdot & \cdot & \cdot & \cdot \\ \cdot & \cdot & \cdot & \cdot & \cdot & \cdot & \cdot \\ a_{n1} & a_{n2} & a_{n3} & \cdot & \cdot & \cdot & a_{nn} \end{pmatrix} \quad X = \begin{pmatrix} x_1 \\ x_2 \\ x_3 \\ \cdot \\ \cdot \\ \cdot \\ x_n \end{pmatrix} \quad B = \begin{pmatrix} b_1 \\ b_2 \\ b_3 \\ \cdot \\ \cdot \\ \cdot \\ b_n \end{pmatrix} \qquad \text{... (5.17)}$$

Note that X and B are column matrices and represent the required solution and the constants on the right hand side of the equations. The column matrix X is also known as the *solution vector.*

If the given system of equation, Eqs.(5.1), has a unique solution, the coefficient matrix A is non-singular, and thus there exists A^{-1} for A.

Pre-multiply the matrix equation by A^{-1} on both sides, we get

$$A^{-1}AX = A^{-1}B$$

Since $A^{-1}A = I$, therefore $IX = A^{-1}B$ which gives $X = A^{-1}B$.

If we know the inverse of the matrix A, we can obtain the solution vector X by post-multiplying it by B. The procedure to find the inverse of a square matrix is described in *Chapter 4*.

Example 5.3: Solve the following system of equation

$$2x_1 - 2x_2 + 5x_3 = 13$$
$$2x_1 + 3x_2 + 4x_3 = 20$$
$$3x_1 - x_2 + 3x_3 = 10$$

Solution:

Given $A = \begin{pmatrix} 2 & -2 & 5 \\ 2 & 3 & 4 \\ 3 & -1 & 3 \end{pmatrix} \quad X = \begin{pmatrix} x_1 \\ x_2 \\ x_3 \end{pmatrix} \quad B = \begin{pmatrix} 13 \\ 20 \\ 10 \end{pmatrix}$

Finding A^{-1} using Gauss-Jordon elimination procedure discussed in Chapter 4, and retaining three decimal digits, we get

$$A^{-1} = \begin{pmatrix} -0.317 & -0.024 & 0.561 \\ -0.146 & 0.220 & -0.049 \\ -0.268 & 0.098 & -0.244 \end{pmatrix}$$

Therefore

$$\begin{pmatrix} x_1 \\ x_2 \\ x_3 \end{pmatrix} = \begin{pmatrix} -0.317 & -0.024 & 0.561 \\ -0.146 & 0.220 & -0.049 \\ -0.268 & 0.098 & -0.244 \end{pmatrix} \times \begin{pmatrix} 13 \\ 20 \\ 10 \end{pmatrix} = \begin{pmatrix} 1.009 \\ 2.012 \\ 3.004 \end{pmatrix}.$$

Thus

$x_1 = 1.009$ $x_2 = 2.012$ $x_3 = 3.004$ (retaining only three decimal digits)

Algorithm 5.6: Matrix Inversion Method

The coefficient matrix A of order $n \times n$ is stored using a two-dimensional array named a of size $n \times n$, the right hand side values are stored in one-dimensional array b of size n, the solution of the system of linear equations is stored in one-dimensional array named x of size n.

```
Begin
    read: n
    for i = 1 to n by 1 do
        for j = 1 to (n+1) by 1 do
            read: a_ij
        endfor
        read: b_i
    endfor
    for i = 1 to n by 1 do
        for j = (n+1) to (n+n) by 1 do
            if ((j -i ) = n) then
                set a_ij = 1
            else
                set a_ij = 0
            endif
        endfor
    endfor
    for k = 1 to n by 1 do
        set temp = a_kk
        for j = k to (n+n) by 1 do
            set a_kj = a_kj / temp
        endfor
        for i = 1 to n by 1 do
```

```
            if ( i ≠ k ) then
               set comm = a_ik
               for j = k to (n+n) by 1 do
                   set a_ij = a_ij - comm × a_kj
                endfor
            endif
        endfor
    endfor
    for i = 1 to n by 1 do
        for j = (n+1) to (n+n) by 1 do
            set a_i(j-n) = a_ij
        endfor
    endfor
    for i = 1 to n by 1 do
        set x_i = 0
        for j = 1 to n by 1 do
           set x_i = x_i +a_ij × b_j
        endfor
    endfor
    write: "Solution of the given system is"
    for i = 1 to n by 1 do
        write: x_i
    endfor
End.
```

5.4.4 Crout's Method

Crout's method, also known as method of *LU* decomposition, the method of matrix factorization, is more economical than other methods. This method has been extensively used in solving large structural analysis problems.

To illustrate this method, we will consider system of three linear equations expressed in the matrix from as shown below:

$$\begin{pmatrix} a_{11} & a_{12} & a_{13} \\ a_{21} & a_{22} & a_{23} \\ a_{31} & a_{32} & a_{33} \end{pmatrix} \begin{pmatrix} x_1 \\ x_2 \\ x_3 \end{pmatrix} = \begin{pmatrix} b_1 \\ b_2 \\ b_3 \end{pmatrix}$$

The above system of equation can be written as

$$AX - B = 0 \qquad \ldots (5.18)$$

where *A* represents the coefficient matrix, *X* the column matrix of unknowns and *B* the column matrix of the constants on the right hand side.

Let us suppose that the above system of equation is reduced to

$$\begin{pmatrix} 1 & u_{12} & u_{13} \\ 0 & 1 & u_{23} \\ 0 & 0 & 1 \end{pmatrix}\begin{pmatrix} x_1 \\ x_2 \\ x_3 \end{pmatrix} = \begin{pmatrix} d_1 \\ d_2 \\ d_3 \end{pmatrix}$$

Then this system of equations can be solved by the back substitution method. Let this upper triangular matrix is denoted by U. This system of equation can be written in the matrix form as

$$UX - D = 0 \qquad \ldots (5.19)$$

where D is the column matrix of the constants on the right hand side of the reduced system of equations.

Suppose there exist a lower triangular matrix

$$\begin{pmatrix} l_{11} & 0 & 0 \\ l_{21} & l_{22} & 0 \\ l_{31} & l_{32} & l_{33} \end{pmatrix}$$

denoted by L and satisfies the condition that if the left hand side of Eq. (5.19) is multiplied by L, we get the right hand side. Therefore, in accordance with the above condition

$$L(UX - D) = AX - B \qquad \ldots (5.20)$$

$$\Rightarrow \quad LU = A \quad \text{and} \quad LD = C \qquad \ldots (5.21)$$

Therefore

$$\begin{pmatrix} l_{11} & 0 & 0 \\ l_{21} & l_{22} & 0 \\ l_{31} & l_{32} & l_{33} \end{pmatrix}\left(\begin{array}{ccc|c} 1 & u_{12} & u_{13} & d_1 \\ 0 & 1 & u_{23} & d_2 \\ 0 & 0 & 1 & d_3 \end{array}\right) = \left(\begin{array}{ccc|c} a_{11} & a_{12} & a_{13} & b_1 \\ a_{21} & a_{22} & a_{23} & b_2 \\ a_{31} & a_{32} & a_{33} & b_3 \end{array}\right)$$

For the sake of convenience, let d's be represented by u_{i4} and b's be represented by a_{i4} to obtain

$$\begin{pmatrix} l_{11} & 0 & 0 \\ l_{21} & l_{22} & 0 \\ l_{31} & l_{32} & l_{33} \end{pmatrix}\begin{pmatrix} 1 & u_{12} & u_{13} & u_{14} \\ 0 & 1 & u_{23} & u_{24} \\ 0 & 0 & 1 & u_{34} \end{pmatrix} = \begin{pmatrix} a_{11} & a_{12} & a_{13} & a_{14} \\ a_{21} & a_{22} & a_{23} & a_{24} \\ a_{31} & a_{32} & a_{33} & a_{34} \end{pmatrix}$$

$$\begin{pmatrix} l_{11} & l_{11}u_{12} & l_{11}u_{13} & l_{11}u_{14} \\ l_{21} & l_{21}u_{12} + l_{22} & l_{21}u_{13} + l_{22}u_{23} & l_{21}u_{14} + l_{22}u_{24} \\ l_{31} & l_{31}u_{12} + l_{32} & l_{31}u_{13} + l_{32}u_{23} + l_{33} & l_{31}u_{14} + l_{32}u_{24} + l_{33}u_{34} \end{pmatrix} = \begin{pmatrix} a_{11} & a_{12} & a_{13} & a_{14} \\ a_{21} & a_{22} & a_{23} & a_{24} \\ a_{31} & a_{32} & a_{33} & a_{34} \end{pmatrix}$$

By comparing the first column on both sides, we get

$$l_{i1} = a_{i1} \qquad \text{for } i = 1, 2, 3$$

By comparing the first row on both sides, we get

$$u_{1j} = a_{1j}/l_{11} \qquad \text{for } j = 2, 3, 4$$

So we have determined the first column of L and the first row of U. We can now proceed to find the second column of L and the second row of U as shown below:

By comparing the second column on both sides, we get

$$l_{22}u_{12}+l_{22} = a_{22} \quad \Rightarrow \quad l_{22} = a_{22} - l_{21}u_{12}$$

$$l_{31}u_{12}+l_{32} = a_{32} \quad \Rightarrow \quad l_{32} = a_{32} - l_{31}u_{12}$$

Now after having the second column of L, we proceed to obtain the second row of U by comparing the second row on both sides, and using the known values of l_{ij}'s as

$$l_{21}u_{13}+l_{22}u_{23} = a_{23} \quad \Rightarrow \quad u_{23} = (a_{23} - l_{21}u_{13})/l_{22}$$

$$l_{21}u_{14}+l_{22}u_{24} = a_{24} \quad \Rightarrow \quad u_{24} = (a_{24} - l_{21}u_{14})/l_{22}$$

Now we get the third column of L followed by the third row of U. The third column of L is obtained by comparing the last element of the third column on both sides as

$$l_{31}u_{13}+l_{32}u_{23}+l_{33} = a_{33} \Rightarrow \quad l_{33} = a_{33} - (l_{31}u_{13}+l_{32}u_{23})$$

Now, we are left with u_{34}, which is found by comparing the last element of the fourth column on both sides as

$$l_{31}u_{14}+l_{32}u_{24}+l_{33}u_{34} = a_{34} \Rightarrow \quad u_{34} = (a_{34} - (l_{31}u_{14}+l_{32}u_{24}))/l_{33}$$

Therefore, when all the elements of U are known, then solution can be obtained by the back substitution method as

$$x_3 = u_{34}$$

$$x_i = u_{i4} - \sum_{j=i+1}^{3} u_{ij}x_j \qquad \text{for } i = 2, 1$$

The general equations to determine the elements of the matrix L and the augmented matrix U for a system of n linear equations in n unknowns are determined as

$$\left.\begin{aligned} & l_{i1} = a_{i1} && \text{for } i = 1 \text{ to } n \\ & u_{1j} = a_{1j}/a_{11} && \text{for } j = 2 \text{ to } n{+}1 \\ & l_{ij} = a_{ij} - \sum_{k=1}^{j-1} l_{ik}u_{kj} && \text{for } j = 2, n \\ & && \text{for } i = j \text{ to } n \\ & && \text{for each value of } j \end{aligned}\right\} \qquad \ldots (5.22a)$$

$$u_{ij} = \left[a_{ij} - \sum_{k=1}^{i-1} l_{ik} u_{kj} \right] / l_{ii} \qquad \left. \begin{array}{l} \text{for } i = 2, n \\ \\ \text{for } j = i+1 \text{ to } n+1 \\ \quad \text{for each value of } i \end{array} \right\} \qquad \ldots (5.22b)$$

And the general back substitution formula is

$$x_n = u_{n(n+1)}$$

$$x_i = u_{i(n+1)} - \sum_{j=i+1}^{n} u_{ij} x_j \qquad \text{for } i = n-1, n-2, \ldots, 3, 2, 1 \qquad \ldots (5.23)$$

Note that if the decomposition

$$LU = A$$

exist, then the elements l_{ii} will be non-zero. This can be seen from the fact that if the above decomposition equation holds, the product of determinant of L and determinant of U equals the determinant of A, i.e.

$$|L||U| = |A|$$

Since we are assuming independent equations, the determinant of A is therefore non-zero. Hence the determinant of L must be non-zero. Recall that the determinant of the triangular matrix is equal to the product of the elements on its principal diagonal, which means that the l_{ii} elements will be non-zero.

Example 5.4: Solve the following system of equations

$$\begin{aligned} 2x_1 + 8x_2 + 2x_3 &= 14 \\ x_1 + 6x_2 - x_3 &= 13 \\ 2x_1 - x_2 + 2x_3 &= 5 \end{aligned}$$

Solution: Using Eqs. (5.22), we get

$$L = \begin{pmatrix} 2 & 0 & 0 \\ 1 & 2 & 0 \\ 2 & -9 & -9 \end{pmatrix} \qquad U = \begin{pmatrix} 1 & 4 & 1 & 7 \\ 0 & 1 & -1 & 3 \\ 0 & 0 & 1 & -2 \end{pmatrix}$$

By applying the back substitution formula given Eqs. (5.23), we get

$$x_3 = -2$$

$$x_2 = 3 - (-1) \times (-2) = 1$$

$$x_1 = 7 - 4 \times 1 - 1 \times (-2) = 5$$

as the solution of the given system of equations.

Algorithm 5.7: Crout's Method

The augmented matrix A of order $n\times(n+1)$ is stored using a two-dimensional array named a of size $n\times(n+1)$, the lower triangular matrix is stored in one-dimensional array l of size $n\times n$, the augmented upper triangular matrix is stored in one-dimensional array u of size $n\times(n+1)$, and the solution vector is stored in one-dimensional array named x of size n.

```
Begin
    read: n
    for i = 1 to n by 1 do
        for j = 1 to (n+1) by 1 do
            read: a_ij
        endfor
    endfor
    for i = 1 to n by 1 do
        set l_i1 = a_i1
    endfor
    for j = 2 to (n+1) by 1 do
        set u_1j = a_1j /a_11
    endfor
    for j = 2 to n by 1 do
        for i = j to n by 1 do
            set sum = 0
            for k = 1 to (j -1) by 1 do
                set sum = sum + l_ik × u_kj
            endfor
            set l_ij = a_ij - sum
        endfor
    endfor
    for i = 2 to n by 1 do
        for j = (i+1) to (n+1) by 1 do
            set sum = 0
            for k = 1 to (i -1) by 1 do
                set sum = sum + l_ik × u_kj
            endfor
            set u_ij = (a_ij - sum)/l_ii
        endfor
    endfor
    set x_n = u_n(n+1)
    for i = (n -1) to 1 by  1 do
        set sum = 0
        for j = (i+1) to n by 1 do
            set sum = sum +u_ij × x_j
        endfor
        set x_i = u_i(n+1) - sum
    endfor
    write: "Solution of the given system is"
    for i = 1 to n by 1 do
```

```
        write: x_i
    endfor
End.
```

The Crout's method requires fewer arithmetic operations than the Gauss elimination and Gauss-Jordon elimination methods. It further can be made more economical of storage by overlapping the L and U matrices on the matrix A itself. This can be easily done, as there is no need to store 0's and 1's of L and U matrices. But the question arises - *How is this done*? The procedure is very simple. Calling element l_{ij} and u_{ij} by simply a_{ij}, the old values of a_{ij}, are replaced by new a_{ij}, values. The first equation is automatically satisfied with no computer execution. The other equations becomes

$$\left.\begin{array}{ll} a_{1j} = a_{1j}/a_{11} & \text{for } j = 2 \text{ to } n+1 \\ a_{ij} = a_{ij} - \sum_{k=1}^{j-1} a_{ik}a_{kj} & \text{for } j = 2, n \\ & \text{for } i = j \text{ to } n \\ & \quad \text{for each value of } j \\ a_{ij} = \left[a_{ij} - \sum_{k=1}^{i-1} a_{ik}a_{kj} \right] / a_{ii} & \text{for } i = 2, n \\ & \text{for } j = i+1 \text{ to } n+1 \\ & \quad \text{for each value of } i \end{array}\right\} \quad \ldots (5.24)$$

And the general back substitution formula will become

$$x_n = a_{n(n+1)}$$

$$x_i = a_{i(n+1)} - \sum_{j=i+1}^{n} a_{ij}x_j \qquad \text{for } i = n-1, n-2, \ldots, 3, 2, 1 \qquad \ldots (5.25)$$

Algorithm 5.8: Modified Crout's Method

The augmented matrix A of order $n\times(n+1)$ is stored using a two-dimensional array named a of size $n\times(n+1)$, lower triangular and augmented upper triangular matrix is stored over the matrix A itself, and solution vector is stored in one-dimensional array named x of size n.

```
Begin
    read: n
    for i = 1 to n by 1 do
        for j = 1 to (n+1) by 1 do
            read: a_ij
        endfor
    endfor
    for j = 2 to (n+1) by 1 do
        set a_1j = a_1j /a_11
    endfor
    for j = 2 to n by 1 do
        for i = j to n by 1 do
            set sum = 0
```

```
            for k = 1 to (j -1) by 1 do
                set sum = sum + aik × akj
            endfor
            set aij = aij - sum
        endfor
    endfor
    for i = 2 to n by 1 do
        for j = (i+1) to (n+1) by 1 do
            set sum = 0
            for k = 1 to (i -1) by 1 do
                set sum = sum + aik × akj
            endfor
            set aij = (aij - sum)/aii
        endfor
    endfor
    set xn = an(n+1)
    for i = (n -1) to 1 by -1 do
        set sum = 0
        for j = (i+1) to n by 1 do
            set sum = sum +aij × xj
        endfor
        set xi = ai(n+1) - sum
    endfor
    write: "Solution of the given system is"
    for i = 1 to n by 1 do
        write: xi
    endfor
End.
```

5.5 ITERATIVE METHODS

In certain cases the iterative methods are preferred over direct methods particularly when the coefficient matrix is sparse i.e. have many zeros. These methods are more rapid. They are more economical in memory requirements of a computer. Sometimes these methods are also used to reduce round-off error in the solutions computed by the direct methods. They can also be used to find the solution of system of non-linear equations. These methods are also called the methods of successive approximations. The values of unknown are improved by substituting directly the previous values of unknowns in the equation. The necessary and sufficient condition for the use of these methods is that the diagonal elements of the coefficient matrix must be dominant i.e. the element a_{11}, a_{22}, a_{33}, . . ., a_{nn} must be sufficiently dominant. Such a system of linear equations is known as a *diagonal system*. Note that it is very difficult to define the exact minimum margin by which these coefficients must dominate the other coefficients to ensure convergence and it is even more difficult to predict the rate of convergence for some combinations of coefficient values when the convergence is assured.

5.5.1 Jacobi's Method

The sequence of steps constituting the Jacobi's method is as follows:

Initialisation Step:

To begin the procedure, assign an initial value to each unknown, if you can make a reasonable guess. These initial values will not affect the convergence, but will effect the number of iterations required for convergence. Usually, we assign value zero to all unknowns if no better initial estimates are at a hand.

Iteration 1

i) Find the value of x_1 from the first equation by substituting the initial values of other unknowns.

ii) Find the value of x_2 from the second equation by substituting the initial values of other unknowns.

iii) Find the value of x_3 from the third equation by substituting the initial values of other unknowns.

⋮

And so on till the value of x_n is computed from the nth equation using the initial values of $x_1, x_2, \ldots, x_{n-1}$.

Iteration 2

i) Find the value of x_1 from the first equation by substituting the values of other unknowns obtained in the first iteration.

ii) Find the value of x_2 from the second equation by substituting the values of other unknowns obtained in the first iteration.

iii) Find the value of x_3 from the third equation by substituting the values of other unknowns obtained in the first iteration.

⋮

And so on till the value of x_n is computed from the nth equation using the values of $x_1, x_2, \ldots, x_{n-1}$ obtained in the first iteration.

Iteration 3

i) Find the value of x_1 from the first equation by substituting the values of other unknowns obtained in the second iteration.

ii) Find the value of x_2 from the second equation by substituting the values of other unknowns obtained in the second iteration.

iii) Find the value of x_3 from the third equation by substituting the values of other unknowns obtained in the second iteration.

⋮

And so on till the value of x_n is computed from the nth equation using the values of $x_1, x_2, \ldots, x_{n-1}$ obtained in the second iteration.

This iterative procedure is continued until the successive value of each unknown differs only within the permissible limits. In other words, the error between the successive values of each unknown is less than or equal to the prescribed tolerance.

Example 5.5: Solve the following system of equations

$$20x_1 + x_2 - 2x_3 = 17$$
$$3x_1 + 20x_2 - x_3 = -18$$
$$2x_1 - 3x_2 + 20x_3 = 25$$

accurate to four significant digits.

Solution: As you can see the system is diagonal system, therefore the convergence is assured. Since we want the solution correct to four significant digits, therefore the iterative process will be terminated as soon as we find that successive iteration do not produce any change at first four significant positions.

We rewrite the given system of equations as

$$x_1 = \frac{1}{20}(17 - x_2 + 2x_3) \qquad x_2 = \frac{1}{20}(-18 - 3x_1 + x_3) \qquad x_3 = \frac{1}{20}(25 - 2x_1 + 3x_2)$$

We start with initial approximation as

$$x_1 = x_2 = x_3 = 0$$

Iteration 1: Substituting the initial values in the above equations, we obtain

$$x_1 = 0.85 \qquad x_2 = -0.9 \qquad x_3 = 1.25$$

Iteration 2: Substituting the values obtained in the first iteration in above equations, we obtain

$$x_1 = 1.02 \qquad x_2 = -0.965 \qquad x_3 = 1.03$$

Iteration 3: Substituting the values obtained in the second iteration in above equations, we obtain

$$x_1 = 1.0012 \qquad x_2 = -1.0015 \qquad x_3 = 1.0033$$

Iteration 4: Substituting the values obtained in the third iteration in above equations, we obtain

$$x_1 = 1.0004 \qquad x_2 = -1.0000 \qquad x_3 = 0.9996$$

Iteration 5: Substituting the values obtained in the fourth iteration in above equations, we obtain

$$x_1 = 1.0000 \qquad x_2 = -1.0000 \qquad x_3 = 1.0000$$

Iteration 6: Substituting the values obtained in the fifth iteration in above equations, we obtain

$$x_1 = 1.0000 \qquad x_2 = -1.0000 \qquad x_3 = 1.0000$$

By comparing the approximations obtained in fifth and sixth iterations, we find that there is no change in first four significant digits, therefore we take the solution obtained in the sixth iteration as the desired solution. Hence, solution correct to four significant digits is

$$x_1 = 1.0000 \qquad x_2 = -1.0000 \qquad x_3 = 1.0000$$

Algorithm 5.9: Jacobi's Method

The augmented matrix A of order $n\times(n+1)$ is stored using a two-dimensional array named a of size $n\times(n+1)$. The solution vectors in two successive iterations are stored in one-dimensional arrays named *old_x* and *new_x* each of size n. The variable *epsilon* represents the prescribed tolerance and variable *maxiter* represents the number of permitted iterations.

```
Begin
    read: n
    for i = 1 to n by 1 do
        for j = 1 to (n+1) by 1 do
            read: a_ij
        endfor
    endfor
    read: maxiter, epsilon
    for i = 1 to n by 1 do        // initialize all variables to zero
        set old_x_i = 0
    endfor
    for k= 1 to maxiter by 1 do
        set bigrelerror = 0
        for i = 1 to n by 1 do
            set sum = 0
            for j = 1 to n by 1 do
                if ( i ≠ j ) then
                    set sum = sum + a_ij × old_x_j
                endif
            endfor
            set new_x_j = (a_i(n+1) - sum)/a_ii
            set relerror = |((new_x_i  - old_x_i)/new_x_i)|
            if ( relerror > bigrelerror ) then
               set bigrelerror = relerror
            endif
        endfor
        if ( bigrelerror ≤ epsilon) then
            write: "Solution converges in ",k," iterations"
            for i = 1 to n by 1 do
                write: x_i
            endfor
            exit
        endif
        for i = 1 to n by 1 do
            set old_x_i = new_x_i
```

```
        endfor
    endfor
    write: "Solution does not converges in ", maxiter," iterations"
End.
```

5.5.2 Gauss-Seidel Method

In the Jacobi's method, even though the new values of unknowns are computed in each iteration, but the values of unknowns in the previous iterations are used in the subsequent iterations. That is, although new value of x_1 is computed from the first equation in a current iteration, but it is not used to compute the new values of other unknowns in the current iteration. Note that the new values of the unknowns are better than the old values, and should be used in preference to the poorer values. This was the main objective of Gauss-Seidel method.

The sequence of steps constituting the Gauss-Seidel method is as follows:

Initialisation Step:

To begin the procedure, assign an initial value to each unknown, if you can make a reasonable guess. These initial values will not effect the convergence, but will effect the number of iterations required for convergence. Usually, we assign value zero to all unknowns if no better initial estimates are at a hand.

Iteration 1

i) Find the value of x_1 from the first equation by substituting the initial values of other unknowns.

ii) Find the value of x_2 from the second equation by substituting current value of x_1 and the initial values of other unknowns.

iii) Find the value of x_3 from the third equation by substituting the current values of x_1 and x_2 and the initial values of other unknowns.

⋮

And so on, till the value of x_n is computed from the nth equation using current values of $x_1, x_2, \ldots, x_{n-1}$.

Iteration 2

i) Find the value of x_1 from the first equation by substituting the values of other unknowns obtained in the first iteration.

ii) Find the value of x_2 from the second equation by substituting current values of other unknowns.

iii) Find the value of x_3 from the third equation by substituting the current values other unknowns.

⋮

And so on till the value of x_n is computed from the nth equation using current values of $x_1, x_2, \ldots, x_{n-1}$.

Iteration 3 *i)* Find the value of x_1 from the first equation by substituting the values of other unknowns obtained in the second iteration.

ii) Find the value of x_2 from the second equation by substituting the current values of other unknowns.

iii) Find the value of x_3 from the third equation by substituting the current values of other unknowns.

⋮ And so on till the value of x_n is computed from the *n*th equation using current values of $x_1, x_2, \ldots, x_{n-1}$.

This iterative procedure is continued until the successive values of each unknown differ only within the permissible limits. In other words, the error between the successive values of each unknown is less than or equal to the prescribed tolerance.

Example 5.6: Solve the following system of equations, accurate to four significant digits.

$$10x_1 + x_2 + 2x_3 = 44$$
$$2x_1 + 10x_2 + x_3 = 51$$
$$x_1 + 2x_2 + 10x_3 = 61$$

Solution: As you can see the system is diagonal system, therefore the convergence is assured. Since we want the solution correct to four significant digits, therefore the iterative process will be terminated as soon as we find that successive iteration do not produce any change at first four significant positions. We rewrite the given system of equations as

$$x_1 = \frac{1}{10}(44 - x_2 - 2x_3) \qquad x_2 = \frac{1}{10}(51 - 2x_1 - x_3) \qquad x_3 = \frac{1}{10}(61 - x_1 - 2x_2)$$

We start with initial approximation as

$$x_1 = x_2 = x_3 = 0$$

Iteration 1: Substituting $x_2 = x_3 = 0$ in the first equation, we obtain

$$x_1 = 4.4$$

Substituting $x_1 = 4.4$ and $x_3 = 0$ in the second equation, we obtain

$$x_2 = 4.22$$

Substituting $x_1 = 4.4$ and $x_2 = 4.22$ in the third equation, we obtain

$$x_3 = 4.816$$

Thus, we obtain

$$x_1 = 4.4 \qquad x_2 = 4.22 \qquad x_3 = 4.816$$

as new approximations at the end of the first iteration.

Iteration 2: Now substituting $x_2 = 4.22$ and $x_3 = 4.816$ in the first equation, we obtain

$x_1 = 4.0154$

Next substituting $x_1 = 4.0154$ and $x_3 = 4.816$ in the second equation, we obtain

$x_2 = 3.0148$

Next substituting $x_1 = 4.0154$ and $x_2 = 3.0148$ in the third equation, we obtain

$x_3 = 5.0955$

Thus, we obtain

$x_1 = 4.0154 \qquad x_2 = 3.0148 \qquad x_3 = 5.0955$

as new approximations at the end of the second iteration.

Iteration 3: Now substituting $x_2 = 3.0148$ and $x_3 = 5.0955$ in the first equation, we obtain

$x_1 = 3.0794$

Next substituting $x_1 = 3.0794$ and $x_3 = 5.0955$ in the second equation, we obtain

$x_2 = 3.9746$

Next substituting $x_1 = 3.0794$ and $x_2 = 3.9746$ in the third equation, we obtain

$x_3 = 4.9971$

Thus, we obtain

$x_1 = 3.0794 \qquad x_2 = 3.9746 \qquad x_3 = 4.9971$

as new approximations at the end of the third iteration.

Iteration 4: Now substituting $x_2 = 3.0794$ and $x_3 = 4.9971$ in the first equation, we obtain

$x_1 = 3.0031$

Next substituting $x_1 = 3.0031$ and $x_3 = 4.9971$ in the second equation, we obtain

$x_2 = 3.9997$

Next substituting $x_1 = 3.0031$ and $x_2 = 3.9997$ in the third equation, we obtain

$x_3 = 4.8001$

Thus, we obtain

$x_1 = 3.0031 \qquad x_2 = 3.9997 \qquad x_3 = 4.8001$

as new approximations at the end of the fourth iteration.

Iteration 5: Now substituting $x_2 = 3.9997$ and $x_3 = 4.8001$ in the first equation, we obtain

$$x_1 = 3.0400$$

Next substituting $x_1 = 3.0400$ and $x_3 = 4.8001$ in the second equation, we obtain

$$x_2 = 4.0120$$

Next substituting $x_1 = 3.0400$ and $x_2 = 4.0120$ in the third equation, we obtain

$$x_3 = 4.8360$$

Thus, we obtain

$$x_1 = 3.0400 \qquad x_2 = 4.0120 \qquad x_3 = 4.8360$$

as new approximations at the end of the fifth iteration.

Iteration 6: Now substituting $x_2 = 4.0120$ and $x_3 = 4.8360$ in the first equation, we obtain

$$x_1 = 3.0316$$

Next substituting $x_1 = 3.0316$ and $x_3 = 4.8360$ in the second equation, we obtain

$$x_2 = 4.0101$$

Next substituting $x_1 = 3.0316$ and $x_2 = 4.0101$ in the third equation, we obtain

$$x_3 = 4.9948$$

Thus, we obtain

$$x_1 = 3.0316 \qquad x_2 = 4.0101 \qquad x_3 = 4.9948$$

as new approximations at the end of the sixth iteration.

Iteration 7: Now substituting $x_2 = 4.0101$ and $x_3 = 4.9948$ in the first equation, we obtain

$$x_1 = 3.0000$$

Next substituting $x_1 = 3.0000$ and $x_3 = 4.9948$ in the second equation, we obtain

$$x_2 = 4.0001$$

Next substituting $x_1 = 3.0000$ and $x_2 = 4.0001$ in the third equation, we obtain

$$x_3 = 5.0000$$

Thus, we obtain

$$x_1 = 3.0000 \qquad x_2 = 4.0001 \qquad x_3 = 5.0000$$

as new approximations at the end of the seventh iteration.

Iteration 8: Now substituting $x_2 = 4.0001$ and $x_3 = 5.0000$ in the first equation, we obtain

$$x_1 = 3.0000$$

Next substituting $x_1 = 3.0000$ and $x_3 = 5.0000$ in the second equation, we obtain

$$x_2 = 4.0000$$

Next substituting $x_1 = 3.0000$ and $x_2 = 4.0000$ in the third equation, we obtain

$$x_3 = 5.0000$$

Thus, we obtain

$$x_1 = 3.0000 \qquad x_2 = 4.0000 \qquad x_3 = 5.0000$$

as new approximations at the end of the eighth iteration.

By comparing the approximations of the seventh and eighth iterations, we find that there is no variation in first four significant digits; therefore we take the solution obtained at the end of the eighth iteration as the desired solution.

Hence, the solution correct to four significant digits (with truncation) is

$$x_1 = 3.000 \qquad x_2 = 4.000 \qquad x_3 = 5.000$$

Algorithm 5.10: Gauss-Seidel Method

The augmented matrix A of order $n \times (n+1)$ is stored using a two-dimensional array named a of size $n \times (n+1)$. The solution vector is stored in one-dimensional arrays named x of size n. The variable *epsilon* represents the prescribed tolerance and variable *maxiter* represents the number of permitted iterations.

```
Begin
    read: n
    for i = 1 to n by 1 do
        for j = 1 to (n+1) by 1 do
            read: aij
        endfor
    endfor
    read: maxiter, epsilon
    for i = 1 to n by 1 do
        set xi = 0
    endfor
    for k = 1 to maxiter by 1 do
        set bigrelerror = 0
        for i = 1 to n by 1 do
            set sum = 0
            for j = 1 to n by 1 do
                if ( i ≠ j ) then
                    set sum = sum + aij × xj
                endif
            endfor
```

```
            set temp = (ai(n+1) - sum)/aii
            set relerror = |((temp - xi)/temp)|
            set xi = temp
            if ( relerror > bigrelerror ) then
                set bigrelerror = relerror
            endif
        endfor
        if ( bigrelerror ≤ epsilon) then
            write: "Solution converges in ", k, " iterations"
            for i = 1 to n by 1 do
                write: xi
            endfor
            exit
        endif
    endfor
    write: "Solution does not converges in ", maxiter, " iterations"
End.
```

5.6 SOLUTION OF HOMOGENEOUS SYSTEM OF LINEAR EQUATIONS: *The Eigen Value Problem*

Consider the following system of linear equations

$$a_{11}x_1 + a_{12}x_2 + a_{13}x_3 + \ldots + a_{1n}x_n = 0$$

$$a_{21}x_1 + a_{22}x_2 + a_{23}x_3 + \ldots + a_{2n}x_n = 0$$

$$a_{31}x_1 + a_{32}x_2 + a_{33}x_3 + \ldots + a_{3n}x_n = 0$$

$$\vdots$$

$$a_{n1}x_1 + a_{n2}x_2 + a_{n3}x_3 + \ldots + a_{nn}x_n = 0$$

Such a system of equations is known as the system of homogeneous linear equations. We already know from the linear algebra that any system of m linear equations in n unknowns has a solution if and only if the coefficient matrix and the augmented matrix of the system have the same *rank*. Hence, the system of homogeneous linear equations always has a solution since the rank of the coefficient matrix and the augmented matrix is always same. The problem of finding the solution of such a system of linear equations is also known as the *eigen value problem*.

Any system of homogeneous linear equations always have a trivial solution $x_1=x_2=x_3= \ldots =x_n= 0$. If the rank r of the coefficient matrix is equal to the order n then only the trivial solution exists. Non-trivial solution exists if and only if the rank r of the coefficient matrix is less than n. For such a system of linear equations, the determinant of the coefficient matrix vanishes and the system will consists of r linearly independent equations and $(n-r)$ dependent equations which are linear combinations of the independent equations. In order to find the solution of the system of homogeneous linear equations, the values of the unknowns are not determined rather a relationship between them is established. Any combination of x_i's that satisfies this relationship constitutes the solution.

In developing the solution of the above system of homogeneous linear equations, the equations are written in the form

$$\begin{aligned}
(a_{11}-\lambda)x_1 + a_{12}x_2 + a_{13}x_3 + \ldots + a_{1n}x_n &= 0 \\
a_{21}x_1 + (a_{22}-\lambda)x_2 + a_{23}x_3 + \ldots + a_{2n}x_n &= 0 \\
a_{31}x_1 + a_{32}x_2 + (a_{33}-\lambda)x_3 + \ldots + a_{3n}x_n &= 0 \\
&\vdots \\
a_{n1}x_1 + a_{n2}x_2 + a_{n3}x_3 + \ldots + (a_{nn}-\lambda)x_n &= 0
\end{aligned} \qquad \ldots (5.27)$$

where λ is a eigen value of the system having unknown values.

In the matrix form the above system of linear equations can be expressed as

$$\begin{pmatrix} a_{11}-\lambda & a_{12} & a_{13} & \cdots & a_{1n} \\ a_{21} & a_{22}-\lambda & a_{23} & \cdots & a_{2n} \\ a_{31} & a_{32} & a_{33}-\lambda & \cdots & a_{3n} \\ \vdots & \vdots & \vdots & & \vdots \\ a_{n1} & a_{n2} & a_{n3} & \cdots & a_{nn}-\lambda \end{pmatrix} \begin{pmatrix} x_1 \\ x_2 \\ x_3 \\ \vdots \\ x_n \end{pmatrix} = 0 \qquad \ldots (5.28)$$

or simply as

$$(A - \lambda I)X = 0 \qquad \ldots (5.29)$$

where the use of identity matrix I allows to use $(A - \lambda I)$ as the coefficient matrix. The column vector X is frequently referred to as *eigen vector* with $x_1, x_2, x_3, \ldots, x_n$ as its components.

Now to find the solution, set determinant of the coefficient matrix equal to zero i.e.

$$| A - \lambda I | = 0 \qquad \ldots (5.30)$$

As discussed in *Chapter 4*, the expansion of this determinant results in n^{th} degree polynomial of type

$$\lambda^n + p_1\lambda^{n-1} + p_2\lambda^{n-2} + p_3\lambda^{n-3} + \ldots + p_n = 0 \qquad \ldots (5.31)$$

which can be solved to find λ_i's. Then substituting these values one at a time into the given system of equations, we can find the corresponding set of relationships between the unknowns.

In *Chapter* 3, we have discussed a method that can be used to compute p_i's, the coefficient of the characteristic polynomial. Then the characteristic polynomial can be solved to find its root, which are actually λ_i's. Then for each λ, we can find the corresponding eigen vector.

But in most of the problems, the emphasis is on finding either the largest or the smallest eigen value and its associated eigen vector. We will discuss a very efficient method, the *power method*, which is an iterative method to find the largest or the smallest eigen value and its associated eigen vector one at a time.

5.6.1 Power Method

To illustrate the power method, we first use it to find the largest eigen value. To find the largest λ, rewrite the Eq. (5.29) as

$$AX = \lambda X \qquad \ldots (5.32)$$

and carry out the following steps:

1. Assign some arbitrary values (usually unity is satisfactory) to the eigen vector components and let this vector be denoted by X_0.
2. Substitute the components of X_0 in place of X on the left-hand side of Eq. (5.32) and carry out the matrix multiplication. This will give the first approximation to the right hand side, where

$$\lambda X = \begin{pmatrix} x_1 \\ x_2 \\ x_3 \\ \cdot \\ \cdot \\ \cdot \\ x_n \end{pmatrix}$$

3. Normalize the vector λX obtained in step 2. Dividing the vector does this by the magnitude of the first component, reducing that component to unity.
4. Use the components of the normalized vector as the improved values of x_i's to be substituted on the left hand of Eq. (5.32) and carry out the matrix multiplication to get another approximation of the right hand side.
5. Repeat steps 3 and 4 until Eq. (5.32) is essentially satisfied i.e. the variation in the eigen vector components from the previous iteration is less than the prescribed tolerance. The normalization factor will be the largest eigen value and the elements of X will be the components of the eigen vector associated with it.

Example 5.6: Solve the following system of equations

$$\begin{aligned} 10x_1 + 4x_2 - x_3 &= 0 \\ 4x_1 + 2x_2 + 3x_3 &= 0 \\ -x_1 + 3x_2 + x_3 &= 0 \end{aligned}$$

for the largest eigen value and its associated eigen vector. The solution must be accurate to four significant digits.

Solution:

The coefficient matrix $A = \begin{pmatrix} 10 & 4 & -1 \\ 4 & 2 & 3 \\ -1 & 3 & 1 \end{pmatrix}$

We begin with $x_1 = x_2 = x_3 = 1$

Iteration 1: $AX = \begin{pmatrix} 10 & 4 & -1 \\ 4 & 2 & 3 \\ -1 & 3 & 1 \end{pmatrix} \times \begin{pmatrix} 1 \\ 1 \\ 1 \end{pmatrix} = \begin{pmatrix} 13 \\ 9 \\ 3 \end{pmatrix} = \begin{pmatrix} 1.0 \\ 0.629307 \\ 0.2307692 \end{pmatrix}$

Iteration 2: Taking $x_1 = 1.0$, $x_2 = 0.6293097$, $x_3 = 0.2307692$

$$AX = \begin{pmatrix} 10 & 4 & -1 \\ 4 & 2 & 3 \\ -1 & 3 & 1 \end{pmatrix} \times \begin{pmatrix} 1.0 \\ 0.629307 \\ 0.2307692 \end{pmatrix} = \begin{pmatrix} 12.53846 \\ 6.076925 \\ 1.307962 \end{pmatrix} = 12.53846 \begin{pmatrix} 1.0 \\ 0.4846626 \\ 0.1042945 \end{pmatrix}$$

Iteration 3: Taking $x_1 = 1.0$, $x_2 = 0.4846626$, $x_3 = 0.1042945$

$$AX = \begin{pmatrix} 10 & 4 & -1 \\ 4 & 2 & 3 \\ -1 & 3 & 1 \end{pmatrix} \times \begin{pmatrix} 1.0 \\ 0.4846626 \\ 0.1042945 \end{pmatrix} = \begin{pmatrix} 11.83436 \\ 5.282208 \\ 0.5582822 \end{pmatrix} = 11.83436 \begin{pmatrix} 1.0 \\ 0.4463453 \\ 0.0471747 \end{pmatrix}$$

Iteration 4: Taking $x_1 = 1.0$, $x_2 = 0.4463453$, $x_3 = 0.0471747$

$$AX = \begin{pmatrix} 10 & 4 & -1 \\ 4 & 2 & 3 \\ -1 & 3 & 1 \end{pmatrix} \times \begin{pmatrix} 1.0 \\ 0.4463453 \\ 0.0471747 \end{pmatrix} = \begin{pmatrix} 11.73831 \\ 5.034215 \\ 0.3862105 \end{pmatrix} = 11.73821 \begin{pmatrix} 1.0 \\ 0.4288743 \\ 0.032902 \end{pmatrix}$$

Iteration 5: Taking $x_1 = 1.0$, $x_2 = 0.4288743$, $x_3 = 0.032902$

$$AX = \begin{pmatrix} 10 & 4 & -1 \\ 4 & 2 & 3 \\ -1 & 3 & 1 \end{pmatrix} \times \begin{pmatrix} 1.0 \\ 0.4288743 \\ 0.032902 \end{pmatrix} = \begin{pmatrix} 11.6826 \\ 4.956454 \\ 0.3195249 \end{pmatrix} = 11.6826 \begin{pmatrix} 1.0 \\ 0.4242597 \\ 0.0273505 \end{pmatrix}$$

Iteration 6: Taking $x_1 = 1.0$, $x_2 = 0.4242597$, $x_3 = 0.0273505$

$$AX = \begin{pmatrix} 10 & 4 & -1 \\ 4 & 2 & 3 \\ -1 & 3 & 1 \end{pmatrix} \times \begin{pmatrix} 1.0 \\ 0.4242597 \\ 0.0273505 \end{pmatrix} = \begin{pmatrix} 11.66969 \\ 4.930571 \\ 0.3001296 \end{pmatrix} = 11.66969 \begin{pmatrix} 1.0 \\ 0.4225109 \\ 0.0257187 \end{pmatrix}$$

Iteration 7: Taking $x_1 = 1.0$, $x_2 = 0.4225109$, $x_3 = 0.0257187$

$$AX = \begin{pmatrix} 10 & 4 & -1 \\ 4 & 2 & 3 \\ -1 & 3 & 1 \end{pmatrix} \times \begin{pmatrix} 1.0 \\ 0.4225109 \\ 0.0257187 \end{pmatrix} = \begin{pmatrix} 11.66432 \\ 4.922178 \\ 0.2932515 \end{pmatrix} = 11.66432 \begin{pmatrix} 1.0 \\ 0.4219857 \\ 0.0251409 \end{pmatrix}$$

Iteration 8: Taking $x_1 = 1.0$, $x_2 = 0.4219857$, $x_3 = 0.0251409$

$$AX = \begin{pmatrix} 10 & 4 & -1 \\ 4 & 2 & 3 \\ -1 & 3 & 1 \end{pmatrix} \times \begin{pmatrix} 1.0 \\ 0.4219857 \\ 0.0251409 \end{pmatrix} = \begin{pmatrix} 11.6628 \\ 4.919394 \\ 0.2910978 \end{pmatrix} = 11.6628 \begin{pmatrix} 1.0 \\ 0.4218021 \\ 0.0249595 \end{pmatrix}$$

Iteration 9: Taking $x_1 = 1.0$, $x_2 = 0.4218021$, $x_3 = 0.0249595$

$$AX = \begin{pmatrix} 10 & 4 & -1 \\ 4 & 2 & 3 \\ -1 & 3 & 1 \end{pmatrix} \times \begin{pmatrix} 1.0 \\ 0.4218021 \\ 0.0249595 \end{pmatrix} = \begin{pmatrix} 11.66225 \\ 4.918483 \\ 0.2903657 \end{pmatrix} = 11.66225 \begin{pmatrix} 1.0 \\ 0.421774 \\ 0.0248979 \end{pmatrix}$$

In iteration number 8 and 9, we find that the relative error between successive values of components of eigen vector is less than the prescribed tolerance, hence approximate solution is

Largest eigen value = 11.66225, and

$$\text{the associated eigen vector} = \begin{pmatrix} 1.0 \\ 0.421774 \\ 0.0248979 \end{pmatrix}$$

Algorithm 5.11: Power Method for largest eigen value

The coefficient matrix *A* of order *n*×*n* is stored using a two-dimensional array named *a* of size *n*×*n*. The *lambda* is the largest eigen value, and the associated eigen vector is stored in one-dimensional arrays named *x* of size *n*. The variable *epsilon* represents the prescribed tolerance.

```
Begin
    read: n
    for i = 1 to n by 1 do
        for j = 1 to n by 1 do
            read: aij
```

```
        endfor
    endfor
    read: epsilon
    for i = 1 to n by 1 do
        set xi = 1
    endfor
    do
       set big_error = 0
       for i = 1 to n by 1 do
           set si = 0
           for j = 1 to n by 1 do
               set si  = si + aij × xj
           endfor
       endfor
       set lambda = | s1 |
       for i = 1 to n by 1 do
           set temp = si / lambda
           set error = |((temp  - xi)/temp)|
           set xi = temp
           if ( error > big_error ) then
               set big_error = error
           endif
       endfor
    while ( big_error > epsilon )
    write: "Largest eigen value =  ", lambda
    write: "Associated eigen vector is"
    for i = 1 to n by 1 do
        write: xi
    endfor
End.
```

To find the smallest eigen value, the matrix equations are to be arranged so that the above stated procedure converges to the smallest eigen value. This is illustrated below:

Pre-multiply Eq. (5.32) by A^{-1} to obtain

$$A^{-1}AX = A^{-1}\lambda X \qquad \Rightarrow \qquad X = \lambda A^{-1}X$$

Dividing it by λ, we get

$$A^{-1}X = \frac{1}{\lambda}X \qquad \ldots (5.33)$$

This equation is in the form which will converge to the largest value of $\frac{1}{\lambda}$ which corresponds to the smallest value of λ. Now apply the same steps as discussed for the largest eigen value.

5.7 MORE SOLVED PROBLEMS

Example 5.7: Investigate the consistency of the following system of equations.

$$6x - 9y - 20z = -4$$
$$4x - 15y + 10z = -1$$
$$2x - 3y - 5z = -1$$

If the system of equations is consistent, find the desired solution.

Solution: For this system of equation, we have

$$A = \begin{pmatrix} 6 & -9 & -20 \\ 4 & -15 & 10 \\ 2 & -3 & -5 \end{pmatrix} \quad B = \begin{pmatrix} -4 \\ -1 \\ -1 \end{pmatrix} \text{ and } [A \vdots B] = \begin{pmatrix} 6 & -9 & -20 & -4 \\ 4 & -15 & 10 & -1 \\ 2 & -3 & -5 & -1 \end{pmatrix}$$

Apply transformation $R_1 \leftrightarrow R_3$ on $[A \vdots B]$, we get

$$[A \vdots B] = \begin{pmatrix} 2 & -3 & -5 & -1 \\ 4 & -15 & 10 & -1 \\ 6 & -9 & -20 & -4 \end{pmatrix}$$

Now apply transformations $R_2 \to R_2 - 2R_1$ and $R_3 \to R_3 - 3R_1$

$$[A \vdots B] = \begin{pmatrix} 2 & -3 & -5 & -1 \\ 0 & -9 & 20 & 1 \\ 0 & 0 & -5 & -1 \end{pmatrix}$$

This matrix is in echelon form containing three non-zero rows, and hence $R([A \vdots B]) = 3$.

Also $R(A) = 3$ as the echelon form of coefficient matrix A is $\begin{pmatrix} 2 & -3 & -5 \\ 0 & -9 & 20 \\ 0 & 0 & -5 \end{pmatrix}$.

Thus, $R(A) = R([A \vdots B]) = 3 =$ number of unknowns. Therefore, the system of equations is consistent and has a unique solution.

Therefore, using back substitution, we get $x = 1/2, y = 1/3, y = 1/5$.

Example 5.8: Is the system of following equations is consistent? If yes, find the desired solution.

$$x - y + z = 3$$
$$2x + y - z = 2$$
$$x + 2y - 2z = 3$$

Solution: For this system of equation, we have

$$A=\begin{pmatrix}1 & -1 & 1\\ 2 & 1 & -1\\ 1 & 2 & -2\end{pmatrix}\quad B=\begin{pmatrix}3\\ 2\\ -1\end{pmatrix}\quad \text{and}\quad [A\vdots B]=\begin{pmatrix}1 & -1 & 1 & 3\\ 2 & 1 & -1 & 2\\ 1 & 2 & -2 & -1\end{pmatrix}$$

Apply elementary transformations on augmented matrix, i.e., $[A\vdots B]$

$R_2 \to R_2 - 2R_1$ and $R_3 \to R_3 - R_1$

$$[A\vdots B]=\begin{pmatrix}1 & -1 & 1 & 3\\ 0 & 3 & -3 & -4\\ 0 & 3 & -3 & -4\end{pmatrix}$$

Now apply elementary transformation $R_3 \to R_3 - R_2$

$$[A\vdots B]=\begin{pmatrix}1 & -1 & 1 & 3\\ 0 & 3 & -3 & -4\\ 0 & 0 & 0 & 0\end{pmatrix}$$

This matrix is in echelon form containing two non-zero rows, and hence $R([A\vdots B])=2$.

Also $R(A)=2$ as the echelon form of coefficient matrix A is $\begin{pmatrix}1 & -1 & 1\\ 0 & 3 & -3\\ 0 & 0 & 0\end{pmatrix}$.

Thus, $R(A)=R([A\vdots B])=2<$ number of unknowns. Therefore, the system of equations is consistent and has infinite solutions.

Let $z=k\in R$, the using back substitution, we get $x = 5/3$, $y = k - 4/3$.

Example 5.9: Given the following system of equations

$$2x+3y+5z=9$$
$$7x+3y-2z=8$$
$$2x+3y+\lambda z=\mu$$

Investigate the values of λ and μ so that the system has (*i*) unique solution, (*ii*) no solution, and (*iii*) infinite solutions.

Solution: For this system of equation, we have

$$A = \begin{pmatrix} 2 & 3 & 5 \\ 7 & 3 & -2 \\ 2 & 3 & \lambda \end{pmatrix} \quad B = \begin{pmatrix} 9 \\ 8 \\ \mu \end{pmatrix} \text{ and } [A \vdots B] = \begin{pmatrix} 2 & 3 & 5 & 9 \\ 7 & 3 & -2 & 8 \\ 2 & 3 & \lambda & \mu \end{pmatrix}$$

For unique solution, $R(A)$ should be equal to 3. This is possible, iff $\lambda \neq 5$. Because, if $\lambda = 5$, then the first row and last row of determinant

$$\begin{vmatrix} 2 & 3 & 5 \\ 7 & 3 & -2 \\ 2 & 3 & \lambda \end{vmatrix}$$

will become same, and from the property of determinants, if two rows of a determinant are identical, then the determinant vanishes.

Thus, for unique solution $\lambda \neq 5$ and μ can have any value.

For no solution, $\lambda = 5$ and μ can take a value for which $R(A) \neq R([A \vdots B])$.

$$A = \begin{pmatrix} 2 & 3 & 5 \\ 7 & 3 & -2 \\ 2 & 3 & \lambda \end{pmatrix} \quad [A \vdots B] = \begin{pmatrix} 2 & 3 & 5 & 9 \\ 7 & 3 & -2 & 8 \\ 2 & 3 & \lambda & \mu \end{pmatrix}$$

That is, $R(A) = 2$ and $R([A \vdots B]) = 2$ only if $\mu = 9$. Therefore, if $\lambda = 5$ and $\mu \neq 9$, the system will have no solution as $R(A) < R([A \vdots B])$.

If $\lambda = 5$ and $\mu = 9$, then the system will have infinite solutions.

EXERCISE . . .

1. Which of the following system of linear equations are consistent? Find the desired solution for the consistent system of equations.

   (*a*) $3x - y + 2z = 3$
   $2x + y - 3z = 5$
   $x - 2y - z = 1$

   (*b*) $3x - y + 2z = 2$
   $2y + z = -1$
   $3x - 5y = 3$

2. Investigate the values of λ and μ so that the following system of linear equations has (*i*) unique solution, (*ii*) no solution, and (*iii*) infinite solutions.

   $7x + 3y - 2z = 8$
   $2x + 3y + 4z = 9$
   $2x + 3y + \lambda z = \mu$

3. Solve the following system of equations using Cramer's rule.

$$3x_1 + x_2 + 2x_3 = 3$$
$$2x_1 - 3x_2 - x_3 = -3$$
$$x_1 - 2x_2 + x_3 = 4$$

4. Solve the following system of equations using Gauss elimination method.

$$2x_1 + x_2 + x_3 = 10$$
$$3x_1 + 2x_2 + 3x_3 = 18$$
$$x_1 + 4x_2 + 9x_3 = 16$$

5. Solve the following system of equations using Gauss elimination method with pivoting.

$$2x_1 + 3x_2 + 4x_3 = 20$$
$$4x_1 + 2x_2 + 3x_3 = 17$$
$$x_1 + 4x_2 + 2x_3 = 17$$

6. Solve the following system of equations using Gauss-Jordon method.

$$2x_1 - 3x_2 + x_3 = -1$$
$$x_1 + 4x_2 + 5x_3 = 25$$
$$3x_1 - 4x_2 + x_3 = 2$$

7. Solve the following system of equations using Matrix inversion method.

$$3x_1 + x_2 + x_3 = 8$$
$$2x_1 - 3x_2 - 2x_3 = -5$$
$$7x_1 + 2x_2 - 5x_3 = 0$$

8. Solve the following system of equations using Crout's method.

$$3x_1 + 2x_2 + 7x_3 = 4$$
$$2x_1 + 3x_2 + x_3 = 5$$
$$3x_1 + 4x_2 + x_3 = 7$$

9. Solve the following system of equations using Jacobi's and Gauss-Seidel method.

$$5x_1 + 2x_2 + x_3 = 12$$
$$x_1 + 4x_2 + 2x_3 = 15$$
$$x_1 + 2x_2 + 5x_3 = 20$$

10. Compare and contrast between
    (*a*) Jacobi's method and Gauss-Seidel method
    (*b*) Gauss elimination method and Gauss-Jordon method
    (*c*) Non-homogeneous and homogeneous system of linear equations.

❑ ❑ ❑

3. Solve the following system of equations using Cramer's rule.

$$3x_1 + x_2 + 2x_3 = 3$$
$$2x_1 - 3x_2 - x_3 = -3$$
$$x_1 + 2x_2 + x_3 = 4$$

4. Solve the following system of equations using Gauss elimination method.

$$2x_1 + x_2 + x_3 = 10$$
$$3x_1 + 2x_2 + 3x_3 = 18$$
$$x_1 + 4x_2 + 9x_3 = 16$$

5. Solve the following system of equations using Gauss elimination method with pivoting.

$$2x_1 + 3x_2 + 4x_3 = 20$$
$$4x_1 + 2x_2 + 3x_3 = 17$$
$$x_1 + x_2 + 2x_3 = 1$$

6. Solve the following system of equations using Gauss-Jordon method.

$$2x_1 + 3x_2 - x_3 = 1$$
$$x_1 + 4x_2 + 5x_3 = 25$$
$$3x_1 - 4x_2 + x_3 = 2$$

7. Solve the following system of equations using Matrix inversion method.

$$x_1 + x_2 + x_3 = 6$$
$$2x_1 + 3x_2 - 2x_3 = -5$$
$$7x_1 + 2x_2 - 5x_3 = 0$$

8. Solve the following system of equations using Crout's method.

$$2x_1 + 2x_2 + 7x_3 = 4$$
$$2x_1 + 3x_2 + x_3 = 5$$
$$3x_1 + 4x_2 + x_3 = 7$$

9. Solve the following system of equations using Jacobi's and Gauss-Seidel method.

$$5x_1 + 2x_2 + x_3 = 12$$
$$x_1 + 4x_2 + 2x_3 = 15$$
$$x_1 + 2x_2 + 5x_3 = 20$$

10. Compare and contrast between
    (a) Jacobi's method and Gauss-Seidel method
    (b) Gauss elimination method and Gauss-Jordon method
    (c) Non-homogeneous and homogeneous system of linear equations.

□ □ □

6 Interpolation and Approximation

Learning Outcomes

After reading this chapter, students will have the ability to

- differentiate between interpolation and extrapolation
- describe the criteria for choosing an interpolation method
- describe various methods of interpolation
- explain the use of difference tables in determining the best order of interpolation polynomial
- write programs for various interpolation methods
- estimate the errors in interpolation
- explain the process of estimating a function using truncated Taylor's series and Chebyshev polynomials

6.1 INTRODUCTION

Suppose x and y are two variable and their relations can be expressed as

$$y = f(x) \qquad x_1 \le x \le x_n \qquad \ldots (6.1)$$

Then we say that x is an independent variable and y is a dependent variable.

When the form of $f(x)$ is known, then the value of y can be computed directly corresponding to any value of x in the range $x_1 \le x \le x_n$.

However, if the form of $f(x)$ is not known, and only a set of values (x_1, y_1), (x_2, y_2), (x_3, y_3),. . ., (x_n, y_n) satisfying the relation $y = f(x)$ are known, then the process of estimating (approximating) the value of dependent variable y for a given value of independent variable x in the range $x_1 \le x \le x_n$ is known as *interpolation*.

However, if we move in opposite direction i.e. estimate the value of dependent variable x for a given value of independent variable y, the process is known as *inverse interpolation*.

The process of estimating the value of independent variable y for a given value of x outside the range $x_1 \le x \le x_n$ is known as *extrapolation*.

In this chapter, our topic of discussion is interpolation. It is important to note that interpolation is a very important topic in numerical analysis. It provides the theoretical foundation for the derivation of formulae for numerical differentiation, numerical integration and the solution of differential equations.

6.2 WHAT GOES IN THE INTERPOLATION?

In the interpolation, we attempt to find a simple function, say $\phi(x)$ such that $f(x)$ and $\phi(x)$ agrees at a set of specified values. The function $\phi(x)$ can take variety of forms. When $\phi(x)$ is polynomial, the process is called *parabolic* or *polynomial interpolation*; and when $\phi(x)$ is a finite trigonometric series, the process is known as *trigonometric interpolation*. Similarly, $\phi(x)$ may be a series of *exponential functions, Legendre polynomials, Bessel functions*, etc.. In practical problems, we always choose $\phi(x)$ to be the simplest function that represents the given function over the given interval. Since polynomials are the simplest functions, we usually take a polynomial for $\phi(x)$, and nearly all the standard formulas for interpolation are polynomial formulas.

6.3 METHODS OF INTERPOLATION

There are a variety of methods available for interpolation, each having its own characteristics. The decision of using a particular method further depends in tabulation of the function. The tabulated points (x_i, y_i), $i = 1, 2, 3,\ldots, n$, of function $y = f(x)$, can be equally spaced or unequally spaced. These methods are discussed one by one, but not in specific order.

Methods for Equally Spaced Functions

- Newton's forward interpolation formula
- Newton's backward interpolation formula
- Gauss's formula
- Stirling's formula
- Bessel's formula
- Laplace-Everett's formula

Methods for Unequally Spaced Functions

- Lagrangian interpolation
- Newton's divided difference interpolation formula

These methods also work well for equally spaced function.

6.4 CHOOSING AN INTERPOLATION METHOD

If the function is tabulated at equal intervals, then we can use either forward difference interpolation formula or backward interpolation formula. Further, if the point for which the value of the function is to be interpolated lies in the upper half of the table then the forward difference interpolation formula will give better approximation to the value of the function.

And if the point lies in the lower half then the backward difference interpolation formula will give better approximation. If the point lies somewhere in the middle of the table, then any one will work.

But if the function is tabulated at unequal intervals, then we use Lagrangian interpolation and Newtons's divided difference interpolation formula.

If the interpolation is required near the centre/middle of the tabulated points, any of the Sterling's formula, Bessel's formula and Laplace-Everett's formula can be used.

6.5 LAGRANGIAN INTERPOLATION

In order to derive a general formula for Lagrangian interpolation, we consider a second order polynomial of type

$$y(x) = a_1(x-x_2)(x-x_3) + a_2(x-x_1)(x-x_3) + a_3(x-x_1)(x-x_2) \quad \ldots (6.2)$$

passing through the points (x_1,y_1), (x_2,y_2) and (x_3,y_3), where a_1, a_2 and a_3 are unknown constants whose values are determined as shown overleaf.

At $x = x_1$ $\quad y(x_1) = a_1(x_1-x_2)(x_1-x_3) \quad \Rightarrow \quad a_1 = \dfrac{y_1}{(x_1 - x_2)(x_1 - x_3)}$

At $x = x_2$ $\quad y(x_2) = a_2(x_2-x_1)(x_2-x_3) \quad \Rightarrow \quad a_2 = \dfrac{y_2}{(x_2 - x_1)(x_2 - x_3)}$

At $x = x_3$ $\quad y(x_3) = a_3(x_3-x_1)(x_3-x_2) \quad \Rightarrow \quad a_3 = \dfrac{y_3}{(x_3 - x_1)(x_3 - x_2)}$

Substituting the values of a_1, a_2 and a_3 in Eq. (6.2), we get

$$y(x) = y_1 \frac{(x - x_2)(x - x_3)}{(x_1 - x_2)(x_1 - x_3)} + y_2 \frac{(x - x_1)(x - x_3)}{(x_2 - x_1)(x_2 - x_3)} + y_3 \frac{(x - x_1)(x - x_2)}{(x_3 - x_2)(x_3 - x_2)}$$

Using the product and summation notation, the above expression can be written as

$$y(x) = \sum_{i=1}^{3} y_i \prod_{\substack{j=1 \\ j \neq i}}^{3} \frac{(x - x_j)}{(x_i - x_j)}$$

which is a second order polynomial passing through (x_1,y_1), (x_2,y_2) and (x_3,y_3).

In general, for n points, we have polynomial of degree (n-1) as follows:

$$y(x) = \sum_{i=1}^{n} y_i \prod_{\substack{j=1 \\ j \neq i}}^{n} \frac{(x - x_j)}{(x_i - x_j)} \qquad \ldots (6.3)$$

This polynomial is known as the *Lagrangian polynomial*, and is very simple to implement on computer.

Note that the Lagrangian polynomial is merely a relation between two variables; therefore by taking y as the independent variable, we can write a formula giving x as a function of y as follows:

$$x(y) = \sum_{i=1}^{n} x_i \prod_{\substack{j=1 \\ j \neq i}}^{n} \frac{(y - y_j)}{(y_i - y_j)} \qquad \ldots (6.4)$$

This polynomial is known as the *Lagrangian inverse polynomial*.

The main uses of Lagrangian polynomial are:

- To find any value of a function when the tabulated values of the function are not equidistant.
- To find the value of the independent variable corresponding to a given value of a function.

For given set of points, we use Lagrangian polynomial of highest order. Note that using a highest order polynomial does not guarantee that the interpolated value will be more accurate. The interpolated value will be best if the degree of the polynomial used in interpolation formula happens to be of the same degree as the inherent degree of relationship among the independent variable and the function values. The main disadvantage of Lagrangian interpolation is that it is difficult to find the order of the polynomial to be fitted. This problem is well taken care of in the Newton's methods of interpolation, which makes use of finite differences.

Example 6.1: Given the table of values as

x	0	1	2	3
$y(x)$	0	2	8	27

Find $y(2.5)$.

Solution: Lagrangian polynomial of degree three and passing through four points is

$$y(x) = y_1 \frac{(x-x_2)(x-x_3)(x-x_4)}{(x_1-x_2)(x_1-x_3)(x_1-x_4)} + y_2 \frac{(x-x_1)(x-x_3)(x-x_4)}{(x_2-x_1)(x_2-x_3)(x_2-x_4)}$$

$$+ y_3 \frac{(x-x_1)(x-x_2)(x-x_4)}{(x_3-x_1)(x_3-x_2)(x_3-x_4)} + y_4 \frac{(x-x_1)(x-x_2)(x-x_3)}{(x_4-x_1)(x_4-x_2)(x_4-x_3)}$$

Therefore

$$y(2.5) = 0 \times \frac{(2.5-1)(2.5-2)(2.5-3)}{(0-1)(0-2)(0-3)} + 2 \times \frac{(2.5-0)(2.5-2)(2.5-3)}{(1-0)(1-2)(1-3)}$$

$$+ 8 \times \frac{(2.5-0)(2.5-1)(2.5-3)}{(2-0)(2-1)(2-3)} + 27 \times \frac{(2.5-0)(2.5-1)(2.5-2)}{(3-0)(3-1)(3-2)}$$

$$= 15.313$$

Example 6.2: Given the table of values as

x	20	25	30	35
$y(x)$	0.342	0.423	0.500	0.650

Find $x(0.390)$.

Solution: This is a problem of inverse interpolation. The Lagrangian inverse polynomial of degree three and passing through four points is

$$x(y) = x_1 \frac{(y-y_2)(y-y_3)(y-y_4)}{(y_1-y_2)(y_1-y_3)(y_1-y_4)} + x_2 \frac{(y-y_1)(y-y_3)(y-y_4)}{(y_2-y_1)(y_2-y_3)(y_2-y_4)}$$

$$+ x_3 \frac{(y - y_1)(y - y_2)(y - y_4)}{(y_3 - y_1)(y_3 - y_2)(y_3 - y_4)} + x_4 \frac{(y - y_1)(y - y_2)(y - y_3)}{(y_4 - y_1)(y_4 - y_2)(y_4 - y_3)}$$

Therefore

$$x(0.390) = 20 \times \frac{(0.390 - 0.423)(0.390 - 0.500)(0.390 - 0.650)}{(0.342 - 0.423)(0.342 - 0.500)(0.342 - 0.650)} +$$

$$25 \times \frac{(0.390 - 0.342)(0.390 - 0.500)(0.390 - 0.650)}{(0.423 - 0.342)(0.423 - 0.500)(0.423 - 0.650)} +$$

$$30 \times \frac{(0.390 - 0.342)(0.390 - 0.423)(0.390 - 0.650)}{(0.500 - 0.342)(0.500 - 0.423)(0.500 - 0.650)} +$$

$$35 \times \frac{(0.390 - 0.342)(0.390 - 0.423)(0.390 - 0.500)}{(0.650 - 0.342)(0.650 - 0.423)(0.650 - 0.500)} +$$

$$= 22.841 \quad \text{(correct to three decimal places)}$$

Algorithm 6.1: Lagrangian Interpolation

Two-dimensional arrays named *x* and *y* each of size *n* are used to store the set of values for the function *f*(*x*).

```
Begin
   read: n, x
   for i = 1 to n by 1 do
       read: xi, yi
   endfor
   set sum = 0
   for i = 1 to n by 1 do
       set prod = 1
       for j = 1 to n by 1 do
           if ( i ≠ j ) then
               set prod = prod × (x  - xj)/(xi -xj)
           endif
       endfor
       set sum = sum + yi × prod
   endfor
   write: "Interpolated value = ", sum
End.
```

6.6 FINITE DIFFERENCES

The *finite differences* as its name implies are either differences between the values of the function or the differences between the past differences.

There are four types of differences:

- Forward differences
- Backward differences
- Divided differences
- Central differences

6.6.1 Forward Differences

If $y_1, y_2, y_3, \ldots, y_n$ denote the values of the function of type $y = f(x)$ at $x = x_1, x_2, x_3, \ldots, x_n$, then $y_2 - y_1, y_3 - y_2, y_4 - y_3, \ldots, y_n - y_{n-1}$ are called the *forward differences* of y. These differences are denoted as $\Delta y_1, \Delta y_2, \Delta y_3, \ldots, \Delta y_{n-1}$. Therefore

$$\Delta y_1 = y_2 - y_1,\ \Delta y_2 = y_3 - y_2,\ \Delta y_3 = y_4 - y_3, \ldots, \Delta y_{n-1} = y_n - y_{n-1}$$

where Δ is called the *forward difference operator*, and $\Delta y_1, \Delta y_2, \Delta y_3, \ldots, \Delta y_{n-1}$ are called *first order forward differences*. The differences of the first order forward differences are called *second order forward differences* and are denoted as $\Delta^2 y_1, \Delta^2 y_2, \Delta^2 y_3, \ldots$ etc.

$$\Delta^2 y_1 = \Delta y_2 - \Delta y_1 \qquad \Delta^2 y_2 = \Delta y_3 - \Delta y_2 \qquad \ldots$$

In the similar manner, the *third order forward differences* are

$$\Delta^3 y_1 = \Delta^2 y_2 - \Delta^2 y_1 \qquad \Delta^3 y_2 = \Delta^2 y_3 - \Delta^2 y_2 \qquad \ldots$$

In general, the *first order forward difference* at the ith point is

$$\Delta y_i = y_{i+1} - y_i$$

and the jth *order forward difference* at the ith point is

$$\Delta^j y_i = \Delta^{j-1} y_{i+1} - \Delta^{j-1} y_i$$

6.6.2 Backward Differences

If $y_1, y_2, y_3, \ldots, y_n$ denote the values of the function of type $y = f(x)$ at $x = x_1, x_2, x_3, \ldots, x_n$, then $y_2 - y_1, y_3 - y_2, y_4 - y_3, \ldots, y_n - y_{n-1}$ are called the *backward differences* of y. These differences are denoted as $\nabla y_2, \nabla y_3, \nabla y_4, \ldots, \Delta y_n$.

Therefore

$$\nabla y_2 = y_2 - y_1,\ \nabla y_3 = y_3 - y_2,\ \nabla y_4 = y_4 - y_3, \ldots, \nabla y_n = y_n - y_{n-1}$$

where ∇ is called the *backward difference operator*, and $\nabla y_2, \nabla y_3, \nabla y_4, \ldots, \Delta y_n$ are called *first order backward differences*.

The differences of the first order forward differences are called *second order backward differences* and are denoted as $\nabla^2 y_3, \nabla^2 y_4, \nabla^2 y_5, \ldots$, etc.

$$\nabla^2 y_3 = \nabla y_3 - \nabla y_2 \qquad \nabla^2 y_4 = \nabla y_4 - \nabla y_3 \qquad \ldots$$

In the similar manner, the *third order backward differences* are

$$\nabla^3 y_4 = \nabla^2 y_4 - \nabla^2 y_3 \qquad \nabla^3 y_5 = \nabla^2 y_5 - \nabla^2 y_4 \ldots$$

In general, the *first order backward difference* at the ith point is

$$\nabla y_i = y_i - y_{i-1}$$

and the jth *order backward difference* at the ith point is

$$\nabla^j y_i = \nabla^{j-1} y_i - \nabla^{j-1} y_{i-1}$$

6.6.3 Divided Differences

If $y_1, y_2, y_3, \ldots, y_n$ denote the values of the function of type $y = f(x)$ at $x = x_1, x_2, x_3, \ldots, x_n$, then

$$\frac{y_2 - y_1}{x_2 - x_1}, \frac{y_3 - y_2}{x_3 - x_2}, \frac{y_4 - y_3}{x_4 - x_3}, \ldots, \frac{y_n - y_{n-1}}{x_n - x_{n-1}}$$

are called the *divided differences* of y, and are denoted as $\Delta_d y_1, \Delta_d y_2, \Delta_d y_3, \ldots, \Delta_d y_{n-1}$. Therefore

$$\Delta_d y_1 = \frac{y_2 - y_1}{x_2 - x_1}, \ \Delta_d y_2 = \frac{y_3 - y_2}{x_3 - x_2}, \ \Delta_d y_3 = \frac{y_4 - y_3}{x_4 - x_3}, \ldots, \Delta_d y_{n-1} = \frac{y_n - y_{n-1}}{x_n - x_{n-1}}$$

where Δ_d is called the *divided difference operator*, and $\Delta_d y_1, \Delta_d y_2, \Delta_d y_3, \ldots, \Delta_d y_{n-1}$ are called *first order divided differences*. The differences of the first order divided differences are called *second order divided differences* and are denoted as $\Delta_d^2 y_1, \Delta_d^2 y_2, \Delta_d^2 y_3, \ldots$, etc.

$$\Delta_d^2 y_1 = \frac{\Delta_d y_2 - \Delta_d y_1}{x_3 - x_1} \qquad \Delta_d^2 y_2 = \frac{\Delta_d y_3 - \Delta_d y_2}{x_4 - x_2} \qquad \ldots$$

In the similar manner, the *third divided forward differences* are

$$\Delta_d^3 y_1 = \frac{\Delta_d^2 y_2 - \Delta_d^2 y_1}{x_4 - x_1} \qquad \Delta_d^3 y_2 = \frac{\Delta_d^2 y_3 - \Delta_d^2 y_2}{x_5 - x_2} \qquad \ldots$$

In general, the *first order divided difference* at the ith point is

$$\Delta_d y_i = \frac{y_{i+1} - y_i}{x_{i+1} - x_i}$$

and the jth *order divided difference* at the ith point is

$$\Delta_d^j y_i = \frac{\Delta_d^{j-1} y_{i+1} - \Delta_d^{j-1} y_i}{x_{i+j} - x_i}$$

6.6.4 Central Differences

For the discussion of central differences and interpolation based on central differences, we will consider method the function of type $y = f(x)$ is tabulated at $x = x_0, x_1, x_2, \ldots, x_n$.

We introduce central difference operator δ defined by

$$\delta f(x) = f\left(x + \frac{h}{2}\right) - f\left(x - \frac{h}{2}\right)$$

Let $y_i = f(x_i)$, where $x_i = x_0 + ih$ and i is not restricted to integral values, then

$$\delta y_r = y_{r+1/2} - y_{r-1/2}$$

If we take $r = \frac{1}{2}$, then $\delta y_{1/2} = y_1 - y_0$

Like wise, other first order central differences can be expressed as

$$\delta y_{1/2} = y_1 - y_0 \quad \delta y_{3/2} = y_2 - y_1 \quad \delta y_{5/2} = y_3 - y_2 \quad \ldots \quad \delta y_{n-1/2} = y_n - y_{n-1}$$

where $\delta y_{1/2}, \delta y_{3/2}, \delta y_{5/2}, \ldots, \delta y_{n-1/2}$ are called *first order central differences.*

The differences of the first order forward differences are called *second order central differences* and are denoted as $\delta y_1^2, \delta y_2^2, \delta y_3^2, \ldots$, etc.

$$\delta y_1^2 = \delta y_{3/2} - \delta y_{1/2} \quad \delta y_2^2 = \delta y_{5/2} - \delta y_{3/2} \quad . \quad . \quad .$$

In the similar manner, the *third order central differences* are

$$\delta y_{3/2}^3 = \delta y_2^2 - \delta y_1^2 \quad \delta y_{5/2}^3 = \delta y_3^2 - \delta y_2^2 \quad . \quad . \quad .$$

In general, the *first order central difference* at the ith point is

$$\delta y_{(2i+1)/2} = y_{i+1} - y_i$$

and the jth *order central difference* at the ith point is

$$\delta^j y_{(2i+1)/2} = \delta^{j-1} y_{i+1} - \delta^{j-1} y_i$$

6.7 DIFFERENCE TABLES

A *difference table* is a table that lists the differences of the function values and the differences of differences in succession. We can construct difference tables for each type of differences. Let us consider these tables one by one.

6.7.1 Forward Difference Table

Let us consider the values y_1, y_2, y_3, y_4 of the function of type $y = f(x)$ tabulated at equally spaced points x_1, x_2, x_3, x_4. Forward difference table along with tabulated points will look like

Table 6.1: Forward difference table

i	x_i	y_i	Δy_i	$\Delta^2 y_i$	$\Delta^3 y_i$
1	x_1	y_1	$\Delta y_1 = y_2 - y_1$	$\Delta^2 y_1 = \Delta y_2 - \Delta y_1$	$\Delta^3 y_1 = \Delta^2 y_2 - \Delta^2 y_1$
2	x_2	y_2	$\Delta y_2 = y_3 - y_2$	$\Delta^2 y_2 = \Delta y_3 - \Delta y_2$	
3	x_3	y_3	$\Delta y_3 = y_4 - y_3$		
4	x_4	y_4			

As we can see from the above table that the forward difference table for a function tabulated at four equally a matrix D of size 3×3 can represent spaced points. In general, the forward difference table for a function tabulated at n equally spaced points can be represented by a matrix of size $(n-1)\times(n-1)$ where the jth order forward difference at the ith point $(\Delta^j y_i)$ is represented by the element d_{ij} of the matrix D. Note that only the elements in the column 1 to $(n-i)$, for rows $i = 1, 2, 3, \ldots, n-1$, are of interest.

Algorithm 6.2: Construction of forward difference table

Two-dimensional arrays named x and y each of size n are used to store the set of values for the function $f(x)$, and two-dimensional array named d of size $(n-1)\times(n-1)$ is used to store the forward difference table.

```
Begin
    for j = 1 to (n-1) by 1 do
        for i = 1 to (n-j) by 1 do
            if ( j = 1 ) then
                set dij = yi+1 - yi
            else
                set dij = d(i+1)(j-1) - di(j-1)
            endif
        endfor
    endfor
End.
```

6.7.2 Backward Difference Table

Let us consider the values y_1, y_2, y_3, y_4 of the function $y = f(x)$ tabulated at equally spaced points x_1, x_2, x_3, x_4. The backward difference table along with tabulated points will look like

Table 6.2: Backward difference table

i	x_i	y_i	∇y_i	$\nabla^2 y_i$	$\nabla^3 y_i$
1	x_1	y_1			
2	x_2	y_2	$\nabla y_2 = y_2 - y_1$		
3	x_3	y_3	$\nabla y_3 = y_3 - y_2$	$\nabla^2 y_3 = \nabla y_3 - \nabla y_2$	
4	x_4	y_4	$\nabla y_4 = y_4 - y_3$	$\nabla^2 y_4 = \nabla y_4 - \Delta y_3$	$\nabla^3 y_4 = \nabla^2 y_3 - \nabla^2 y_2$

Algorithm 6.3: Construction of backward difference table

Two-dimensional arrays named x and y each of size n are used to store the set of values for the function $f(x)$, and two-dimensional array named d of size $(n-1)\times(n-1)$ is used to store the backward difference table.

```
Begin
   for j = 1 to (n-1) by 1 do
      for i = (j+1) to n by 1 do
         if ( j = 1 ) then
            set dij = yi - yi-1
         else
            set dij = di(j-1) - d(i-1)(j-1)
         endif
      endfor
   endfor
End.
```

6.7.3 Divided Difference Table

Let us consider the values y_1, y_2, y_3, y_4 of the function $y = f(x)$ tabulated at points x_1, x_2, x_3, x_4, not necessarily equally spaced. The divided difference table along with tabulated points is shown in Table 6.3.

Like forward and backward difference tables, the divided difference table for a function tabulated at n points, not necessarily equally spaced, by a matrix D of size $(n-1)\times(n-1)$.

The *j*th order divided difference at the *i*th point ($\Delta_d^j y_i$) is represented by the element d_{ij} of the matrix. Note that only the elements in the column 1 to *n-i*, for row *i* = 1, 2, 3, . . ., *n*–1, are of interest.

Table 6.3: Divided difference table

i	x_i	y_i	$\Delta_d y_i$	$\Delta_d^2 y_i$	$\Delta_d^3 y_i$
1	x_1	y_1	$\Delta_d y_1 = \dfrac{y_2 - y_1}{x_2 - x_1}$	$\Delta_d^2 y_1 = \dfrac{\Delta_d y_2 - \Delta_d y_1}{x_3 - x_1}$	$\Delta_d^3 y_1 = \dfrac{\Delta_d^2 y_2 - \Delta_d^2 y_1}{x_4 - x_1}$
2	x_2	y_2	$\Delta_d y_2 = \dfrac{y_3 - y_2}{x_3 - x_2}$	$\Delta_d^2 y_2 = \dfrac{\Delta_d y_3 - \Delta_d y_2}{x_4 - x_2}$	
3	x_3	y_3	$\Delta_d y_3 = \dfrac{y_4 - y_3}{x_4 - x_3}$		
4	x_4	y_4			

Algorithm 6.4: Construction of divided difference table

Two-dimensional arrays named *x* and *y* each of size *n* are used to store the set of values for the function *f*(*x*), and two-dimensional array named *d* of size (*n*–1)×(*n*–1) is used to store the divided difference table.

```
Begin
   for j = 1 to (n -1) by 1 do
      for i = 1 to (n-j) by 1 do
         if ( j = 1 ) then
            set d_ij = (y_i+1 - y_i)/(x_i+1 - x_i)
         else
            set d_ij = (d_(i+1)(j-1) - d_i(j-1))/(x_i+j - x_i)
         endif
      endfor
   endfor
End.
```

6.7.4 Central Difference Table

Let us consider the values y_0, y_1, y_2, y_3 of the function $y = f(x)$ tabulated at equally spaced points x_0, x_1, x_2, x_3. The central difference table along with tabulated points is shown in Table 6.4.

Table 6.4: Central difference table

x	y	δy	δ^2	δ^3
x_0	y_0			
		$\delta y_{1/2} = y_1 - y_0$		
x_0+h	y_1		$\delta y_1^2 = \delta y_{3/2} - \delta y_{1/2}$	
		$\delta y_{3/2} = y_2 - y_1$		$\delta y_{3/2}^3 = \delta y_2^2 - \delta y_1^2$
			$\delta y_2^2 = \delta y_{5/2} - \delta y_{3/2}$	
x_0+2h	y_2			
		$\delta y_{5/2} = y_3 - y_2$		
x_0+3h	y_3			

6.7.5 Some Observations about Difference Tables

The following observations are made regarding interpolation using difference tables:

1. The higher order differences become smaller in size. Further, in the Newton's forward and backward interpolation formulae (described in the next section), the nth order difference is divided by $n!$, thereby further reducing its contribution to the value of the interpolated function.
2. If the function $f(x)$ happens to be a polynomial of degree n, then the nth order difference would be constant and the $(n+1)^{th}$ and higher differences would be zero.

   To prove this fact, consider assume that

   $$f(x) = a_1 + a_2x + a_3x^2 + \ldots + a_nx^n$$

   Now $\Delta f_1 = f(x_1+h) - f(x_1)$

   $$= [a_1 + a_2(x+h) + a_3(x+h)^2 + \; + \ldots + a_n(x+h)^n]$$

   $$- [a_1 + a_2x + a_3x^2 + \ldots + a_nx^n]$$

   $$= a_2h + 2a_3hx_1 + \ldots + a_{n-1}\, x_1^{n-1}$$

   This is a polynomial of degree $(n-1)$. By repeated application of above rule, we get

   $$\Delta^2 f_1 = \text{A polynomial of degree } (n-2)$$

   $$\Delta^3 f_1 = \text{A polynomial of degree } (n-3)$$

   $$\vdots$$

   A polynomial of degree $(n-n)$, *i.e.*, constant, which means that is not a function of x.

Therefore

$$\Delta^{n+1} f_1 = \Delta^n f_2 - \Delta^n f_1 = \text{constant} - \text{constant} = 0$$

Thus, it easy to decide the order of the polynomial to be used by inspecting the difference tables.

Algorithm 6.5: Finding order of polynomial

The difference table is represented by a two-dimensional array named d of size $(n-1)\times(n-1)$.

```
Begin
    for j = 1 to (n-1) by 1 do
        set temp = d1j
        for i = 1 to (n-j) by 1 do
            if (temp ≠ dij ) then
                set temp = dij
            endif
        endfor
        if (temp = d1j) then
            write: "Order of Polynomial ", j
            exit
        endif
    endfor
    write: "Order of Polynomial ", n-1
End.
```

Example 6.3: Given the table of values as

x	2.0	2.1	2.2	2.3	2.4	2.5
$y(x)$	13.00	14.261	15.648	17.167	18.824	20.625

Find the order of the interpolating polynomial, which might be suitable for above set of values.

Solution: Since the points are equally spaced, we can use the forward difference table which is constructed below:

i	x_i	y_i	Δy_i	$\Delta^2 y_i$	$\Delta^3 y_i$	$\Delta^4 y_i$	$\Delta^5 y_i$
1	2.0	13.000	1.261	0.126	0.6	0.0	0.0
2	2.1	14.261	1.387	0.132	0.6	0.0	
3	2.2	15.648	1.519	0.138	0.6		
4	2.3	17.167	1.657	0.144			
5	2.4	18.824	1.801				
6	2.5	20.625					

By inspecting the table, we find that third order differences become constant and the higher order differences are zero. From these facts we conclude that the required order of interpolating polynomial is three.

6.8 NEWTON'S METHODS OF INTERPOLATION

Newton's methods of interpolation are further divided into the following methods depending on the type of differences being used:

- Newton's Forward Difference Interpolation Formula
- Newton's Backward Difference Interpolation Formula
- Newton's Divided Difference Interpolation Formula

If the function is tabulated at equal intervals, then we can use either forward difference interpolation formula or backward interpolation formula. However, if the point for which the value of the function is to be interpolated lies in the upper half of the table, then the forward difference interpolation formula will give better approximation to the value of the function; and if the point lies in the lower half, then the backward difference interpolation formula will give better approximation. If the point lies somewhere in the middle of the table, then any one will work. But if the function is tabulated at unequal intervals, then we use the divided difference interpolation formula, which also works well for equally spaced points.

If the interpolation is required near the centre/middle of the tabulated points, then a difference which is average of the forward and backward differences, called *central difference,* would be useful. A number of formulae using central differences like Sterling's formula, Bessel's formula, Gauss's formula, etc., are derived in literature.

6.8.1 Newton's Forward Difference Interpolation Formula

Let us assume that the function $y(x)$ is tabulated at $(n+1)$ equally spaced (interval size h) points. In order to derive the formula for Newton's forward difference interpolation assume a polynomial of type

$$y(x) = a_1 + a_2(x-x_1) + a_3(x-x_1)(x-x_2) + \ldots + a_{n+1}(x-x_1)(x-x_2)\ldots(x-x_n) \qquad \ldots (6.5)$$

It is the nth order polynomial in x. If we assume that it is an interpolation polynomial for the table values x_i and y_i, then the polynomial must pass through all points. Therefore, we can obtain y_i by substituting the corresponding x_i for x.

At $x = x_1$ $\quad y_1 = a_1$

At $x = x_2$ $\quad y_2 = a_1 + a_2(x_2-x_1)$

At $x = x_3$ $\quad y_3 = a_1 + a_2(x_3-x_1) + a_3(x_3-x_1)(x_3-x_2)$

and so on.

Since x_i's are equally spaced, therefore we can write

$$x_{i+1} - x_i = h \quad \text{and} \quad x_{i+m} - x_i = mh$$

then we can express the function values y_i's in terms of interval values as

$$y_1 = a_1$$

$$y_2 = a_1 + a_2 h$$

$$y_3 = a_1 + a_2(2h) + a_3(2h)(h)$$

$$\vdots$$

$$y_{n+1} = a_1 + a_2(nh) + a_3(nh)((n-1)h) + \dots + a_{n+1}(nh)((n-1)h) \dots (h)$$

Now solving for $a_1, a_2, a_3, \dots, a_{n+1}$, we get

$$a_1 = y_1$$

$$a_2 = \frac{y_2 - a_1}{h} = \frac{y_2 - y_1}{h}$$

$$a_3 = \frac{y_3 - a_2(2h) - a_1}{2h^2} = \frac{\left[y_3 - \frac{y_2 - y_1}{h} 2h - y_1 \right]}{2h^2}$$

$$= \frac{y_3 - 2y_2 + 2y_1 - y_1}{2h^2} = \frac{y_3 - 2y_2 + y_1}{2!h^2}$$

$$\vdots$$

$$a_{n+1} = \frac{y_{n+1} - ny_n + \dots + y_1}{n!h^n}$$

Using the forward difference table, we get

$$a_1 = y_1$$

$$a_2 = \frac{\Delta y_1}{h}$$

$$a_3 = \frac{\Delta^2 y_1}{2!h^2}$$

$$\vdots$$

$$a_{n+1} = \frac{\Delta^n y_1}{n!h^n}$$

Therefore substituting these values of $a_1, a_2, a_3, \dots, a_{n+1}$ in Eq.(6.5), we get

$$y(x) = y_1 + \frac{\Delta y_1}{h}(x - x_1) + \frac{\Delta^2 y_1}{2!h}(x - x_1)(x - x_2) + ... + \frac{\Delta^n y_1}{n!h^n}(x - x_1)(x - x_2)...(x - x_n) \quad ...(6.6)$$

If we use the relation $\dfrac{x - x_1}{h} = u$ then $x - x_1 = hu$

$$x - x_2 = x - (x_1 + h) = hu - h = h(u-1)$$
$$x - x_3 = x - (x_2 + h) = h(u-1) - h = h(u-2)$$
$$\vdots$$
$$x - x_n = h(u-(n-1))$$

Substituting these values $(x - x_1)$, $(x - x_2)$, $(x - x_3)$, ..., $(x - x_n)$ in Eq. (6.6), we get

$$y(x) = y_1 + \frac{\Delta y_1}{h} hu + \frac{\Delta^2 y_1}{2!h} hu \times h(u-1) + ... + \frac{\Delta^n y_1}{n!h^n} hu \times h(u-1) \times ... \times h(u-(n-1)) \quad ...(6.7)$$

Simplifying, we get

$$y(x) = y_1 + \Delta y_1 u + \frac{\Delta^2 y_1}{2!} u(u-1) + ... + \frac{\Delta^n y_1}{n!} u(u-1)...(u-(n-1)) \quad ...(6.8)$$

The above derivation assumes that $x_1 < x < x_2$

If we want to interpolate at any point between x_k and x_{k+1}, i.e. $x_k < x < x_{k+1}$, then the formula takes the form

$$\boxed{y(x) = y_k + \Delta y_k u + \frac{\Delta^2 y_k}{2!} u(u-1) + ... + \frac{\Delta^{n-k} y_k}{(n-k)!} u(u-1)...(u-((n-k)-1))} \quad ...(6.9)$$

This polynomial is called the *Newton's forward difference interpolation formula* and is particularly suitable if one wants to interpolate near the top of the table.

Example 6.4: Given the table of values as

x	2.0	2.25	2.50	2.75	3.0
$y(x)$	9.00	10.06	11.25	12.56	14.00

Find $y(2.35)$.

Solution: The forward difference table for the given set of values is

i	x_i	y_i	Δy_i	$\Delta^2 y_i$	$\Delta^3 y_i$	$\Delta^4 y_i$
1	2.00	9.00	1.06	0.13	–0.01	0.02
2	2.25	10.06	1.19	0.12	0.01	
3	2.50	11.25	1.31	0.13		
4	2.75	12.56	1.44			
5	3.00	14.00				

Since the point $x = 2.35$ lies between second and third point, therefore we consider the forward differences at the second point. The Newton's forward difference interpolating polynomial for above case becomes

$$y(x) = y_2 + \Delta y_2 u + \frac{\Delta^2 y_2}{2} u(u-1) + \frac{\Delta^3 y_2}{6} u(u-1)(u-2)$$

Now $u = \dfrac{x - x_2}{h} = \dfrac{2.35 - 2.25}{0.25} = 0.4$

Therefore

$$y(2.35) = 10.06 + 1.19 \times 0.4 + \frac{0.12}{2} \times 0.4 \times (0.4-1) + \frac{0.01}{6} \times 0.4 \times (0.4-1) \times (0.4-2)$$

$$= 10.522 \qquad \text{(truncated to three decimal digits)}$$

Algorithm 6.6: Newton's forward interpolation formula

Two-dimensional arrays named x and y each of size n are used to store the set of values for the function $f(x)$, and two-dimensional array named d of size $(n-1)\times(n-1)$ is used to store the forward difference table.

```
Begin
    read: n, x
    for i = 1 to n by 1 do
        read: x_i, y_i
    endfor
    if (( x < x_1 ) or ( x > x_n ) )then
        write: "Value lies outside tabulated range"
        exit
    endif
    set i = 2
    while ( x > x_i ) do
        set i = i + 1
    endwhile
    set k = i - 1
    set u = (x-x_k)/(x_{k+1}-x_k)
```

```
    for j = 1 to (n-1) by 1 do   // Construct forward difference table
        for i = 1 to (n-j) by 1 do
            if ( j = 1 ) then
                set d_ij = y_i+1 - y_i
            else
                set d_ij = d_(i+1)(j-1) - d_i(j-1)
            endif
        endfor
    endfor
    set sum = y_k
    for i = 1 to (n-k) by 1 do
        set prod = 1.0
        for j = 0 to (i-1) by 1 do
            set prod = prod × (u-j)
        endfor
        Find i ! and let its value be m
        set sum = sum + (d_ki × prod)/m
    endfor
    write: "Interpolated value = ",  sum
End.
```

6.8.2 Newton's Backward Difference Interpolation Formula

Let us assume that the function $y(x)$ is tabulated at $(n+1)$ equally spaced (interval size h) points. In order to derive the formula for Newton's backward difference interpolation assume a polynomial of type

$$y(x) = a_1 + a_2(x-x_n) + a_3(x-x_n)(x-x_{n-1}) + \dots + a_{n+1}(x-x_n)(x-x_{n-1}) \dots (x-x_1) \qquad \dots (6.10)$$

It is the nth order polynomial in x. If we assume that it is an interpolation polynomial for the table values x_i and y_i, then the polynomial must pass through all points. Therefore, we can obtain y_i by substituting the corresponding x_i for x.

At $x = x_n$ $\quad y_n = a_1$

At $x = x_{n-1}$ $\quad y_{n-1} = a_1 + a_2(x_{n-1}-x_n)$

At $x = x_{n-2}$ $\quad y_{n-2} = a_1 + a_2(x_{n-2}-x_n) + a_3(x_{n-2}-x_n)(x_{n-2}-x_{n-1})$

and so on.

Since x_i's are equally spaced, therefore we can write

$$x_i - x_{i-1} = h$$

$$\Rightarrow \quad x_{i-1} - x_i = -h \quad \text{and} \quad x_{i-m} - x_i = -mh$$

then we can express the function values y_i's in terms of interval values as

$$y_n = a_1$$
$$y_{n-1} = a_1 + a_2(-h)$$
$$y_{n-2} = a_1 + a_2(-2h) + a_3(-2h)(-h)$$
$$\vdots$$
$$y_1 = a_1 + a_2(-nh) + a_3(-nh)(-(n-1)h) + \ldots + a_{n+1}(-nh)(-(n-1)h)\ldots(-h)$$

Now solving for $a_1, a_2, a_3, \ldots, a_{n+1}$, we get

$$a_1 = y_n$$

$$a_2 = \frac{y_{n-1} - a_1}{h} = \frac{y_{n-1} - y_n}{h}$$

$$a_3 = \frac{y_{n-2} - a_2(-2h) - a_1}{2h^2}$$

$$= \frac{\left[y_{n-2} - \frac{y_{n-1} - y_n}{h}(-2h) - y_n\right]}{2h^2} = \frac{y_{n-2} - 2y_{n-1} + 2y_n - y_n}{2h^2} = \frac{y_{n-2} - 2y_{n-1} + y_n}{2!h^2}$$

$$\vdots$$

$$a_{n+1} = \frac{y_1 - ny_2 + \ldots + y_n}{n!h^n}$$

Using the backward difference table, we get

$$a_1 = y_n$$

$$a_2 = \frac{\nabla y_n}{h}$$

$$a_3 = \frac{\nabla^2 y_n}{2!h^2}$$

$$\vdots$$

$$a_{n+1} = \frac{\nabla^n y_n}{n!h^n}$$

Therefore substituting these values of $a_1, a_2, a_3, \ldots, a_{n+1}$ in Eq. (6.10), we get

$$y(x) = y_n + \frac{\nabla y_n}{h}(x - x_n) + \frac{\nabla^2 y_n}{2!h}(x - x_n)(x - x_{n-1}) + \ldots \qquad \ldots (6.11)$$
$$+ \frac{\nabla^n y_n}{n!h^n}(x - x_n)(x - x_{n-1})\ldots(x - x_1)$$

If we use the relation $\frac{x - x_n}{h} = u$, then

$x - x_n = hu$

$x - x_{n-1} = x - (x_n - h) = x - x_n + h = hu + h = h(u+1)$

$x - x_{n-2} = x - (x_{n-1} - h) = x - x_{n-1} + h = h(u+1) + h = h(u+2)$

$\vdots$

$x - x_1 = h(u+(n-1))$

Substituting these values of $(x - x_n)$, $(x - x_{n-1})$, $(x - x_{n-2})$, ..., $(x - x_1)$ in Eq. (6.11), we get

$$y(x) = y_n + \frac{\nabla y_n}{h} hu + \frac{\nabla^2 y_n}{2!h} hu \times h(u+1) + \ldots + \frac{\nabla^n y_n}{n!h^n} hu \times h(u+1) \times \ldots \times h(u+(n-1))$$

Simplifying, we get

$$y(x) = y_n + \nabla y_n u + \frac{\nabla^2 y_n}{2!} u(u+1) + \ldots + \frac{\nabla^n y_n}{n!} u(u+1)\ldots(u+(n-1)) \qquad \ldots (6.12)$$

The formula in Eq. (6.12) assumes that

$x_{n-1} < x < x_n$

If we want to interpolate at any point laying between x_{k-1} and x_k, i.e. $x_{k-1} < x < x_k$, then the formula takes the form

$$y(x) = y_k + \nabla y_k u + \frac{\nabla^2 y_k}{2!} u(u+1) + \ldots + \frac{\nabla^{k-1} y_k}{(k-1)!} u(u+1)\ldots(u+((k-1)-1)) \qquad \ldots (6.13)$$

This polynomial is called the *Newton's backward difference interpolation formula* and is particularly suitable if one wants to interpolate near the bottom of the table.

Example 6.5: Given the table of values as

x	2.5	3.0	3.5	4.0	4.5
$y(x)$	9.75	12.45	15.70	19.52	23.75

Find $y(4.25)$.

Solution: The backward difference table for the given set of values is

i	x_i	y_i	∇y_i	$\nabla^2 y_i$	$\nabla^3 y_i$	$\nabla^4 y_i$
1	2.5	9.75				
2	3.0	12.45	2.70			
3	3.5	15.70	3.25	0.55		
4	4.0	19.52	3.82	0.57	0.02	
5	4.5	23.75	4.23	0.41	–0.16	–0.18

Since the point $x = 4.25$ lies between the fourth and the fifth point, therefore we consider the backward differences at the fifth point. The Newton's backward difference interpolating polynomial for above case becomes

$$y(x) = y_5 + \nabla y_5 u + \frac{\nabla^2 y_5}{2} u(u+1) + \frac{\nabla^3 y_5}{6} u(u+1)(u+2) + \frac{\nabla^4 y_5}{24} u(u+1)(u+2)(u+3)$$

Now $u = \dfrac{x - x_5}{h} = \dfrac{4.25 - 4.5}{0.5} = -0.5$

Therefore

$$\begin{aligned} y(4.25) &= 23.75 + 4.23 \times (-0.5) + \frac{0.41}{2} \times (-0.5) \times (-0.5+1) \\ &\quad + \frac{-0.16}{6} \times (-0.5) \times (-0.5+1) \times (-0.5+2) \\ &\quad + \frac{-0.18}{24} \times (-0.5) \times (-0.5+1) \times (-0.5+2) \times (-0.5+3) \\ &= 21.601 \qquad \text{(truncated to three decimal digits)} \end{aligned}$$

Algorithm 6.7: Newton's backward interpolation formula

Two-dimensional arrays named x and y each of size n are used to store the set of values for the function $f(x)$, and two-dimensional array named d of size $(n–1)\times(n–1)$ is used to store the backward difference table.

```
Begin
    read: n, x
    for i = 1 to n by 1 do
        read: xi, yi
    endfor
    if (( x < x1 ) or ( x > xn ) )then
        write: "Value lies outside tabulated range"
        exit
    endif
    set i = 2
    while ( x > xi ) do
        set i = i + 1
    endwhile
    set k = i
    set u = (x -xk)/(xk -xk-1)
    for j = 1 to (n-1) by 1 do   //generate backward difference table
        for i = j+1 to n by 1 do
            if ( j = 1 ) then
                set dij = yi  - yi-1
            else
                set dij = di(j-1) - d(i-1)(j-1)
            endif
```

```
        endfor
    endfor
    set sum = yk
    for i = 1 to (k-1) by 1 do
        set prod = 1.0
        for j = 0 to (i-1) by 1 do
            set prod = prod × (u+j)
        endfor
        Find i! and let its value be m
        set sum = sum + dki × prod/m
    endfor
    write: "Interpolated value = ", sum
End.
```

6.8.3 Newton's Divided Difference Interpolation Formula

Let us assume that the function $y(x)$ is tabulated at $(n+1)$ equally spaced (interval size h) points. In order to derive the formula for Newton's divided difference interpolation assume a polynomial of type

$$y(x) = a_1 + a_2(x-x_1) + a_3(x-x_1)(x-x_2) + \ldots + a_{n+1}(x-x_1)(x-x_2)\ldots(x-x_n) \qquad \ldots(6.14)$$

It is the nth order polynomial in x. If we assume that it is an interpolation polynomial for the table values x_i and y_i, then the polynomial must pass through all points. Therefore, we can obtain y_i by substituting the corresponding x_i for x.

At $x = x_1$ $\quad y_1 = a_1$

At $x = x_2$ $\quad y_2 = a_1 + a_2(x_2-x_1)$

At $x = x_3$ $\quad y_3 = a_1 + a_2(x_3-x_1) + a_2(x_3-x_1)(x_3-x_2)$ $\quad$ and so on.

Solving for $a_1, a_2, a_3, \ldots, a_{n+1}$, we get

$$a_1 = y_1$$

$$a_2 = \frac{y_2 - a_1}{x_2 - x_1} = \frac{y_2 - y_1}{x_2 - x_1}$$

$$a_3 = \frac{y_3 - a_2(x_3 - x_1) - a_1}{(x_3 - x_1)(x_3 - x_2)} = \frac{\left[y_3 - \frac{y_2 - y_1}{x_2 - x_1}(x_3 - x_1) - y_1\right]}{(x_3 - x_1)(x_3 - x_2)}$$

$$= \frac{y_3 - y_1}{(x_3 - x_1)(x_3 - x_2)} - \frac{y_2 - y_1}{(x_3 - x_1)(x_3 - x_2)} = \frac{1}{(x_3 - x_2}\left[\frac{y_3 - y_1}{(x_3 - x_1)} - \frac{y_2 - y_1}{(x_3 - x_1)}\right]$$

$$\vdots$$

and so on.

Using the divided difference table, we get

$$a_1 = y_1$$

$$a_2 = \Delta_d y_1$$

$$a_3 = \Delta_d^2 y_1$$

$$\vdots$$

$$a_{n+1} = \Delta_d^n y_1$$

Therefore substituting these values of $a_1, a_2, a_3, \ldots, a_{n+1}$ in Eq. (6.14), we get

$$y(x) = y_1 + \Delta_d y_1(x - x_1) + \Delta_d^2 y_1(x - x_1)(x - x_2) + \ldots + \Delta_d^n y_1(x - x_1)(x - x_2)\ldots(x - x_n)$$

The above derivation assumes that

$$x_1 < x < x_2$$

If we want to interpolate at any point between x_k and x_{k+1}, i.e. $x_k < x < x_{k+1}$, then the formula takes the form

$$\boxed{\begin{aligned} y(x) = y_k &+ \Delta_d y_k (x - x_k) + \Delta_d^2 y_k (x - x_k)(x - x_{k+1}) + \ldots \\ &+ \Delta_d^{n-k} y_k (x - x_k)(x - x_{k+1})\ldots(x - x_{n-1}) \end{aligned}} \quad \ldots (6.15)$$

This polynomial is called the *Newton's divided difference interpolation formula* and is particularly suitable if the function is tabulated at *unequal intervals.*

Example 6.6: Given the table of values as

x	2.5	3.0	4.5	4.75	6.0
$y(x)$	8.85	11.45	20.66	22.85	38.60

Find $y(3.5)$.

Solution: The divided difference table for the given set of values is

i	x_i	y_i	$\Delta_d y_i$	$\Delta_d^2 y_i$	$\Delta_d^3 y_i$	$\Delta_d^4 y_i$
1	2.50	8.85	5.20	0.470	0.456	–0.029
2	3.00	11.45	6.14	1.497	0.354	
3	4.50	20.66	8.76	2.560		
4	4.75	22.85	12.6			
5	6.00	38.60				

Since the point x = 3.5 lies between second and third point, therefore we consider the divided differences at the second point. The Newton's divided difference interpolating polynomial for above case becomes

$$y(x) = y_2 + \Delta_d y_2(x - x_2) + \Delta_d^2 y_2(x - x_2)(x - x_3) + \Delta_d^3 y_2(x - x_2)(x - x_3)(x - x_4)$$

Therefore

$$y(3.5) = 11.45 + 6.14 \times (3.5 - 3) + 1.497 \times (3.5 - 3)(3.5 - 4.5)$$
$$+ 0.354 \times (3.5 - 3)(3.5 - 4.5)(3.5 - 4.75)$$
$$= 13.993 \qquad \text{(rounded to three decimal digits)}$$

Algorithm 6.8: Newton's divided interpolation formula

Two-dimensional arrays named x and y each of size n are used to store the set of values for the function $f(x)$, and two-dimensional array named d of size $(n-1)\times(n-1)$ is used to store the divided difference table.

```
Begin
    read: n, x
    for i = 1 to n by 1 do
        read: xi, yi
    endfor
    if (( x < x1 ) or ( x > xn ) ) then
        write: "Value lies outside tabulated range"
        exit
    endif
    set i = 2
    while ( x > xi ) do
        set i = i + 1
    endwhile
    set k = i - 1
    for j = 1 to (n-1) by 1 do    // generate divided difference table
        for i = 1 to (n-j) by 1 do
            if ( j = 1 ) then
                set dij = (yi+1 - yi)/(xi+1 - xi)
            else
                set dij = (d(i+1)(j-1) - di(j-1))/(xi+j - xi)
            endif
        endfor
    endfor
    set sum = yk
    for i = 1 to n by 1 do
        set prod = 1.0
        for j = 0 to (i-1) by 1 do
            set prod = prod × (x -xk+j)
        endfor
        set sum = sum + dki × prod
    endfor
    write: "Interpolated value = ", sum
End.
```

6.9 GAUSS'S INTERPOLATION FORMULA

If x takes the values x_0-2h, x_0-h, x_0, x_0+h, x_0+2h and corresponding values of $y = f(x)$ are $y_{-2}, y_{-1}, y_0, y_1, y_2$, then central difference table in two notations will take the form

Table 6.5: Central difference table with alternate notations

x	y	δy	δ^2	δ^3	δ^4
x_0-2h	y_{-2}				
		$\delta y_{-3/2}(\Delta y_{-2})$			
x_0-h	y_{-1}		$\delta^2 y_{-1}(\Delta^2 y_{-2})$		
		$\delta y_{-1/2}(\Delta y_{-1})$		$\delta^3 y_{-1/2}(\Delta^3 y_{-2})$	
x_0	y_0		$\delta^2 y_0(\Delta^2 y_{-1})$		$\delta^4 y_0(\Delta^4 y_{-2})$
		$\delta y_{1/2}(\Delta y_0)$		$\delta^3 y_{1/2}(\Delta^3 y_{-1})$	
x_0+h	y_1		$\delta^2 y_1(\Delta^2 y_0)$		
		$\delta y_{3/2}(\Delta y_1)$			
x_0+2h	y_2				

Consider the Newton's forward interpolation formula at first tabulated point is

$$y(x) = y_0 + \Delta y_0 u + \frac{\Delta^2 y_0}{2!} u(u-1) + \frac{\Delta^3 y_0}{3!} u(u-1)(u-2) + \ldots$$

From Table 6.5, we have

$$\Delta^3 y_{-1} = \Delta^2 y_0 - \Delta^2 y_{-1} \qquad \Rightarrow \qquad \Delta^2 y_0 = \Delta^3 y_{-1} + \Delta^2 y_{-1}$$

Similarly

$$\Delta^3 y_0 = \Delta^4 y_{-1} + \Delta^3 y_{-1}$$
$$\Delta^4 y_0 = \Delta^5 y_{-1} + \Delta^4 y_{-1}$$

Substituting these values in Newton's forward interpolation method, we get

$$y(x) = y_0 + \Delta y_0 u + \frac{(\Delta^3 y_{-1} + \Delta^2 y_{-1})}{2!} u(u-1) + \frac{(\Delta^4 y_{-1} + \Delta^3 y_{-1})}{3!} u(u-1)(u-2) + \ldots$$

On simplifying, it gives

$$y(x) = y_0 + \Delta y_0 u + \frac{\Delta^2 y_{-1}}{2!} u(u-1) + \frac{\Delta^3 y_{-1}}{3!} (u+1)u(u-1) + \frac{\Delta^4 y_{-2}}{4!} (u+1)u(u-1)(u-2) + \ldots \qquad \ldots (6.16)$$

This is called *Gauss's forward interpolation formula.* In the other notation, it takes the form

$$y(x) = y_0 + \delta y_{1/2} u + \frac{\delta^2 y_0}{2!} u(u-1) + \frac{\delta^3 y_{1/2}}{3!}(u+1)u(u-1) + \frac{\delta^4 y_0}{4!}(u+1)u(u-1)(u-2) + \ldots \qquad \ldots (6.17)$$

From Table 6.5, we also have

$$\Delta^2 y_{-1} = \Delta y_0 - \Delta y_{-1} \quad \Rightarrow \quad \Delta y_0 = \Delta y_{-1} + \Delta^2 y_{-1}$$

Similarly

$$\Delta^2 y_0 = \Delta^2 y_{-1} + \Delta^3 y_{-1}$$

$$\Delta^3 y_0 = \Delta^3 y_{-1} + \Delta^4 y_{-1}$$

Substituting these values in Newton's forward interpolation method, we get

$$y(x) = y_0 + (\Delta y_{-1} + \Delta^2 y_{-1})u + \frac{(\Delta^2 y_{-1} + \Delta^3 y_{-1})}{2!} u(u-1) + \frac{(\Delta^3 y_{-1} + \Delta^4 y_{-i})}{3!} u(u-1)(u-2) + \ldots$$

On simplifying, it gives

$$y(x) = y_0 + \Delta y_{-1} u + \frac{\Delta^2 y_{-1}}{2!} u(u+1) + \frac{\Delta^3 y_{-2}}{3!}(u+1)u(u-1) + \frac{\Delta^4 y_{-2}}{4!}(u+2)(u+1)u(u-1) + \ldots \qquad \ldots (6.18)$$

This is called *Gauss's backward interpolation formula.* In the other notation, it takes the form

$$y(x) = y_0 + \delta y_{-1/2} u + \frac{\delta^2 y_0}{2!} u(u+1) + \frac{\delta^3 y_{-1/2}}{3!}(u+1)u(u-1) + \frac{\delta^4 y_0}{4!}(u+2)(u+1)u(u-1) + \ldots \qquad \ldots (6.19)$$

6.10 STIRLING'S FORMULA

Gauss's forward interpolation formula is

$$y(x) = y_0 + \Delta y_{-1} u + \frac{\Delta^2 y_{-1}}{2!} u(u+1) + \frac{\Delta^3 y_{-2}}{3!}(u+1)u(u-1) + \frac{\Delta^4 y_{-2}}{4!}(u+2)(u+1)u(u-1) + \ldots$$

Taking the average of these, we obtain

$$y(x) = y_0 + \left(\frac{\Delta y_0 + \Delta y_{-1}}{2}\right)u + \Delta^2 y_{-1}\frac{u^2}{2!} + \left(\frac{\Delta^3 y_{-1} + \Delta^3 y_{-2}}{2}\right)\left(\frac{u(u^2-1)}{3!}\right) + \Delta^4 y_{-2}\frac{u^2(u^2-1) + \ldots}{4!} \quad \ldots (6.20)$$

This is known as *Stirling's formula.*

6.11 BESSEL'S FORMULA

Gauss's forward interpolation formula is

$$y(x) = y_0 + \Delta y_0 u + \frac{\Delta^2 y_{-1}}{2!}u(u-1) +$$

$$\frac{\Delta^3 y_{-1}}{3!}(u+1)u(u-1) + \frac{\Delta^4 y_{-2}}{4!}(u+1)u(u-1)(u-2) + \ldots$$

We have $\Delta^2 y_0 - \Delta^2 y_{-1} = \Delta^3 y_{-1}$

$\Rightarrow \quad \Delta^2 y_{-1} = \Delta^2 y_0 - \Delta^3 y_{-1}$

Similarly $\Delta^4 y_{-2} = \Delta^4 y_{-1} - \Delta^5 y_{-2}$ etc.

Now Gauss's forward interpolation formula can be written as

$$y(x) = y_0 + \Delta y_0 u + \left(\frac{1}{2}\Delta^2 y_{-1} + \frac{1}{2}\Delta^2 y_{-1}\right)\frac{u(u-1)}{2!} + \frac{u(u^2-1)}{3!}\Delta^3 y_{-1} +$$

$$\frac{u(u^2-1)(u-2)}{4!}\left(\frac{1}{2}\Delta^4 y_{-2} + \frac{1}{2}\Delta^4 y_{-2}\right) + \ldots$$

$$= y_0 + \Delta y_0 u + \frac{1}{2}\frac{u(u-1)}{2!}\Delta^2 y_{-1} + \frac{1}{2}\frac{u(u-1)}{2!}(\Delta^2 y_0 - \Delta^3 y_{-1}) +$$

$$\frac{u(u^2-1)}{3!}\Delta^3 y_{-1} + \frac{1}{2}\frac{u(u^2-1)(u-2)}{4!}\Delta^4 y_{-2} +$$

$$\frac{1}{2}\frac{u(u^2-1)(u-2)}{4!}(\Delta^4 y_{-1} - \Delta^5 y_{-2}) + \ldots$$

$$y(x) = y_0 + \Delta y_0 u + \frac{u(u-1)}{2!}\frac{\Delta^2 y_{-1} + \Delta^2 y_0}{2} + \frac{(u-1/2)u(u-1)}{3!}\Delta^3 y_{-1} + \frac{(u+1)u(u-1)(u-2)}{4!}\frac{\Delta^4 y_{-2} + \Delta^4 y_{-1}}{2} + \ldots \quad \ldots (6.21)$$

This is known as *Bessel's formula.*

6.12 LAPLACE-EVERETT'S FORMULA

Again consider Gauss's forward interpolation formula is

$$y(x) = y_0 + \Delta y_0 u + \frac{\Delta^2 y_{-1}}{2!}u(u-1) + \frac{\Delta^3 y_{-1}}{3!}(u+1)u(u-1) + \frac{\Delta^4 y_{-2}}{4!}(u+1)u(u-1)(u-2) + \ldots$$

We eliminate the odd differences in above by using the relations

$$\Delta y_0 = y_1 - y_0,\ \Delta^3 y_{-1} = \Delta^2 y_0 - \Delta^2 y_{-1},\ \Delta^5 y_{-2} = \Delta^4 y_{-1} - \Delta^4 y_{-2}$$

Then, the Gauss's forward interpolation takes the form

$$y(x) = y_0 + (y_1 - y_0)u + \Delta^2 y_{-1}\frac{u(u-1)}{2!} + \frac{(u+1)u(u-1)}{3!}(\Delta^2 y_0 - \Delta^2 y_{-1}) + \frac{(u+1)u(u-1)(u-2)}{4!}\Delta^4 y_{-2} + \frac{(u+2)(u+1)u(u-1)(u-2)}{5!}(\Delta^2 y_0 - \Delta^2 y_{-1}) + \ldots$$

$$= (1-u)y_0 + uy_1 - \frac{u(u-1)(u-2)}{3!}\Delta^2 y_{-1} + \frac{(u+1)u(u-1)}{3!}\Delta^2 y_0 - \frac{(u+1)u(u-1)(u-2)(u-3)}{5!}\Delta^4 y_{-2} + \frac{(u+2)(u+1)u(u-1)(u-2)}{5!}\Delta^4 y_{-1} - \ldots$$

To change the terms with negative sign, put $v = 1 - u$, we obtain

$$y(x) = vy_0 + \frac{v(v^2-1)}{3!}\Delta^2 y_{-1} + \frac{v(v^2-1)(v^4-4)}{5!}\Delta^4 y_{-2} + \ldots + uy_1 + \frac{u(u^2-1)}{3!}\Delta^2 y_0 + \frac{u(u^2-1)(u^4-4)}{5!}\Delta^4 y_{-1} + \ldots \quad \ldots (6.22)$$

This is known as *Laplace-Everett's formula.*

6.13 ERROR PROPAGATION IN DIFFERENCE TABLES

It must be remembered that the higher order differences may be unreliable due to the loss of significant digits when differences are taken due to the rapid build up errors in difference tables. The errors in differences may also occur due to errors in the original data. The errors in the original data may arise on account of a number of reasons such as wrong measurements. We will demonstrate the error build up in difference tables by taking the example of the forward difference table.

(*i*) Due to identical errors in all data points

y_i	Δy_i	$\Delta^2 y_i$	$\Delta^3 y_i$	$\Delta^4 y_i$
e	0	0	0	0
e	0	0	0	
e	0	0		
e	0			
e				

(*ii*) Due to alternate errors in the data points

y_i	Δy_i	$\Delta^2 y_i$	$\Delta^3 y_i$	$\Delta^4 y_i$
$-e$	$2e$	$-4e$	$8e$	$-16e$
e	$-2e$	$4e$	$-8e$	
$-e$	$2e$	$-4e$		
e	$-2e$			
$-e$				

(*iii*) Due to error in one of the data points

y_i	Δy_i	$\Delta^2 y_i$	$\Delta^3 y_i$	$\Delta^4 y_i$
0	0	e	$-3e$	$6e$
0	e	$-2e$	$3e$	
e	$-e$	e		
0	0			
0				

6.14 ESTIMATE OF ERRORS IN INTERPOLATION

Let us have an estimate of the error in interpolation that occurs by approximating a polynomial of finite order through a set of tabulated points. This error is called the truncation error. The error term is similar to the remainder term in a Taylor series expansion of the function.

The error term for various interpolation formulae is:

(*i*) Lagrangian interpolation formula

$$\text{error} = \frac{h^{n+1} f^{n+1}(\mu)}{(n+1)!}\left[\prod_{i=k}^{n+1}(x - x_i)\right]$$

(*ii*) Newton's forward difference interpolation formula

$$\text{error} = \frac{f^{n-k+1}(\mu)}{(n-k+1)!}\left[\prod_{i=0}^{n-k}(u - i)\right]$$

(*iii*) Newton's backward difference interpolation formula

$$\text{error} = \frac{f^{k}(\mu)}{k!}\left[\prod_{i=0}^{k}(u + i)\right]$$

(*iv*) Newton's divided difference interpolation formula

$$\text{error} = \frac{f^{n-k+1}(\mu)}{(n-k+1)!}\left[\prod_{i=k}^{n}(x - x_i)\right]$$

where $x_1 \leq \mu \leq x_n$ for all of the above cases.

6.15 APPROXIMATIONS OF FUNCTIONS

Our aim is to represent a function with a minimum error as this is a central problem in the software development because it is more economical to compute the values of the common functions using an efficient approximation than to store a table of values and employ interpolation techniques. Remember that computers are essentially arithmetic devices; the most elaborate function they can compute is a rational function, a ratio of polynomials. In this section, we will limit our discussion to representation functions by polynomials. There are two ways to approximate a function by a polynomial. These are

- Using truncated Taylor's series.
- Using Chebyshev polynomials.

The familiar Taylor-series expansion represents the function with very small error near the point of expansion, but the error increases rapidly (proportional to a power) as we employ it at points farther away. In computers, we have no control over where in an interval the approximation will be used, so Taylor series is not usually appropriate.

6.15.1 Taylor Series Representation

Suppose a function $f(x)$ has upto $(n+1)$st derivatives in an interval $[a, b]$, then it may be expressed near $x = x_0$ in $[a, b]$ as

$$f(x) = f(x_0) + f^1(x_0)(x - x_0) + f^2(x_0)\frac{(x - x_0)^2}{2!} + f^3(x_0)\frac{(x - x_0)^2}{3!} + \ldots$$
$$+ f^n(x_0)\frac{(x - x_0)^n}{n!} + f^{n+1}(s)\frac{(x - x_0)^{n+1}}{(n+1)!} \qquad \ldots (6.23)$$

where $f^1(x_0), f^2(x_0), f^3(x_0), \ldots$ are the 1st, 2nd, 3rd order derivatives of $f(x)$ evaluated at x_0.

The term $\quad f^{n+1}(s)\dfrac{(x - x_0)^{n+1}}{(n+1)!}$

is called the *remainder term.* Here s is a number that is a function of x and lies between x and x_0. The remainder term gives the truncation error if only the first n terms in the Taylor series are used to represent the function.

Therefore

$$\text{Truncation error} = \left|f^{n+1}(s)\right|\frac{\left|(x - x_0)^{n+1}\right|}{(n+1)!} \qquad \ldots (6.24)$$

$$\leq \frac{\left|(x - x_0)^{n+1}\right|}{(n+1)!}M \qquad \text{where } M = \max\left|f^{n+1}(s)\right| \text{ for } x \text{ in } [a, b].$$

Observe that the Taylor series is a polynomial with base function 1, $(x-x_0)$, $(x-x_0)^2$, …, $(x-x_0)^n$. The coefficients are constants given by $f(x_0)$, $f^1(x_0)$ $\dfrac{f^2(x_0)}{2!}$, … etc. Thus the series can be written in the nested form.

6.15.2 Chebyshev Polynomials

The Chebyshev Polynomial of degree n over the interval $[-1, 1]$ is defined by the relation

$$T_n(x) = \cos(n\cos^{-1}(x)) \qquad \ldots (6.25)$$

from which follows the relation

$$T_n(x) = T_{-n}(x) \qquad \ldots (6.26)$$

Taking $\cos^{-1}(x) = \theta$, so that $x = \cos(\theta)$. Thus from Eq. (6.25), we obtain

$$T_n(x) = \cos(n\theta)$$

Hence, by putting $n = 0, 1$, we get

$$T_0(x) = 1 \text{ and } T_1(x) = x$$

Now using the relation

$$\cos((n-1)\theta) + \cos((n+1)\theta) = 2\cos(n\theta)cox(\theta)$$

we get

$$T_{n-1}(x) + T_{n+1}(x) = 2xT_n(x)$$

$$\Rightarrow \quad T_{n+1}(x) = 2xT_n(x) - T_{n-1}(x) \qquad \ldots (6.27)$$

This is the recurrence relation that can be used to compute successively all $T_n(x)$ since we know $T_0(x)$ and $T_1(x)$.

The first few Chebyshev polynomials are

$$\begin{aligned} T_0(x) &= 1 \\ T_1(x) &= x \\ T_2(x) &= 2x^2 - 1 \\ T_3(x) &= 4x^2 - 3x \\ T_4(x) &= 8x^4 - 8x^2 + 1 \\ T_5(x) &= 16x^5 - 20x^3 + 5x \\ T_6(x) &= 32x^6 - 48x^4 + 18x^2 - 1 \end{aligned} \qquad \ldots (6.28)$$

Note that the coefficients of x^n in $T_n(x)$ is always 2^{n-1}.

Using the Chebyshev polynomials given in Eq. (6.28), we can express 1, x, x^2, ...,x^6 in terms of Chebyshev polynomials as

$$\begin{aligned} 1 &= T_0(x) \\ x &= T_1(x) \\ x^2 &= (1/2)[T_0(x) + T_1(x)] \\ x^3 &= (1/4)[3T_1(x) + T_3(x)] \\ x^4 &= (1/8)[3T_0(x) + 4T_2(x) + T_4(x)] \\ x^5 &= (1/16)[10T_1(x) + 5T_3(x) + T_5(x)] \\ x^6 &= (1/32)[10T_0(x) + 15T_2(x) + 6T_4(x) + T_6(x)] \end{aligned} \qquad \ldots (6.29)$$

These expressions will be useful in the economization of power series. Thus, it is possible to express the Taylor series expansion of e^{-x} in terms of Chebyshev polynomials as

$$e^{-x} = 1 - x + \frac{x^2}{2!} - \frac{x^3}{3!} + \frac{x^4}{4!} - \frac{x^5}{5!} + \ldots \qquad \ldots (6.30)$$

$$e^{-x} = T_0(x) - T_1(x) + \frac{1}{4}[T_0(x) + T_2(x)] - \frac{1}{24}[3T_1(x) + T_3(x)] \qquad \ldots (6.31)$$

$$+ \frac{1}{195}[3T_0(x) + 4T_2(x) + T_4(x)] - \frac{1}{1920}[10T_1(x) + 5T_3(x) + T_5(x)] + \ldots$$

$$e^{-x} = 1.26606\, T_0(x) - 1.13021\, T_1(x) + 0.27148\, T_2(x) - 0.04427\, T_3(x) \qquad \ldots (6.32)$$

$$+ 0.0054687\, T_4(x) - 0.0005208\, T_5(x) + \ldots$$

If we expand $T_0(x), T_1(x), T_2(x), T_3(x), T_4(x), T_5(x)$ using their polynomial equivalents and truncate after six terms, we obtain

$$e^{-x} \approx 1.00045 - 1.000022x + 0.4991992x^2 - 0.166488x^3 \qquad \ldots (6.33)$$

$$+ 0.043794x^4 - 0.008687x^5$$

Comparing this Taylor series expansion of Eq. (6.30), we observe that there are slight differences in the coefficients of different powers of x. The main advantage of this representation as a sum of Chebyshev polynomials is that, for a given error bound, we can truncate the series of Eq. (6.33) with a smaller number of terms compared to Taylor series. Also, the error is more uniformly distributed for various arguments. The possibility of a series with a lower number of terms is called *economization of power series.* The maximum error in the six term Chebyshev representation of Eq. (6.33) is 0.00045 whereas, that of Taylor series representation of Eq. (6.30) is 0.0014. Thus, we have to add one more term in Taylor series to ensure that the error is less than that in the Chebyshev approximation.

Example 6.7: We want to compute sin(x) correct to three significant digits. Obtain a series with minimum number of terms using Taylor series and Chebyshev series.

Solution: It is required to compute sin(x) correct to three significant digits. The Taylor series representation for sin(x) is

$$\sin(x) = x - \frac{x^3}{3!} + \frac{x^5}{5!} - \frac{x^7}{7!} + \ldots$$

$$\Rightarrow \quad \sin(x) = x - \frac{x^3}{6} + \frac{x^5}{120} - \frac{x^7}{5040} + \ldots \qquad \ldots (6.34)$$

Since the solution is required to be accurate to three significant digits and the truncation error after three terms of Eq. (6.34) is ≤ (1/5040)=0.000198, thus we truncate after three terms to obtain

$$\sin(x) = x - \frac{x^3}{6} + \frac{x^5}{120} \quad (6.35)$$

Now expressing the above series using Chebyshev polynomials, we obtain

$$\sin(x) = T_1(x) - \frac{1}{24}[3T_1(x) + T_3(x)] + \frac{1}{1920}[10T_1 + 5T_3 + T_5]$$

$$= 0.8802\ T_1(x) - 0.03906\ T_3(x) + 0.00052\ T_5(x)$$

As the coefficient of $T_5(x)$ = 0.00052 and as $|T_5(x)| \le 1$ for all x, therefore the truncation error if we omit the last term above still give 3 significant digit accuracy. Thus, if we omit, we will get

$$\sin(x) = 0.8802T_1(x) - 0.03906T_3(x) \quad \ldots (6.36)$$

Substituting $T_1(x) = x$ and $T_1(x) = (4x^3 - 3x)$, we get

$$\sin(x) = 0.9974x - 0.1562Tx^3 \quad \ldots (6.37)$$

which gives sin(x) correct to 3 significant digits with only two terms for any value of x.

EXERCISE . . .

1. What is interpolation? Explain with the help of a suitable example?
2. What are various types of differences? Give some advantages of the difference tables. Also discuss the error build up in the difference tables.
3. Show the derivation of the following interpolating polynomials:
   (*a*) Langrangian polynomial
   (*b*) Newton's backward difference polynomial
4. How will you choose a particular interpolation method?
5. Given a function in the form of a table as

x	2.0	3.0	4.0
$y(x)$	6.6	9.2	8.6

   Interpolate the value of $y(x)$ using *Lagrangian polynomial* at x = 2.8, 3.1.

6. Given a function in the form of a table as

x	2.0	3.0	4.0
$y(x)$	6.6	9.2	8.6

   Interpolate the value x using *Lagrangian inverse polynomial* for $y(x)$ = 7.2, 10.0.

7. Given a function in the form of table as

x	2.5	3.0	4.5	4.75	6.0
$y(x)$	8.85	11.45	20.66	22.85	38.60

Suggest the appropriate order of interpolating polynomial.

8. Given a function in the form of a table as

x	2.0	2.25	2.50	3.0
$y(x)$	9.0	10.06	11.25	14.0

Interpolate the value of x at $y(x) = 10.10$.

9. Given a function in the form of a table as

x	2.0	2.1	2.2	2.3
$y(x)$	11.0	12.261	13.648	15.167

(*a*) Estimate the order of polynomial that might be suitable for interpolating the function.

(*b*) Using polynomial of estimated order in (*a*), interpolate the value of $y(x)$ at $x = 2.05$ using Newton's forward difference interpolating polynomial.

10. In an experiment, the pressure of the gas is recorded as 50, 53, 57, 68, and 82 at intervals of two seconds. The first recording was done at zero seconds. Obtain an estimate of the pressure of the gas at time of 2.5 seconds and 6.2 seconds.
11. The distance (in kilometres) covered a car in a given time (in minutes) is given in the following table:

Time (t)	5	10	13	15	18
Distance(s)	5	12	17	21	29

What will be the approximate distance travelled at $t = 10$ minutes.

12. Express the following as polynomials in x:

(*a*) $T_0(x) + 2T_1(x) + T_2(x)$ (*b*) $2T_0(x) - (1/4)T_2(x) + (1/8)T_4(x)$

(*c*) $5T_0(x) + 2T_1(x) + 4T_2(x) - 8T_3(x)$

13. Express the following as polynomials in x:

(*a*) $1 + x - x^2 + x^3 - x^4$ (*b*) $1 - x^2 + 2x^4 - 4x^6$

(*c*) $1 - x + x^2 - x^3 + x^4$

❑ ❑ ❑

Curve Fitting and Method of Least Squares

Learning Outcomes

After reading this chapter, students will have the ability to

- explain the process of curve fitting
- list various methods for curve fitting
- describe the method of least squares
- explain the procedure for curve fitting
- describe the steps for fitting various types of curves
- use these methods to fit a given curve through a set of given data
- write programs for fitting variety of curves

7.1 INTRODUCTION

Very often, in practice, a relationship is found to exist between two or more variables. It is frequently desirable to express this relationship in a mathematical form by determining an equation connecting these variables. The general problem of finding equations of approximating curves which fit given set of data is called *regression analysis* or *curve fitting*. There are various methods available for approximating curves, but the most popular and useful method is the *method of least squares.*

In general sense, the curve fitting involves the determination of the continuous function

$$y = f(x) \qquad \ldots (7.1)$$

which results in the most reasonable or best fit for the given set of values of (x_1, y_1), (x_2, y_2), (x_3, y_3), . . ., (x_n, y_n). The particular form of $f(x)$ sometimes is known in advance from the physical laws or theory associated with the x and y variables being measured. In other cases, the useful practice is to prepare a scatter diagram, i.e. plot all or some of the points if number of points is large, and the appearance of the plotted points may suggest the particular form of $f(x)$. Figure 7.1 depicts some popular curves drawn in this manner.

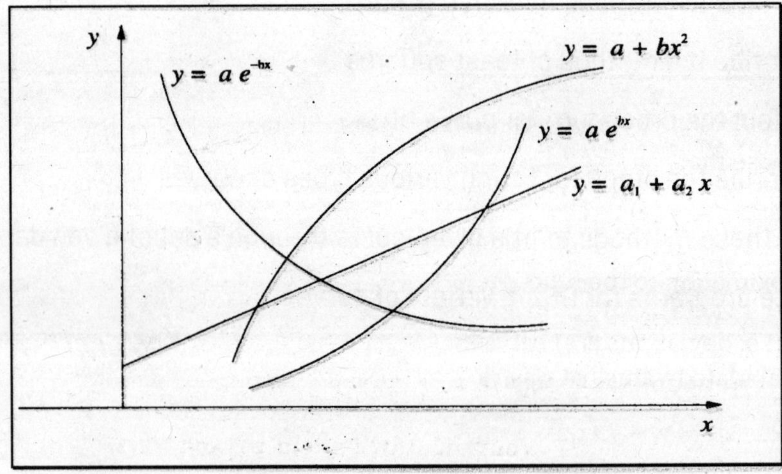

Figure 7.1: Some popular curves

Different methods available in literature to fit a curve to a given data are:

- Graphical Method
- Method of group averages
- Method of least squares
- Method of moments

The method of least squares is the best method to fit a curve to the given data.

In this chapter, we will focus our discussion on method of least squares for fitting variety of curves through a set of points given in the form of a table.

7.2 METHOD OF LEAST SQUARES

Let us suppose that $x_1, x_2, x_3, \ldots, x_n$ are the values of the independent variables x and $y_1, y_2, y_3, \ldots, y_n$ are values of the dependent variable y. The scatter diagram for this given data is:

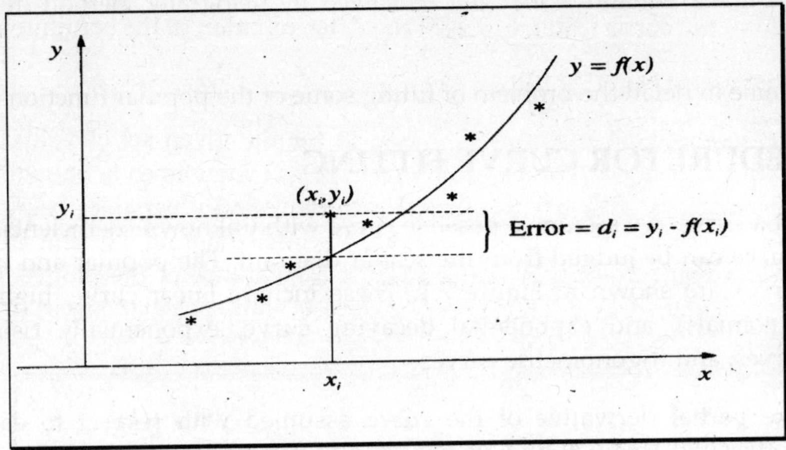

Figure 7.2: Scatter diagram

Let us suppose that

$$y = f(x)$$

be an approximation to the function.

The errors (i.e. deviations d_i's) between y's obtained from the above approximation and the actual tabulated values of y_is are

$$d_1 = y_1 - f(x_1)$$
$$d_2 = y_2 - f(x_2)$$
$$d_3 = y_3 - f(x_3)$$
$$\vdots$$
$$d_n = y_n - f(x_n)$$

The least square method states that

> *A curve is a best fit (i.e. a best curve), if the sum of the squares of deviations of the individual points from the curve is minimum.*

We form the sum, say S, of the squares of the deviations as

$$S = d_1^2 + d_2^2 + d_3^2 + \ldots + d_n^2 = \sum_{i=1}^{n} d_i^2 \qquad \ldots (7.2)$$

This sum will be either maximum or minimum if the partial derivatives of type $\partial S / \partial a = 0$, where a is one of the unknowns assumed in the approximation $y = f(x)$.

This way, by equating all the partial derivatives to zero, we get a system of non-homogeneous linear equations that can be solved by using any method discussed in *Chapter 5*.

We now examine in detail the problem of fitting some of the popular functions.

7.3 PROCEDURE FOR CURVE FITTING

In order to fit a curve, we assume a possible curve with unknown coefficients. The type of possible curve can be judged from the scatter diagram. The popular and most often occurring curves are shown in Figure 7.1. These include linear curve, higher degree curves (polynomials), and exponential decaying curve, exponentially rising curve, geometric curves, and trigonometric curves.

Then we take partial derivative of the curve assumed with respect to each of the unknown coefficients. This will give us a system of non-homogeneous linear equations with number of equations equal to number of unknowns. Then, this system of equation can be solved using any method of finding solution of system of non-homogeneous linear equations.

Once unknown coefficients are determined, we substitute their values in assumed curve, and thus the procedure for curve fitting reaches the logical end.

7.4 FITTING A STRAIGHT LINE

Let us suppose that $x_1, x_2, x_3, \ldots, x_n$ are the values of the independent variable x and $y_1, y_2, y_3, \ldots, y_n$ are values of the dependent variable y. These are plotted as shown in Figure 7.3.

The simplest type of the approximating curve is a straight line, whose equation can be written as

$$y = f(x) = a_2 x + a_1 \qquad \ldots (7.3)$$

where a_2 and a_1 represent the slope and the intercept of the line, respectively, and are known as *regression coefficients*, and the line is known as the regression line of the y on x. Now to find the equation of the best line we have to compute these regression coefficients.

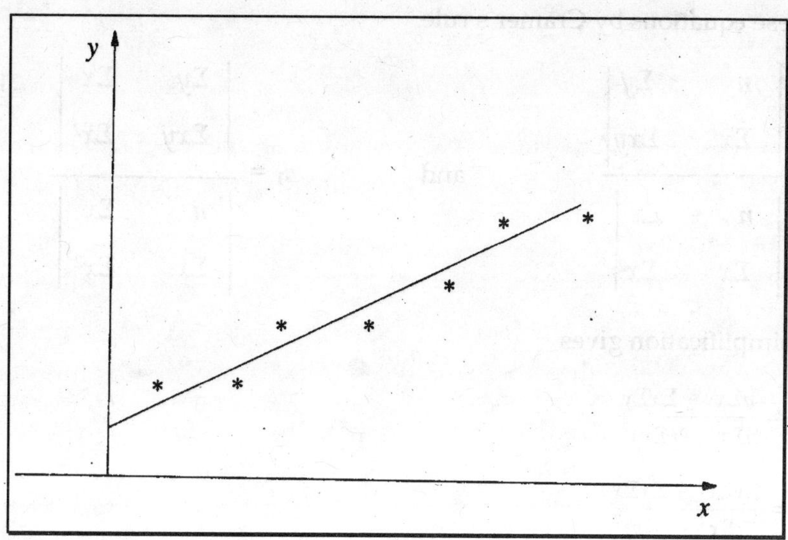

Figure 7.3: Fitting a Straight Line

The procedure used is described below:

The sum of the squares of the deviations is

$$S = \sum_{i=1}^{n} d_i^2 = \sum_{i=1}^{n} (y - f(x))^2 = \sum_{i=1}^{n} (y - a_2 x - a_1)^2$$

Therefore

$$\frac{\partial S}{\partial a_1} = 2\sum_{i=1}^{n} (y - a_2 x - a_1)(-1) \qquad \text{. . . (7.4)}$$

$$\frac{\partial S}{\partial a_2} = 2\sum_{i=1}^{n} (y - a_2 x - a_1)(-x) \qquad \text{. . . (7.5)}$$

According to the principle of least squares, these partial derivatives must be equal to zero. Hence, equating them to zero and simplifying, we get the following linear equations

$$na_1 + a_2 \Sigma x = \Sigma y \qquad \text{. . . (7.6)}$$

$$a_1 \Sigma x + a_2 \Sigma x^2 = \Sigma xy \qquad \text{. . . (7.7)}$$

These equations are called *normal equations* for the least square line. In the above normal equations we have used Σ instead of $\sum_{i=1}^{n}$ for simplicity and will continue to use this notation in rest of the chapter.

Solving these equations by Cramer's rule

$$a_2 = \frac{\begin{vmatrix} n & \Sigma y \\ \Sigma x & \Sigma xy \end{vmatrix}}{\begin{vmatrix} n & \Sigma x \\ \Sigma x & \Sigma x^2 \end{vmatrix}} \quad \text{and} \quad a_1 = \frac{\begin{vmatrix} \Sigma y & \Sigma x \\ \Sigma xy & \Sigma x^2 \end{vmatrix}}{\begin{vmatrix} n & \Sigma x \\ \Sigma x & \Sigma x^2 \end{vmatrix}}$$

which on simplification gives

$$a_2 = \frac{n\Sigma xy - \Sigma x\Sigma y}{n\Sigma x^2 - (\Sigma x)^2} \quad \ldots (7.8)$$

$$a_1 = \frac{\Sigma y\Sigma x^2 - \Sigma x\Sigma xy}{n\Sigma x^2 - (\Sigma x)^2} \quad \ldots (7.9)$$

Once a_2 is computed, the value of a_1 can also be computed as

$$a_1 = \frac{\Sigma y}{n} - a_2 \frac{\Sigma x}{n} \quad \ldots (7.10)$$

Proceeding on the same lines, we can find the regression line of x on y, whose general form may be represented as

$$x = g(y) = b_2 y + b_1 \quad \ldots (7.11)$$

whose normal equations will be

$$nb_1 + b_2\Sigma y = \Sigma x \quad \ldots (7.12)$$

$$b_1\Sigma y + b_2\Sigma y^2 = \Sigma yx \quad \ldots (7.13)$$

Solving these equations by Cramer's rule, we get

$$b_2 = \frac{\begin{vmatrix} n & \Sigma x \\ \Sigma y & \Sigma yx \end{vmatrix}}{\begin{vmatrix} n & \Sigma y \\ \Sigma y & \Sigma y^2 \end{vmatrix}} \quad \text{and} \quad a_1 = \frac{\begin{vmatrix} \Sigma x & \Sigma y \\ \Sigma yx & \Sigma y^2 \end{vmatrix}}{\begin{vmatrix} n & \Sigma y \\ \Sigma y & \Sigma y^2 \end{vmatrix}}$$

which on simplification gives

$$b_2 = \frac{n\Sigma yx - \Sigma y\Sigma x}{n\Sigma y^2 - (\Sigma y)^2} \qquad \text{... (7.14)}$$

$$b_1 = \frac{\Sigma x\Sigma y^2 - \Sigma y\Sigma yx}{n\Sigma y^2 - (\Sigma y)^2} \qquad \text{... (7.15)}$$

Once b_2 is computed, the value of b_1 can also be computed as

$$b_1 = \frac{\Sigma x}{n} - b_2 \frac{\Sigma y}{n} \qquad \text{... (7.16)}$$

An important property of these regression lines is that they intersect at a point

$$\left(\frac{\Sigma x}{n}, \frac{\Sigma y}{n}\right).$$

Example 7.1: Given a table of values for the function as

x :	0.1	0.2	0.3	0.4	0.5	0.6
y :	5.1	5.3	5.6	5.7	5.9	6.1

Determine both the regression lines, and also prove that they intersect at $\left(\frac{\Sigma x}{n}, \frac{\Sigma y}{n}\right)$.

Solution: The values of Σx, Σx^2, Σy, Σy^2, Σxy are computed as shown in the following table.

i	x_i	y_i	x_i^2	y_i^2	$x_i\, y_i$
1	0.1	5.1	0.01	26.01	0.51
2	0.2	5.3	0.04	28.09	1.06
3	0.3	5.6	0.09	31.36	1.68
4	0.4	5.7	0.16	32.49	2.28
5	0.5	5.9	0.25	34.81	2.95
6	0.6	6.1	0.36	37.21	3.66
$n = 6$	$\Sigma x = 2.1$	$\Sigma y = 33.7$	$\Sigma x^2 = 0.91$	$\Sigma y^2 = 189.97$	$\Sigma xy = 12.14$

Regression line of y on x

Let the regression line of y on x be of type

$y = a_2 x + a_1$

Substituting the values computed in the above table in Eqs. (7.8) and (7.9), and simplifying we get

$$a_2 = \frac{n\Sigma xy - \Sigma x\Sigma y}{n\Sigma x^2 - (\Sigma x)^2} = \frac{6\times 12.14 - 2.1\times 33.7}{6\times 0.91 - (2.1)^2} = 1.97$$

$$a_1 = \frac{\Sigma y\Sigma x^2 - \Sigma x\Sigma xy}{n\Sigma x^2 - (\Sigma x)^2} = \frac{33.7\times 0.91 - 2.1\times 12.14}{6\times 0.91 - (2.1)^2} = 4.93$$

The final value of a_2 and a_1 is rounded to two decimal places.

Thus the regression line of y on x is

$$y = 1.97x + 4.93$$

Regression line of x on y

Let the regression line of x on y be of type

$$x = b_2y + b_1$$

Substituting the values computed in the above table in Eqs. (7.8) and (7.9), and simplifying we get

$$b_2 = \frac{n\Sigma xy - \Sigma y\Sigma x}{n\Sigma y^2 - (\Sigma y)^2} = \frac{6\times 12.14 - 33.7\times 2.1}{6\times 189.97 - (33.7)^2} = 0.50$$

$$b_1 = \frac{\Sigma x\Sigma y^2 - \Sigma y\Sigma xy}{n\Sigma y^2 - (\Sigma y)^2} = \frac{2.1\times 189.97 - 33.7\times 12.14}{6\times 189.97 - (33.7)^2} = -2.46$$

The final value of b_2 and b_1 is rounded to two decimal places.

Thus the regression line of x on y is

$$x = 0.50y - 2.46$$

Now we would see that they intersect at $\left(\frac{\Sigma x}{n}, \frac{\Sigma y}{n}\right)$, i.e., (0.35, 5.62). This we can be done if we can prove that both of these regression lines pass through it. Let us first check for the regression line of y on x, *i.e.*,

$$y = 1.97x + 4.93$$

Substituting $x = 0.35$ on the right hand side, we get

$$y = 1.97 * 0.35 + 4.93 = 0.69 + 4.93 = 5.62$$

which shows that the regression line of y on x passes through (0.35, 5.62).

Now check for the regression line of x on y i.e.

$$x = 0.50y - 2.46$$

Substituting $y = 5.62$ on the right hand side, we get

$$x = 0.50 * 5.62 - 2.46 = 2.81 - 2.46 = 0.35$$

which shows that the regression line of x on y also passes through (0.35, 5.62).

Since both the regression lines passes through the point (0.35, 5.62), thus we conclude that they intersect at the point $\left(\frac{\Sigma x}{n}, \frac{\Sigma y}{n}\right)$.

Algorithm 7.1: Fitting a regression line of Y on X

Two one-dimensional arrays named x and y, each of size n, are used to store the data points at which the function is tabulated.

```
Begin
    read: n
    for i = 1 to n by 1 do
        read: xi, yi
    endfor
    set sx  = 0
    set sx2 = 0
    set sy  = 0
    set sxy = 0
    for i = 1 to n by 1 do
        set sx  = sx + xi
        set sx2 = sx2 + xi × xi
        set sy  = sy + yi
        set sxy = sxy + xi × yi
    endfor
    set denom = n×sx2 - sx×sx
    set a2 = (n×sxy - sx×sy)/ denom
    set a1 = (sy×sx2 - sx×sxy)/ denom
    write: "Regression coefficients are"
    write: "Slope of the regression line = ", a2
    write: "Y-intercept for the line = ", a1
End.
```

7.5 FITTING A POLYNOMIAL

In general it may be necessary to fit a higher degree polynomial rather a straight line. In this section, we will see how to fit a polynomial of a specific form.

The actual degree of the polynomial will be described by the physical consideration. However, we can utilize the property of difference tables. Recall that we have proved in

Chapter 6 that if a polynomial happens to be of mth degree, then the mth order differences will be constant and the higher order differences vanishes. Therefore by generating a difference table, we can find the degree of the polynomial.

Let us suppose that $x_1, x_2, x_3, \ldots, x_n$ are the values of the independent variable x and $y_1, y_2, y_3, \ldots, y_n$ are the values of the dependent variable y. And further suppose that we want to fit a polynomial of degree m of the form

$$y = f(x) = a_{m+1}x^m + a_m x^{m-1} + a_{m-1}x^{m-2} + \ldots + a_2 x + a_1 \quad \ldots (7.17)$$

where $a_{m+1}, a_m, a_{m-1}, \ldots, a_2, a_1$ are to be computed to give the best polynomial fit.

The sum of squares of the deviations is given by

$$S = \sum_{i=1}^{n} (y - f(x))^2 = \sum_{i=1}^{n} \left[y - (a_{m+1}x^m + a_m x^{m-1} + a_{m-1}x^{m-2} + \ldots + a_2 x + a_1) \right]^2$$

Differentiating S with respect to $a_1, a_2, a_3, \ldots, a_{m+1}$, and equating each one to zero we obtain the following normal equations

$$\left.\begin{aligned}
a_1 n + a_2 \Sigma x + a_3 \Sigma x^2 + \ldots + a_{m+1} \Sigma x^m &= \Sigma y \\
a_1 \Sigma x + a_2 \Sigma x^2 + a_3 \Sigma x^3 + \ldots + a_{m+1} \Sigma x^{m+1} &= \Sigma xy \\
a_1 \Sigma x^2 + a_2 \Sigma x^3 + a_3 \Sigma x^4 + \ldots + a_{m+1} \Sigma x^{m+2} &= \Sigma x^2 y \\
\vdots \qquad & \qquad \vdots \\
a_1 \Sigma x^m + a_2 \Sigma x^{m+1} + a_3 \Sigma x^{m+2} + \ldots + a_{m+1} \Sigma x^{m+m} &= \Sigma x^m y
\end{aligned}\right\} \quad \ldots (7.18)$$

These are $(m+1)$ non-homogeneous equations in $(m+1)$ unknowns, and are called *normal equations* for a polynomial fit.

$$\begin{pmatrix}
n & \Sigma x & \Sigma x^2 & \cdots & \Sigma x^m \\
\Sigma x & \Sigma x^2 & \Sigma x^3 & \cdots & \Sigma x^{m+1} \\
\Sigma x^2 & \Sigma x^3 & \Sigma x^4 & \cdots & \Sigma x^{m+2} \\
\vdots & \vdots & \vdots & & \vdots \\
\Sigma x^m & \Sigma x^{m+1} & \Sigma x^{m+2} & \cdots & \Sigma x^{m+m}
\end{pmatrix}
\begin{pmatrix} a_1 \\ a_2 \\ a_3 \\ \vdots \\ a_{m+1} \end{pmatrix}
=
\begin{pmatrix} \Sigma y \\ \Sigma xy \\ \Sigma x^2 y \\ \vdots \\ \Sigma x^m y \end{pmatrix}$$

The above system can be written as $CA = B$, where

$$C = \begin{pmatrix} n & \Sigma x & \Sigma x^2 & \cdots & \Sigma x^m \\ \Sigma x & \Sigma x^2 & \Sigma x^3 & \cdots & \Sigma x^{m+1} \\ \Sigma x^2 & \Sigma x^3 & \Sigma x^4 & \cdots & \Sigma x^{m+2} \\ \cdot & \cdot & \cdot & \cdots & \cdot \\ \cdot & \cdot & \cdot & \cdots & \cdot \\ \cdot & \cdot & \cdot & \cdots & \cdot \\ \Sigma x^m & \Sigma x^{m+1} & \Sigma x^{m+2} & \cdots & \Sigma x^{m+m} \end{pmatrix} \quad A = \begin{pmatrix} a_1 \\ a_2 \\ a_3 \\ \cdot \\ \cdot \\ \cdot \\ a_{m+1} \end{pmatrix} \quad B = \begin{pmatrix} \Sigma y \\ \Sigma xy \\ \Sigma x^2 y \\ \cdot \\ \cdot \\ \cdot \\ \Sigma x^m y \end{pmatrix}$$

The elements of the matrix C are computed as

$$c_{ij} = \sum_{k=1}^{n} x_k^{i+j-2} \qquad \text{for } i = 1, 2, 3, \ldots, m+1$$

$$\text{for } j = 1, 2, 3, \ldots, m+1 \quad \text{(for each value of } i) \qquad \ldots (7.19)$$

and the elements of the vector B are computed as

$$b_i = \sum_{k=1}^{n} x_k^{i-1} y_k \qquad \text{for } i = 1, 2, 3, \ldots, m+1 \qquad \ldots (7.20)$$

Before we can solve these equations, we have to compute the above summations, and then using any of the methods to find the solution of a system of non-homogeneous equations discussed in *Chapter 5*, we can find the unknown coefficients and this is how the polynomial relationship is established.

In the following algorithm, we will be using the Gauss elimination method with pivoting. For computer implementation, we will be storing the constants of the normal equations as the last column of the augmented matrix (i.e. as $(m+2)^{th}$ column). The step required to compute these coefficients are summarised in the following algorithm.

Example 7.2: Given a table of values for the function as

x :	1.0	1.5	2.0	2.5	3.1	4.0
y :	1.1	1.3	1.6	2.0	3.4	4.2

Fit a second degree polynomial.

Solution: Let $y = a_3x^2 + a_2x + a_1$ be a second degree polynomial.

The normal equations for this second-degree polynomial are

$$na_1 + a_2\Sigma x + a_3\Sigma x^2 = \Sigma y$$

$$a_1\Sigma x + a_2\Sigma x^2 + a_3\Sigma x^3 = \Sigma xy$$

$$a_1\Sigma x^2 + a_2\Sigma x^3 + a_3\Sigma x^4 = \Sigma x^2 y$$

Computing the values of Σx, Σx^2, Σx^3, Σx^4, Σy, Σxy, and Σx^2y and substituting in the normal equations, we get

$$6a_1 + 14.1a_2 + 39.11a_3 = 13.6$$

$$14.1a_1 + 39.11a_2 + 121.791a_3 = 38.59$$

$$39.11a_1 + 121a_2 + 409.477a_3 = 122.799$$

Solving this system of non-homogeneous equations in three unknowns by Gauss elimination method using pivoting, we get

$a_1 = 0.678$ $\quad a_2 = 0.137$ $\quad a_3 = 0.195$ (rounded off to three decimal places)

Thus the equation of the second-degree polynomial is

$$y = 0.195x^2 + 0.137x + 0.678$$

Algorithm 7.2: To Fit a polynomial of degree *m*

Two one-dimensional arrays named *x* and *y*, each of size *n*, are used to store the data points at which the function is tabulated. Third one-dimensional array named *a* of size (*m*+1) is used to store the computed values of regression coefficients.

```
Begin
    read: m, n
    for i = 1 to n by 1 do
        read: x_i, y_i
    endfor
    for i = 1 to (m+1) by 1 do
        for j = 1 to (m+1) by 1 do
            set c_ij = 0
            for k = 1 to n by 1 do
                set c_ij = c_ij + (x_k)^(i+j-2)
            endfor
        endfor
        set c_i(m+2) = 0
        for k = 1 to n by 1 do
            set c_i(m+2) = c_i(m+2) + (x_k)^(i-1) × y_k
        endfor
    endfor
```

```
    for k = 1 to m by 1 do
        set max = | c_ii |
        set p = k
        for q = (k+1) to (m+1) by 1 do
            if (| c_qk | > max ) then
                set max = | c_qk |
                set p = q
            endif
        endfor
        if ( p ≠ k ) then
            for q = k to (m+1) by 1 do
                set temp = c_kq
                set c_kq = c_pq
                set c_pq = temp
            endfor
        endif
        for i = (k+1) to (m+1) by 1 do
            set temp = c_ik/c_kk
            for j = k to (m+2) by 1 do
                set c_ij = c_ij - temp × c_kj
            endfor
        endfor
    endfor
    set a_m+1 = c_(m+1)(m+2)/c_(m+1)(m+1)
    for i = m to 1 by -1 do
        set sum = 0
        for j = (i+1) to (m+1) by 1 do
            set sum = sum + c_ij × a_j
        endfor
        set a_i = (c_i(m+2) - sum)/c_ii
    endfor
    write: "Regression coefficients are"
    for i = 1 to (m+1) by 1 do
        write: a_i
    endfor
End.
```

7.6 FITTING A NON-LINEAR FUNCTION

In many real life situations, the experimental data follow some other geometric forms which can be deduced from the inspection of the scatter diagram or other physical considerations, such as exponential curves, trigonometric curves, parabola, ellipse, etc.

The normal practice of fitting these functions is to transform them to linear form, which may not always be possible. In this section, we will consider some familiar functions that can be transformed to linear form.

7.6.1 Fitting a Geometric Curve

The geometric curve is described by the equation

$$y = ax^b \qquad \ldots (7.21)$$

where b is a real number.

Taking logarithm on both sides, we get

$$\log y = \log a + b \log x$$

Substituting

$$z = \log y,\ c = \log a, \text{ and } t = \log x$$

in the above equation, we get

$$z = bt + c$$

which is in linear form. The same steps as discussed in the case of straight line fitting are applied here, and we get the following normal equations for the above geometric curve

$$nc + b\Sigma t = \Sigma z$$

$$c\Sigma t + b\Sigma t^2 = \Sigma tz$$

$$\Rightarrow \qquad nc + b\Sigma \log x = \Sigma \log y$$

$$c\Sigma \log x + b\Sigma(\log x)^2 = \Sigma \log x \log y$$

Solving these equations, we get

$$\left.\begin{aligned} b &= \frac{n\Sigma \log x \log y - \Sigma \log x \Sigma \log y}{n\Sigma(\log x)^2 - (\Sigma \log x)^2} \\ c &= \frac{\Sigma \log y \Sigma(\log x)^2 - \Sigma \log x \Sigma \log x \log y}{n\Sigma(\log x)^2 - (\Sigma \log x)^2} \end{aligned}\right\} \qquad \ldots (7.22)$$

Once the values of b and c are computed, the value of a is computed from the equation

$$a = e^c \qquad \ldots (7.23)$$

7.6.2 Fitting an Exponential Curve

The exponential curves are described by the equations of type

$$y = ae^{bx} \quad \text{and} \quad y = ae^{-bx} \qquad \ldots (7.24)$$

We take the case of the exponential curve of form

$$y = ae^{bx}$$

Taking logarithm on both sides, we get

$$\log y = \log a + bx$$

Substituting

$$z = \log y,\ c = \log a$$

in the above equation, we get

$$z = bx + c$$

which is in linear form. The same steps as discussed in the case of straight line fitting are applied here, and we get the following normal equations for the above exponential curve

$$nc + b\Sigma x = \Sigma z$$

$$c\Sigma x + b\Sigma x^2 = \Sigma xz$$

$$\Rightarrow \quad nc + b\Sigma x = \Sigma \log y$$

$$c\Sigma x + b\Sigma x^2 = \Sigma x \log y$$

Solving these equations, we get

$$\left.\begin{aligned} b &= \frac{n\Sigma x \log y - \Sigma x \Sigma \log y}{n\Sigma x^2 - (\Sigma x)^2} \\ c &= \frac{\Sigma \log y \Sigma x^2 - \Sigma x \Sigma x \log y}{n\Sigma x^2 - (\Sigma x)^2} \end{aligned}\right\} \qquad \ldots (7.25)$$

Once the values of b and c are computed, the value of a is computed from the equation

$$a = e^c$$

7.6.3 Fitting a Hyperbola

The hyperbola is described by the equations of type

$$y = \frac{1}{a + bx} \qquad \ldots (7.26)$$

Substituting $t = \dfrac{1}{y}$ in the above equation, we get

$t = bx + a$ which is in linear form.

The same steps as discussed in the case of straight-line fitting are applied here, and we get the following normal equations for the above hyperbola

$$na + b\Sigma x = \Sigma t$$

$$a\Sigma x + b\Sigma x^2 = \Sigma xt$$

$$\Rightarrow \quad na + b\Sigma x = \Sigma \frac{1}{y}$$

$$a\Sigma x + b\Sigma x^2 = \Sigma x \frac{1}{y}$$

Solving these equations, we get

$$\left.\begin{aligned} b &= \frac{n\Sigma \frac{x}{y} - \Sigma x \Sigma \frac{1}{y}}{n\Sigma x^2 - (\Sigma x)^2} \\ a &= \frac{\Sigma \frac{1}{y} \Sigma x^2 - \Sigma x \Sigma \frac{x}{y}}{n\Sigma x^2 - (\Sigma x)^2} \end{aligned}\right\} \quad \ldots (7.27)$$

7.6.4 Fitting a Trigonometric Function

A trigonometric function is described by the equations of type

$$y = A\sin(kt + \theta) \quad \ldots (7.28)$$

Where A, k, and θ are unknowns. We assume that k is known. Expanding the above equation, we get

$$y = A\sin kt \cos\theta + A\sin\theta \cos kt$$

Substituting

$$a = A\cos\theta \qquad b = A\sin\theta$$

in the above equation, we get

$$y = a\sin kt + b\cos kt$$

which is in linear form.

The same steps as discussed in the case of straight line fitting are applied here, and we get the following normal equations for the above trigonometric function

$$a \sin^2 kt + b\Sigma \sin kt \cos kt = \Sigma y \sin kt$$

$$a\Sigma \sin kt \cos kt + b\Sigma \cos^2 kt = \Sigma y \cos kt$$

Solving these equations, we get

$$\left.\begin{aligned} b &= \frac{\Sigma \sin^2 kt\Sigma y \cos kt - \Sigma \sin kt \cos kt\Sigma y \sin kt}{\Sigma \sin^2 kt\Sigma \cos^2 kt - (\Sigma \sin kt\Sigma \cos kt)^2} \\ a &= \frac{\Sigma \cos^2 kt\Sigma y \sin kt - \Sigma y \cos kt\Sigma \sin kt \cos kt}{\Sigma \sin^2 kt\Sigma \cos^2 kt - (\Sigma \sin kt\Sigma \cos kt)^2} \end{aligned}\right\} \quad \ldots (7.29)$$

Once the values of b and a are computed, the value of A and θ are computed as

$$\left.\begin{aligned} A &= \sqrt{a^2 + b^2} \\ \theta &= \tan^{-1}\left(\frac{b}{a}\right) \end{aligned}\right\} \quad \ldots (7.30)$$

Example 7.3: Given a table of values for the function as

x :	600	500	400	350
y :	2	10	26	61

It is known that a relation of type

$$y = ae^{bx}$$

exist. Find the best possible values for a and b.

Solution: The normal equations for curve of type $y = ae^{bx}$ are

$$nc + b\Sigma x = \Sigma \log y$$

$$c\Sigma x + b\Sigma x^2 = \Sigma x \log y$$

where $c = \log a$.

Compute the values of Σx, Σx^2, $\Sigma \log y$, and $\Sigma x \log y$ as shown in the table on next page.

i	x_i	y_i	$\log y_i$	x_i^2	$x_i \log y_i$
1	600	2	0.30103	360000	180.618
2	500	10	1.0	250000	500.0
3	400	26	1.414973	160000	565.9892
4	350	61	1.78533	122500	624.8655
$n = 4$	$\Sigma x = 1850$	$\Sigma y = 99$	$\Sigma \log y = 4.501333$	$\Sigma x^2 = 892500$	$\Sigma x \log y = 1871.473$

Substituting these values in the above normal equations, we get

$$4c + 1850b = 4.501333$$

$$1850c + 892500b = 1871.473$$

Solving these normal equations, we get

$$c = 3.764167$$

$$b = -0.0057056$$

Since

$$c = \log a$$

therefore

$$a = \text{antilog } c = \text{antilog } (3.764167) = 43.12777$$

Hence the relation between x and y is

$$y = 43.12777e^{-0.0057056}$$

7.7 MULTIPLE REGRESSION

Till now we have considered the case where dependent variable is a function of single independent variable i.e. a relation of type $y = f(x)$, we can call it a case of *simple regression*. In this section, we will consider a case where the dependent variable is a function of two or more independent variables i.e. a relation of type $y = f(x_1, x_2, x_3, \ldots, x_k)$, we can call it a case of *multiple regression*.

In multiple regression, we deal with data consisting of $n(r+1)$-tuples $(x_{i1}, x_{i2}, x_{i3}, \ldots, x_{ik}, y_i)$, where x's are assumed to be known with error while y's are values of dependent variable which may contain error.

Some problems where data of this kind arise are enumerated below:

- Studies designed to determine the effect of various climatic conditions on a metal's resistance to corrosion.
- Effect of kiln temperature, humidity, and iron content on the strength of a ceramic coating.
- Effect of factory production, consumption level, and stocks in storage on the price of a product.

For simplicity, we will consider the problem for two independent variables and further consider that the regression equation is linear and is of type

$$y = a_0 + a_1x_1 + a_2x_2 \qquad \ldots (7.31)$$

where a_0, a_1, a_2 are regression coefficients and are to be determined. Applying method of least squares to determine these coefficients, we minimize the sum of squares of the deviations, *i.e.*, we minimize

$$S = \sum_{i=1}^{n}[y_i - (a_0 + a_1x_{i1} + a_2x_{i2})]^2 \qquad \ldots (7.32)$$

Differentiating Eq. (7.31) w. r. t. a_0, a_1 and a_2, we get

$$\left.\begin{aligned} \frac{\partial S}{\partial a_0} &= 2\sum_{i=1}^{n}\left[y_i - (a_0 + a_1x_{i1} + a_2x_{i2})\right](-1) \\ \frac{\partial S}{\partial a_0} &= 2\sum_{i=1}^{n}\left[y_i - (a_0 + a_1x_{i1} + a_2x_{i2})\right](-x_{i1}) \\ \frac{\partial S}{\partial a_0} &= 2\sum_{i=1}^{n}\left[y_i - (a_0 + a_1x_{i1} + a_2x_{i2})\right](-x_{i2}) \end{aligned}\right\} \qquad \ldots (7.33)$$

Equating them with zero and simplifying, we get

$$\left.\begin{aligned} na_0 + a_1\Sigma x_{i1} + a_2\Sigma x_{i2} &= \Sigma y_i \\ a_0\Sigma x_{i1} + a_1\Sigma x_{i1}^2 + a_2\Sigma x_{i1}x_{i2} &= \Sigma x_{i1}\,y_i \\ a_0\Sigma x_{i2} + a_1\Sigma x_{i1}x_{i2} + a_2\Sigma x_{i2}^2 &= \Sigma x_{i2}\,y_i \end{aligned}\right\} \qquad \ldots (7.34)$$

This is a system of three linear equation in three unknowns (a_0, a_1 and a_2), and solving these equations we obtain the values of these unknowns, and hence we determine the equation of regression equation (plane).

The same procedure can be extended to any number of independent variables and regression equations of any order.

Example 7.5: The following table shows the number of twists required breaking a certain kind of forged alloy bar and percentages of two alloying elements present in the metal:

Number of twists (y)	Percent of element A (x_1)	Percent of element B (x_2)	Number of twists (y)	Percent of element A (x_1)	Percent of element B (x_2)
41	1	5	49	2	5
69	3	5	65	4	5
40	1	10	50	2	12
58	3	11	57	4	9
31	1	15	36	2	17

Fit a regression plane and use it to estimate the number of twists required to break one of the bars when $x_1 = 2.5$ and $x_2 = 12$.

Solution:

i	y_i	x_{i1}	x_{i2}	x_{i1}^2	x_{i2}^2	$x_{i1}x_{i2}$	$x_{i1}y_i$	$x_{i2}y_i$
1	41	1	5	1	25	5	41	205
2	49	2	5	2	25	10	98	245
3	69	3	5	9	25	15	207	345
4	65	4	5	16	25	20	260	325
5	40	1	10	1	100	10	40	400
6	50	2	12	4	144	24	100	600
7	58	3	11	9	121	33	174	638
8	57	4	9	16	81	36	228	513
9	31	1	15	1	225	15	31	465
10	36	2	17	4	289	34	72	612
n=10	$\Sigma y_i = 496$	$\Sigma x_{i1} = 23$	$\Sigma x_{i2} = 84$	$\Sigma x_{i1}^2 = 63$	$\Sigma x_{i2}^2 = 1060$	$\Sigma x_{i1}x_{i2} = 202$	$\Sigma x_{i1}y_i = 1251$	$\Sigma x_{i2}y = 4348$

Substituting these computed values in Eq. (7.33), we get

$$10a_0 + 23a_1 + 84a_2 = 496$$

$$23a_0 + 63a_1 + 202a_2 = 1251$$

$$84a_0 + 202a_1 + 1060a_2 = 4348$$

Solving this system of equation, we get

$$a_0 = 22.926, \quad a_1 = 10.696, \quad a_2 = 0.247$$

Thus, estimated regression plane is

$$y = 22.926 + 10.696\,x_1 + 0.247\,x_2$$

Now substituting $x_1 = 2.5$ and $x_2 = 12$, we get

$$\hat{y} = 22.926 + 10.696 \times 2.5 + 0.247 \times 12 = 52.63$$

7.8 CURVE FITTING BY SUM OF EXPONENTIALS

In the field of science and engineering, enormous situations are encountered that requires the fitting of a sum of exponentials of the form

$$y = A_1 e^{\lambda_1 x} + A_2 e^{\lambda_2 x} + A_3 e^{\lambda_3 x} + \ldots + A_n e^{\lambda_n x} \quad \ldots (7.35)$$

where $A_1, A_2, A_3, \ldots, A_n$ and $\lambda_1, \lambda_2, \lambda_3, \ldots, \lambda_n$ are unknowns.

Eq. (7.35) satisfies a differential equation of type

$$\frac{d^n y}{dx^n} + a_1 \frac{d^{n-1} y}{dx^{n-1}} + a_2 \frac{d^{n-2} y}{dx^{n-2}} + \ldots + a_n y = 0 \quad \ldots (7.36)$$

where $a_1, a_2, a_3, \ldots, a_n$ are unknowns

Froberg suggested a method for computing these derivatives numerically, in Eq. (7.36) at the given points and substituting them in Eq. (7.35), thus obtaining a system of n linear equations for n unknowns $a_1, a_2, a_3, \ldots, a_n$ that can be solved.

Again it can be seen that $\lambda_1, \lambda_2, \lambda_3, \ldots, \lambda_n$ are the roots of the polynomial equation

$$\lambda^n + a_1 \lambda^{n-1} + a_2 \lambda^{n-2} + \ldots + a_n = 0 \quad \ldots (7.37)$$

which when solved, enables us to determine $A_1, A_2, A_3, \ldots, A_n$ from Eq. (7.34) by the method of least squares.

An obvious disadvantage of the method is the numerical evaluation of the derivatives whose accuracy deteriorates with their increasing order and therefore leading to unreliable results.

Eq. (7.36) can be solved by **Moore's method** that is described below for $n = 2$.

For $n = 2$, Eq. (7.35) becomes

$$y = A_1 e^{\lambda_1 x} + A_2 e^{\lambda_2 x} \quad \ldots (7.38)$$

This satisfies the differential equation

$$\frac{d^2 y}{dx^2} = a_1 \frac{dy}{dx} + a_2 y \quad \ldots (7.39)$$

Assume that a is the initial value of x and integrate Eq. (7.39) w.r.t. x, we will get

$$y'(x) - y'(a) = a_1 y(x) - a_1 y(a) + a_2 \int_a^x y(x)dx \qquad \text{... (7.40)}$$

where $y'(x) = \dfrac{dy(x)}{dx}$

Integrating Eq. (7.40) again w.r.t. x from a to x, we will get

$$y(x) - y(a) - (x-a)y'(a) = a_1 \int_a^x y(x)dx - a_1(x-a)y(a) + a_2 \int_a^x \int_a^x y(x)dxdx \qquad \text{... (7.41)}$$

Using the result from literature

$$\int_a^x ... \int_a^x f(x)dx....dx = \frac{1}{(n-1)!}\int_a^x (x-t)^{n-1} f(t)dt \qquad \text{... (7.42)}$$

Eq. (7.41) simplifies to

$$y(x) - y(a) - (x-a)y'(a) = a_1 \int_a^x y(x)dx - a_1(x-a)y(a) + a_2 \int_a^x (x-t)y(t)dt \qquad \text{... (7.43)}$$

In order to use Eq. (7.43), to set up a system of linear equations in terms of a_1 and a_2, $y'(a)$ need to be eliminated and this is done in following way:

Choosing two data points x_1 and x_2 such that $a - x_1 = x_2 - a$, then from Eq. (7.41), we will get

$$y(x_1) - y(a) - (x_1 - a)y'(a) = a_1 \int_a^{x_1} y(x)dx - a_1(x_1 - a)y(a) + a_2 \int_a^{x_1} (x_1 - t)y(t)dt$$

$$y(x_2) - y(a) - (x_2 - a)y'(a) = a_1 \int_a^{x_2} y(x)dx - a_1(x_2 - a)y(a) + a_2 \int_a^{x_2} (x_2 - t)y(t)dt$$

Adding these two equations and simplifying by using $a - x_1 = x_2 - a$, we will get

$$y(x_1) + y(x_2) - 2y(a) = a_1\left[\int_a^{x_1} y(x)dx + \int_a^{x_2} y(x)dx\right] + a_2\left[\int_a^{x_1}(x_1 - t)y(t)dt + \int_a^{x_2}(x_2 - t)y(t)dt\right] \qquad \text{... (7.44)}$$

Now Eq. (7.44) can be used to set up a system of linear equations for a_1 and a_2, and then we obtain λ_1 and λ_2 from characteristic equation

$$\lambda^2 = a_1\lambda + a_2 \qquad \text{... (7.45)}$$

Finally, A_1 and A_2 can be obtained by the method of least squares. Thus, we have the required equation of the form.

Once the values of A_1, A_2, λ_1 and λ_2 are determined, we have obtained the required form

$$y = A_1 e^{\lambda_1 x} + A_2 e^{\lambda_2 x}$$

7.9 CURVE FITTING WITH CUBIC SPLINES

Consider the problem of interpolating through a set of equally spaced points (x_i, y_i), $i = 1, 2, 3, \ldots n$ by means of spline fitting. Then the cubic spline is such that

- $f(x)$ is a linear polynomial outside the interval (x_0, x_n),
- $f(x)$ is a cubic polynomial inside each of the subintervals,
- $f'(x)$ and $f''(x)$ are continuous at each point.

Since $f(x)$ is cubic in each of the subintervals, $f''(x)$ will be linear and can be written as

$$f''(x) = \frac{1}{h}[(x_{i+1} - x) f''(x_i) + (x - x_i) f''(x_{i+1})] \qquad \ldots (7.46)$$

Integrating Eq. (7.46) twice, we obtain

$$f(x) = \frac{1}{h}\left[\frac{(x_{i+1} - x)^3}{3!} f''(x_i) + \frac{(x - x_i)^3}{3!} f''(x_{i+1})\right] + a_i(x_{i+1} - x) + b_i(x - x_i) \qquad \ldots (7.47)$$

The constants of integration a_i, b_i are determined by substituting the values of $y = f(x)$ at x_i and x_{i+1}. Thus

$$a_i = \frac{1}{h}\left[y_i - \frac{h^2}{3!} f''(x_i)\right] \quad \text{and} \quad b_i = \frac{1}{h}\left[y_{i+1} - \frac{h^2}{3!} f''(x_{i+1})\right]$$

Substituting these values and writing $f''(x) = M_i$ in Eq. (7.47), we get

$$f(x) = \frac{(x_{i+1} - x)^3}{6h} M_i + \frac{(x - x_i)^3}{6h} M_{i+1} + \frac{x_{i+1} - x}{h}\left(y_i - \frac{h^2}{6} M_i\right) + \frac{x - x_i}{h}\left(y_{i+1} - \frac{h^2}{6} M_{i+1}\right) \qquad \ldots (7.48)$$

Differentiating Eq. (7.48) w.r.t. x, we get

$$f'(x) = -\frac{(x_{i+1} - x)^2}{2h} M_i + \frac{(x - x_i)^2}{2h} M_{i+1} - \frac{h}{6}(M_{i+1} - M_i) + \frac{1}{h}(y_{i+1} - y_i)$$

To impose the condition of continuity of $f'(x)$, we set

$$f'(x - \varepsilon) = f'(x + \varepsilon) \quad \text{as } \varepsilon \to 0$$

$$\frac{h}{6}(2M_i - M_{i-1}) + \frac{1}{h}(y_i - y_{i-1}) = -\frac{h}{6}(2M_i + M_{i+1}) + \frac{1}{h}(y_{i+1} - y_i)$$

$$\Rightarrow \quad M_{i-1} + 4M_i + M_{i+1} = \frac{6}{h^2}(y_{i-1} - 2y_i + y_{i+1}), \quad \text{for } i = 1 \text{ to } n-1 \qquad \ldots (7.49)$$

Since the graph is linear for $x < x_0$ and $x > x_n$, we have

$$M_0 = M_n = 0 \qquad \ldots (7.50)$$

Eqs. (7.49), (7.50) gives $(n+1)$ equations in $(n+1)$ unknowns M_i's ($i = 0, 1, 2, \ldots, n$), which can be solved. Substituting the values in Eq. (7.48), we get the concerned cubic spline.

Example 7.5: Consider the following data

x :	1	2	3	4
y :	1	2	5	11

Find the cubic spline and evaluate $y(1.5)$.

Solution: Since $h = 1$ and $n = 3$, from Eq. (7.49), we obtain

$$M_{i-1} + 4M_i + M_{i+1} = \frac{6}{h^2}(y_{i-1} - 2y_i + y_{i+1}), \quad \text{for } i = 1, 2$$

$$\Rightarrow \quad M_0 + 4M_1 + M_2 = 6(y_0 - 2y_1 + y_2) = 12$$

$$M_1 + 4M_2 + M_3 = 6(y_1 - 2y_2 + y_3) = 18$$

Since $M_0 = M_2 = 0$, therefore

$$4M_1 + M_2 = 12 \quad \text{and} \quad M_1 + 4M_2 = 18$$

which gives $M_1 = 2$ and $M_2 = 4$

Thus Eq. (7.47) gives the splines as

$$f(x) = \begin{cases} \frac{1}{3}(x^3 - 3x^2 + 5x) & 1 \le x \le 2 \\ \frac{1}{3}(x^3 - 3x^2 + 5x) & 2 \le x \le 3 \\ \frac{1}{3}(-2x^3 + 24x^2 - 76x + 81) & 3 \le x \le 4 \end{cases}$$

Therefore, $y(1.5) = f(1.5) = 1.375$

EXERCISE . . .

1. What is regression analysis? How is it different from interpolation?
2. Explain the method of least squares. How method of least squares can be used to fit non-linear functions?
3. The observations from an experiment are obtained for y by varying x as shown in the following table

x:	1.0	1.5	2.0	2.5	3.0	3.5
y:	6.2	7.5	9.0	10.0	11.5	12.0

Determine the suitable regression line.

4. Given the data below

x:	1.2	1.4	1.6	1.8	2.0	2.4
y:	4.2	6.3	8.4	11.6	15.0	24.2

Determine the both the regression lines. Also prove that they intersect at $(\bar{x}, \bar{y})$, where $\bar{x}$ and $\bar{y}$ are the arithmetic means of x and y respectively.

5. The growth rate of bacteria (n) in a culture after t seconds is given below. Fit a curve of the form $n = ab^t$ and estimate n when $t = 2.5$ seconds.

t:	0	1	2	3	4	5
n:	32	45	65	93	125	185

6. Given the data below. Fit a geometric curve of the form $y = ax^b$.

x:	1	2	3	4	5	6
y:	0.5	2.1	4.5	7.5	12.6	19.8

7. Find the normal equations to fit a curve of type

$$y = a + \frac{b}{x} + \frac{c}{x^2}$$

8. An experiment was conducted to determine the value v of the carbon dioxide in a given volume of water at different temperatures t. The results of the experiment are tabulated below:

t:	1	5	10	15	20	25
v:	1.75	1.45	1.16	1.02	0.85	0.60

Obtain a relation of the form $v = a + bt$ which best fits these observations.

9. Consider the values given in the following table. Find the cubic splines for these and hence evaluate $y(1.5)$ and $y'(2)$.

x:	1	2	3
y:	–6	–1	16

10. Consider values given in following table. Using cubic splines, evaluate $y(1.5)$ and $y'(3)$.

x:	1	2	3	4
y:	1	5	11	8

11. Consider the values given in the following table:

x:	1	2	3	4	5
y:	30	15	32	18	25

Find the cubic splines corresponding to the interval [2, 3].

❑ ❑ ❑

8 Numerical Differentiation

Learning Outcomes

After reading this chapter, students will have the ability to

- explain the process and need for numerical differentiation
- method for differentiating a graphical function
- methods for differentiating tabulated functions
- use these methods to find numerical differentiation of given function in tabulated form
- write programs for various methods

8.1 INTRODUCTION

Many problems in science and engineering involve the use of differentiation of various types of functions. If a function is expressed in mathematical form, its derivative can be easily obtained by analytical methods.

The first order derivative of a function $f(x)$ is defined as

$$f^1(x) = \frac{df(x)}{dx}$$

$$= \underset{\delta x \to 0}{\text{limit}} \frac{f(x+\delta x) - f(x)}{\delta x} \qquad \ldots (8.1)$$

so if the derivative is desired at the point $x = x_0$, then we have to compute the ratio

$$\frac{f(x_0 + \delta x) - f(x_0)}{\delta x} \qquad \text{for small value of } \delta x.$$

But in many situations, the values of the independent variable x and the corresponding values of the dependent variable y are available either in graphic form or tabular form. Consider the situation where the distance covered by an athlete as a function of time during his/her run for a 100-meter race is given in the following table:

Time (Secs.) :	0	1	2	3	4	5	6	7	8	9	10
Distance (Mts.) :	0	2.5	10	20	30.5	50	54	65.5	77.2	88.5	100

Here many questions can be asked, for example:

1. What was the speed of the athlete after 5 seconds?
2. What was the speed of the athlete as he approached the tape?
3. What was the acceleration of the athlete after 88 seconds?

Normally, speed and acceleration are calculated analytically. Suppose y represents the distance traveled by the athlete in x seconds, then

$$\text{Speed} = \frac{dy}{dx} \qquad \text{Acceleration} = \frac{d^2y}{dx^2}$$

But the mathematical relationship between y and x is not known in the above example. Therefore, it is not possible to use analytical methods. However, we can use the data given in the table to approximate the values of $\frac{dy}{dx}$ and $\frac{d^2y}{dx^2}$. Technique, which performs such calculations, is known as *numerical differentiation*.

8.2 DIFFERENTIATING A GRAPHICAL FUNCTION

If the function is in graphical form, then we can compute the derivative of the function at $x = x_0$ as illustrated below.

The derivative is given by

$$f^1(x_0) = \frac{f(x_0 + \delta x) - f(x_0)}{\delta x}$$

We know that $f^1(x_0)$ is the slope of the line CB and is a good approximation to the slope of the curve at point C.

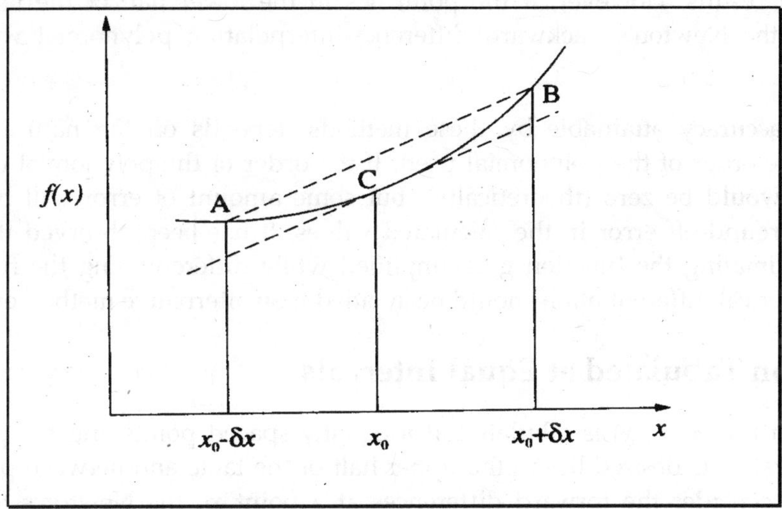

Figure 8.1: Function in graphical form

We can have another approximation to slope (derivative) by using line segment AC, *i.e.*,

$$f^1(x_0) = \frac{f(x_0) - f(x_0 - \delta x)}{\delta x}$$

Even we can have a third approximation to the slope by using the arithmetic mean of the above two approximations, i.e.

$$f^1(x_0) = \frac{f(x_0 + \delta x) - f(x_0 - \delta x)}{2\delta x}$$

This approximation represents the slope of the line AB. Since AC and CB are the segments of line AB, so all of the above three approximations to the slope of the line at $x = x_0$ are equally good.

8.3 DIFFERENTIATING A TABULATED FUNCTION

The real problem in differentiating a function comes when the function is given in the form of a table. The tabulated function cannot be differentiated directly. However, approximating it by a polynomial using the technique of interpolation can differentiate the function. If a function is tabulated at equal intervals, we can use the Newton's forward and backward difference interpolating polynomials, and if the function is tabulated at unequal intervals, we can use the Newton's divided difference-interpolating polynomial.

If the function is tabulated at equal intervals and the point in question lies in the upper half of the tabulated function, the Newton's forward difference interpolating polynomial will give better results. However, if the point lies in the lower half of the tabulated function, then the Newton's backward difference-interpolating polynomial will give better results.

However, the accuracy attainable by these methods depends on the nature of the function and the order of the polynomial used. If the order of the polynomial is exact, then the error would be zero (theoretically), but some amount of error will be there because of the round-off error in the calculated values. It has been observed that any error in approximating the function gets amplified while differentiating the function. Therefore, numerical differentiation should be avoided if an alternative method exists.

8.3.1 Function Tabulated at Equal Intervals

We assume that function $f(x)$ is tabulated at n equally spaced points and the point at which the derivative is desired lies in the upper half of the table and between points x_k and x_{k+1}. If we consider the forward differences at a point x_k, the Newton's forward difference interpolating polynomial takes the form

$$y(x) = y_k + \Delta y_k u + \frac{\Delta^2 y_k}{2!} u(u-1) + \frac{\Delta^3 y_k}{3!} u(u-1)(u-2) + \ldots \qquad \ldots (8.2)$$

where $u = \left(\frac{x - x_k}{h}\right)$, and Δy_k, $\Delta^2 y_k$, ... are forward differences of the function values at a point $x = x_k$.

We can use this polynomial to find the derivative

$$y^1(x) = \frac{dy(x)}{dx}$$

Since y is a function of u, and u in turn is a function of x, therefore we can write

$$\frac{dy}{dx} = \left(\frac{dy}{du}\right)\left(\frac{du}{dx}\right)$$

As $u = \left(\frac{x - x_k}{h}\right)$, therefore $\frac{du}{dx} = \frac{1}{h}$, which gives

$$\frac{dy}{du} = \Delta y_k + \frac{\Delta^2 y_k}{2!}\left[(u-1)+u\right] + \frac{\Delta^3 y_k}{3!}\left[(u-1)(u-2)+u(u-2)+u(u-1)\right] + \ldots$$

Therefore

$$\frac{dy}{dx} = \frac{1}{h}\begin{bmatrix} \Delta y_k + \frac{\Delta^2 y_k}{2!}\left[(u-1)+u\right] \\ + \frac{\Delta^3 y_k}{3!}\left[(u-1)(u-2)+u(u-2)+u(u-1)\right] + \ldots \end{bmatrix} \quad \ldots (8.3)$$

The above formula can be written using summation and product notation as

$$\frac{dy}{dx} = \frac{1}{h}\left[\Delta y_k + \sum_{j=2}^{n-k}\left[\frac{\Delta^j y_k}{j!}\sum_{p=0}^{j-1}\prod_{\substack{m=0 \\ m \neq p}}^{j-1}(u-m)\right]\right] \quad \ldots (8.4)$$

This is the general formula for evaluating $\frac{dy}{dx}$ from the forward difference table. The error term involves the unknown quantity unless x is one of the tabulated points.

This formula will be further simplified if the point at which the derivative is desired to be one of the tabulated points. Suppose, we want to compute the derivative at the tabulated point $x = x_k$, then $u = 0$, and Eq. (8.3), we have

$$\left.\frac{dy}{dx}\right|_{x=x_k} = \frac{1}{h}\left[\Delta y_k + \frac{\Delta^2 y_k}{2!}\left[(-1)+0\right] + \frac{\Delta^3 y_k}{3!}\left[(-1)(-2)+0+0\right]\ldots\right]$$

$$\left.\frac{dy}{dx}\right|_{x=x_k} = \frac{1}{h}\left[\Delta y_k - \frac{\Delta^2 y_k}{2} + \frac{\Delta^3 y_k}{3} - \ldots\right]$$

We have already discussed in Chapter 6 that the forward difference table for a function tabulated at n points can be represented by a matrix D of order $(n-1)\times(n-1)$, where the difference Δ_i^j can be represented by the element d_{ij}, the element in the ith row and the jth column of the matrix D.

Then we can write

$$\left.\frac{dy}{dx}\right|_{x=x_k} = \frac{1}{h}\left[d_{k1} - \frac{d_{k2}}{2} + \frac{d_{k3}}{3} - \ldots\right]$$

Using summation notation, the formula becomes

$$\left.\frac{dy}{dx}\right|_{x=x_k} = \frac{1}{h}\sum_{j=1}^{n-k}\left[(-1)^{j+1}\frac{d_{kj}}{j}\right] \qquad \ldots (8.5)$$

Example 8.1: Given the following table

x :	0.50	0.75	1.00	1.25	1.50
$y = f(x)$:	0.13	0.42	1.00	1.95	2.35

Find $f^1(0.75)$.

Solution: The forward difference table for the given data is

i	x_i	y_i	Δy_i	$\Delta^2 y_i$	$\Delta^3 y_i$	$\Delta^4 y_i$
1	0.50	0.13	0.29	0.29	0.08	–1.00
2	0.75	0.42	0.58	0.37	–0.92	
3	1.00	1.00	0.95	–0.55		
4	1.25	1.95	0.40			
5	1.50	2.35				

Since $h = 0.25$, and the derivative is desired at the $x = 0.75$ (i.e. $k = 2$). Expanding the formula up to third term, we get

$$\left.\frac{dy}{dx}\right|_{x=0.75} = \frac{1}{h}\left[d_{21} - \frac{d_{22}}{2} + \frac{d_{23}}{3}\right]$$

$$= \frac{1}{0.25}\left[0.58 - \frac{0.37}{2} + \frac{-0.92}{3}\right] = \frac{1}{0.25}[0.58 - 0.185 - 0.307] = 0.352$$

Example 8.2: Using the data of Example 8.1, evaluate $f^1(0.85)$.

Solution: Since the point $x = 0.85$ lies between the 2nd and 3rd point i.e. between $x = 0.75$ and $x = 1.0$, therefore we consider the forward differences at the 2nd point i.e. $k = 2$.

Since $n = 5$, $k = 2$, therefore $n–k = 3$, thus expanding the formula of Eq. (8.4) up to three terms, we get

$$\frac{dy}{dx} = \frac{1}{h}\left[\Delta y_2 + \frac{\Delta^2 y_2}{2!}[(u-1)+u] + \frac{\Delta^3 y_2}{3!}[(u-1)(u-2)+u(u-2)+u(u-1)]\right]$$

Now $u = \frac{x - x_2}{0.25} = \frac{0.85 - 0.75}{0.25} = 0.4$

Substituting values, we get

$$\left.\frac{dy}{dx}\right|_{x=0.85} = \frac{1}{0.25}\left[0.58 + \frac{0.37}{2}[(0.4-1)+0.4]\right.$$

$$\left. + \frac{-0.92}{6}[(0.4-1)(0.4-2)+0.4(0.4-2)+0.4(0.4-1)]\right] = 2.124$$

Algorithm 8.1: To compute derivative at the tabulated point

One dimensional arrays named x and y are used to hold the table of n values. Two dimensional array named d is used to hold the forward difference table.

```
Begin
    read: n, x
    for i = 1 to n by 1 do
        read: xi, yi
    endfor
    if (( x < x1 ) or ( x > xn ) )then
        write:"Value lies outside tabulated range"
        exit
    endif
    set i = 2
    while ( x > xi ) do
        set i = i + 1
    endwhile
    set k = i - 1
    set h = xk+1 - xk
    for j = 1 to (n-1) by 1 do
        for i = 1 to (n-j) by 1 do
            if ( j = 1 ) then
                set dij = yi+1 - yi
            else
                set dij = d(i+1)(j-1) - di(j-1)
            endif
        endfor
    endfor
    set sum = 0
    for j = 1 to (n-k) by 1 do
        set sum = sum + (-1)^(j+1) × dkj / j
```

```
    endfor
    set t = s/h
    write: "Value of the derivative = ", t
End.
```

Using Backward Differences

Proceeding on similar lines, we can get the formula for numerical differentiation using Newton's backward difference interpolating polynomial as

$$\frac{dy}{dx} = \frac{1}{h}\left[\nabla y_k + \sum_{j=2}^{k-1}\left[\frac{\nabla^j y_k}{j!}\sum_{p=0}^{j-1}\prod_{\substack{m=0\\m\neq p}}^{j-1}(u+m)\right]\right] \quad \ldots (8.6)$$

for $x_k < x < x_{k+1}$, and

$$\left.\frac{dy}{dx}\right|_{x=x_k} = \frac{1}{h}\sum_{j=1}^{k-1}\frac{d_{kj}}{j} \quad \ldots (8.7)$$

for any tabulated point, say $x = x_k$

Example 8.3: The distance (s) covered as a function of time (t) by an athlete during his/her run for the 50 meter race is given in the following table

Time (Secs.)	0	1	2	3	4	5	6
Distance (Mts.)	0	2.5	8.5	15.5	24.5	36.5	50

Determine the speed of the athlete at $t = 5$ seconds?

Solution: Given $h = 1.0$, and the derivative (speed) is desired at the 6th point, which lies in the lower half of the table, therefore we use the formula for differentiation based on backward differences. The backward difference table for the given data is

i	t_i	s_i	∇s_i	$\nabla^2 s_i$	$\nabla^3 s_i$	$\nabla^4 s_i$	$\nabla^5 s_i$	$\nabla^6 s_i$
1	0	0						
2	1	2.5	2.5					
3	2	8.5	6.0	3.5				
4	3	15.5	7.0	1.0	–2.5			
5	4	24.5	9.0	2.0	1.0	3.5		
6	5	36.5	12.0	3.0	1.0	0.0	–3.5	
7	6	50.0	13.5	1.5	–1.5	–2.5	–2.5	1.0

The total number of data points is seven and we will consider backward differences at point number 6, *i.e.*, $t = 5$ seconds.

As $k-1 = 5$, expanding the formula given by Eq. (8.6) up to 5th term, we get

$$\left.\frac{ds}{dt}\right|_{t=5} = \frac{1}{h}\left[d_{61} + \frac{d_{62}}{2} + \frac{d_{63}}{3} + \frac{d_{64}}{4} + \frac{d_{65}}{5}\right]$$

$$= \left[12.0 + \frac{3.0}{2} + \frac{1.0}{3} + \frac{0.0}{4} + \frac{-3.5}{5}\right]$$

$$= 13.133 \qquad \text{(correct to two decimal digits)}$$

Example 8.4: Use the data of Example 8.3, and find the speed of the athlete at $t = 4.5$ seconds.

Solution: Since the point $t = 4.5$ lies between the 5th and 6th point i.e. between $t=4$ and $t=5$, therefore we consider the backward differences at the 6th point i.e. $k = 6$.

Since $k = 6$, therefore $k - 1 = 5$

Thus expanding the formula of Eq. (8.6) up to three terms, we get

$$\frac{ds}{dt} = \frac{1}{h}\left[\nabla s_6 + \frac{\nabla^2 s_6}{2!}\left[(u+1) + u\right] + \frac{\nabla^3 s_6}{3!}\left[(u+1)(u+2) + u(u+2) + u(u+1)\right]\right.$$

$$+ \frac{\nabla^4 s_6}{4!}\left[(u+1)(u+2)(u+3) + u(u+2)(u+3) + u(u+1)(u+3) + u(u+1)(u+2)\right]$$

$$+ \frac{\nabla^5 s_6}{5!}\left[(u+1)(u+2)(u+3)(u+4) + u(u+2)(u+3)(u+4) + u(u+1)(u+3)(u+4)\right.$$

$$\left.\left. + u(u+1)(u+2)(u+4) + u(u+1)(u+2)(u+3)\right]\right]$$

$$u = \frac{t - t_6}{h} = \frac{4.5 - 5}{1.0} = -0.5$$

Substituting the values of $\nabla s_6, \nabla^2 s_6, \ldots, \nabla^5 s_6$ and that of u and simplifying, we get

Speed of the athlete at $t = 4.5$ seconds = 12.07 m/s.

8.3.2 Function Tabulated at Unequal Intervals

We assume that the function $f(x)$ is tabulated at n unequally spaced points and the point at which the derivative is desired lies between the points x_k and x_{k+1}. The Newton's divided difference interpolating polynomial takes the form

$$y(x) = y_k + \Delta_d y_k (x - x_k) + \Delta_d^2 y_k (x - x_k)(x - x_{k+1}) + \Delta_d^3 y_k (x - x_k)(x - x_{k+1})(x - x_{k+2}) + ... \quad ...(8.8)$$

where $\Delta_d y_k$, $\Delta_d^2 y_k$, ... are divided differences of the function values at a point x(8.8).

We can use this polynomial to find the derivative

$$y^1(x) = \frac{dy(x)}{dx}$$

$$\frac{dy(x)}{dx} = \Delta_d y_k + \Delta_d^2 y_k \left[(x - x_k) + (x - x_{k+1})\right] + \Delta_d^3 y_k \begin{bmatrix} (x - x_{k+1})(x - x_{k+2}) + (x - x_k)(x - x_{k+2}) \\ +(x - x_k)(x - x_{k+1}) \end{bmatrix} + ...$$

Using summation and product notation, the above formula can be written as

$$\frac{dy}{dx} = \Delta_d y_k + \sum_{j=2}^{n-k} \left[\Delta_d^j y_k \sum_{m=0}^{j-1} \prod_{\substack{p=0 \\ p \neq m}}^{j-1} (x - x_{k+p}) \right] \quad ...(8.9)$$

Example 8.5: The distance (s) covered by a car in a given time (t) is given in the following table:

Time (Minutes)	10	12	16	17	22
Distance (Kilometers)	12	15	20	22	32

Find the speed of the car at t = 14 minutes.

Solution: The divided difference table for the given data is

i	t_i	s_i	$\Delta_d s_i$	$\Delta_d^2 s_i$	$\Delta_d^3 s_i$	$\Delta_d^4 s_i$
1	10	12	1.50	–0.042	0.027	–0.004
2	12	15	1.25	0.150	–0.015	
3	16	20	2.00	0.000		
4	17	22	2.00			
5	22	32				

The given data point lies between 2nd and 3rd point, therefore we consider the differences at the 2nd point i.e. $k = 2$.

As $n = 5$, $n - k = 3$, thus expanding the formula given by Eq. (8.9) up to third term, we get

$$\frac{ds}{dt} = \Delta_d s_2 + \Delta_d^2 s_2\left[(t - t_3) + (t - t_2)\right]$$

$$+ \Delta_d^3 s_2\left[(t - t_3)(t - t_4) + (t - t_2)(t - t_4) + (t - t_2)(t - t_3)\right]$$

$$\left.\frac{ds}{dt}\right|_{t=14} = 1.25 + 0.15 \times \left[(14 - 16) + (14 - 12)\right]$$

$$+ (-0.015) \times \left[(14 - 16)(14 - 17) + (14 - 12)(14 - 17) + (14 - 12)(14 - 16)\right]$$

$$= 1.310$$

Hence, the speed of the car at $t = 14$ minutes is 1.310 Kilometer per minute.

Algorithm 8.2: To compute derivative when function is tabulated at unequal intervals

One dimensional arrays named x and y are used to hold the table of n values. Two-dimensional array named d is used to hold the divided difference table.

```
Begin
    read: n, x
    for i = 1 to n by 1 do
        read: xi, yi
    endfor
    if (( x < x1 ) or ( x > xn ) )then
        write: "Value lies outside tabulated range"
        exit
    endif
    set i = 2
    while ( x > xi ) do
        set i = i + 1
    endwhile
    set k = i -1
    for j = 1 to (n-1) by 1 do
        for i = 1 to (n-j) by 1 do
            if ( j = 1 ) then
                set dij = (yi+1 - yi)/(xi+1 - xi)
            else
                set dij = (d(i+1)(j-1) - di(j-1))/(xi+j - xi)
            endif
        endfor
    endfor
    set sum = dk1
```

```
    for j = 2 to (n-k) by 1 do
        set sum2 = 0
        for m = 0 to (j-1) by 1 do
            set prod = 1.0
            for p = 0 to (j-1) by 1 do
                if ( p ≠ m ) then
                    set prod = prod × (x - x_{k+p})
                endif
            endfor
            set sum2 = sum2 + prod
        endfor
        set sum1 = sum1 + d_{kj} × sum2
    endfor
    write: "Value of derivative = ", sum1
End.
```

8.4 HIGHER ORDER DERIVATIVES

Higher order derivative such as $\frac{d^2y}{dx^2}, \frac{d^3y}{dx^3}$ etc. can be obtained by repeated numerical differentiation. We consider again that the function is tabulated at n equally spaced points and the point in which the derivative is desired lies between the points x_k and x_{k+1}. If we consider forward difference at point x_k, Newton's forward difference interpolating polynomial takes the form

$$y(x) = y_k + \Delta y_k u + \frac{\Delta^2 y_k}{2!} u(u-1) + \frac{\Delta^3 y_k}{3!} u(u-1)(u-2)$$
$$+ \frac{\Delta^4 y_k}{4!} u(u-1)(u-2)(u-3) + \ldots$$

Therefore

$$\frac{dy}{dx} = \frac{1}{h}\left[\Delta y_k + \frac{\Delta^2 y_k}{2!}\left[(u-1)+u\right] + \frac{\Delta^3 y_k}{3!}\left[(u-1)(u-2)+u(u-2)+u(u-1)\right]\right.$$
$$\left. + \frac{\Delta^4 y_k}{4!}\begin{bmatrix}(u-1)(u-2)(u-3)+u(u-2)(u-3)\\ +u(u-1)(u-3)+u(u-1)(u-2)\end{bmatrix} + \ldots\right]$$

Differentiating again, we get the second derivative as

$$\frac{d^2y}{dx^2} = \frac{1}{h^2}\left[\Delta^2 y_k + \Delta^3 y_k (u-1) + \Delta^4 y_k \frac{6u^2 - 18u + 11}{12} + \ldots\right]$$

Similarly

$$\frac{d^3y}{dx^3} = \frac{1}{h^3}\left[\Delta^3 y_k + \Delta^4 y_k \frac{2u-3}{4} + ...\right]$$

At table point, say $x = x_k$, $u = 0$

$$\left.\frac{d^2y}{dx^2}\right|_{x=x_k} = \frac{1}{h^2}\left[\Delta^2 y_k - \Delta^3 y_k + \frac{11}{12}\Delta^4 y_k + ...\right]$$

$$\left.\frac{d^3y}{dx^3}\right|_{x=x_k} = \frac{1}{h^3}\left[\Delta^3 y_k - \frac{3}{2}\Delta^4 y_k + ...\right]$$

The formulae for computing higher order derivatives based on backward and divided differences can be derived by proceeding on similar lines.

Example 8.6: The distance (*s*) covered by a car in a given time (*t*) is given in the following table:

Time (Minutes)	10	12	14	16	18
Distance (Kilometers)	12	15	20	27	37

Find the acceleration of the car at $t = 13$ minutes.

Solution: We know that acceleration = $\frac{d^2s}{dt^2}$. Forward difference table for given data is

i	t_i	s_i	Δs_i	$\Delta^2 s_i$	$\Delta^3 s_i$	$\Delta^4 s_i$
1	10	12	3	2	0	1
2	12	15	5	2	1	
3	14	20	7	3		
4	16	27	10			
5	18	37				

The given data point lies between 2nd and 3rd point, therefore we consider the differences at the 2nd point, *i.e.*, $k = 2$.

$$\text{Acceleration} = \frac{d^2s}{dt^2} = \frac{1}{h^2}\left[\Delta^2 s_2 + \Delta^3 s_2 (u-1)\right]$$

Since $h = 2$, therefore $u = \frac{t - t_2}{h} = \frac{13-12}{2} = 0.5$

Substituting the values, we get acceleration = $\left.\frac{d^2s}{dt^2}\right|_{t=13} = \frac{1}{2^2}\left[2 + 1\times(0.5-1)\right] = 0.375$

Hence, the acceleration of the car at $t = 14$ minutes is 0.375 Kilometer per minute.

EXERCISE . . .

1. From the data given in the following table

x	0	1	2	3	4	5	6
y	0	2	2.75	3.4	3.2	2.5	2.2

Use forward difference formula to compute approximate value of dy/dx at $x = 2, 2.5$.

2. Consider the data of problem 1. Can the forward difference formula be used to estimate dy/dx at $x = 6$. If not, what could be done if it was required to estimate dy/dx?
3. The pressure of a certain gas was recorded at intervals of two seconds, and the results are listed in the following table:

Time (t)	0	2	4	6	8
Pressure(p)	50	53	57	68	82

Estimate the rate of change of pressure at $t = 2.5$ and $t = 4$.

4. From the data given in the following table

x	3	4	5	6	7	8.
y	4	6.6	7.7	9.0	10.5	12.2

Use backward difference formula to compute approximate value of dy/dx at $x = 7, 7.5$.

5. Consider the data of problem 4. Can the backward difference formula be used to estimate dy/dx at $x = 3$. If not, what could be done if it was required to estimate dy/dx?
6. The distance (in kilometers) covered a car in a given time (in minutes) is given in the following table:

Time (t)	5	10	13	15	18
Distance(s)	5	12	17	21	29

Estimate the speed and acceleration of the car at $t = 10$ minutes.

7. Consider the data given in the following table:

x	10	11	12	13	14
y	15	12.8	10.6	8.5	6.4

Estimate the value of dy/dx and d^2y/dx^2 at $x = 10, 10.5, 11, 11.2$ using the forward difference formula.

❑ ❑ ❑

9 Numerical Integration

Learning Outcomes

After reading this chapter, students will have the ability to

- explain the process and need for numerical integration
- various methods for integrating tabulated functions
- explain the process of adaptive integration
- explain the process double integration
- use these methods to find numerical integration of given function in tabulated form
- write programs for various methods
- explain the handling of improper integrals

9.1 INTRODUCTION

Numerical integration is the process of computing the value of a definite integral from a set of numerical values of the function referred to as *integrand*. When applied to the integration of a function of a single variable, the process is sometimes called *numerical quadrature*. When applied to compute double integral of a function of two independent variables, the process is called *numerical cubature*. If a function is defined as a mathematical expression, then its integral is usually determined using the technique of calculus.

We usually encounter three kinds of problems in numerical integration:

1. Certain functions do not possess closed form solutions. For example, the integral of $\frac{\sin x}{x}$ cannot be expressed in a closed form.
2. Closed form solutions do exist, but these solutions are too complex and difficult to use for computations.
3. Functions that are available in the form of tables and no mathematical relationship between the variables are known.

In all such situations, the values of the integral can be obtained by numerical technique whose aim is to provide effective procedures for the approximate evaluation of definite integrals.

In practice, given set of values for a function $f(x)$, the following table of data

x :	a	x_1	x_2	x_3	...	b
$f(x)$:	$f(a)$	$f(x_1)$	$f(x_2)$	$f(x_3)$	...	$f(b)$

is used to compute the value of the integral $\int_a^b f(x)dx$.

The principal idea behind all these techniques is:

(i) Replace the function to be integrated $f(x)$ by approximating polynomial $p(x)$, and

(ii) Integrate $p(x)$ and take $\int_a^b p(x)dx$ to be the required approximation of $\int_a^b f(x)dx$.

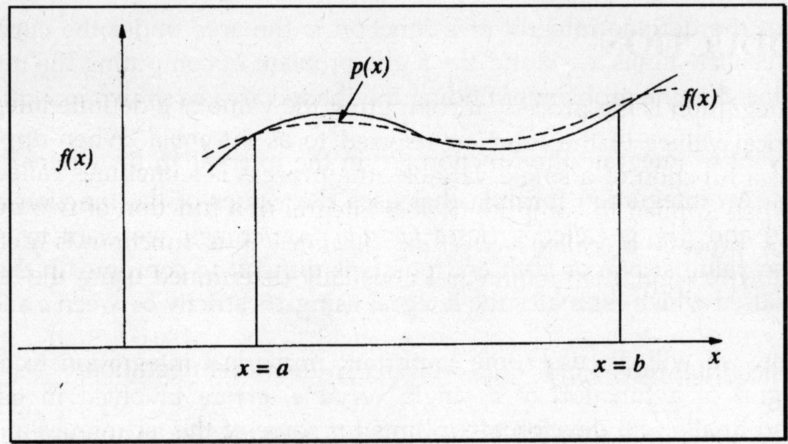

Figure 9.1: Illustrating approximating polynomial $p(x)$ for function $f(x)$

The value of the definite integral of a function $f(x)$, which is a continuous function of x in the interval $[a,b]$, is given by

$$I = \int_a^b f(x)dx = F(b) - F(a) \qquad \dots (9.1)$$

where $F(x)$ is the anti-derivative of $f(x)$.

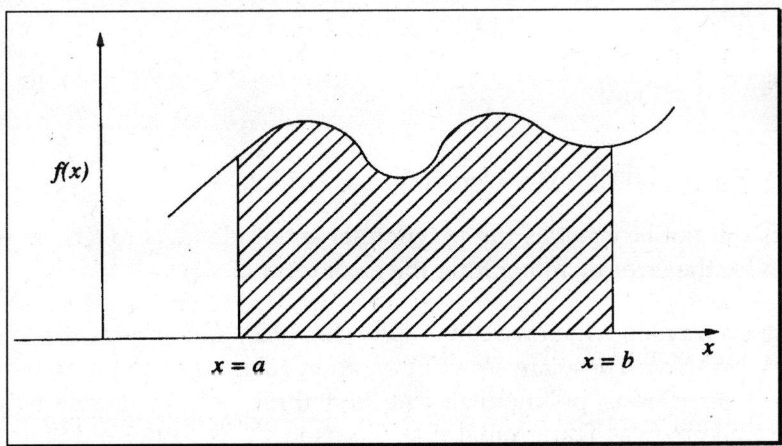

Figure 9.2: Definite integral represented by the shaded area

Remember that the definite integral of a function is the area under the curve $y = f(x)$ enclosed between the limits $x = a$ and $x = b$. the problem of computing the integral of a function is reduced to the problem of finding the shaded area as shown in Figure 9.2.

Suppose we want to integrate the function $f(x)$ in the interval $[a,b]$, where a and b are one of the x_i's. An integration formula that uses the values of the function at the end points, *i.e.*, $f(a)$ and $f(b)$, is called a *closed formula.* Sometimes, we want to integrate a function whose value at one or both end points is difficult to compute, in this case the *open formula* is used which estimates the integral using x_i's strictly between a and b.

In this chapter, we will discuss some important numerical integration techniques to compute integral of a function of a single variable, errors involved in using these techniques, and finally will develop algorithms for some of the techniques and the rest will be left as an exercise for the readers. Note that we will consider closed formulae only.

9.2 NEWTON-COTES INTEGRATION FORMULAE

As mentioned earlier, the formulae for numerical integration are based on fitting a polynomial through the given number of points, *i.e.*, approximating the given function by an interpolating polynomial function, and then integrating this approximating function. When the points are equi-spaced, the *Newton's forward difference interpolating polynomial* is a convenient starting point.

Therefore, we can write

$$\int_a^b f(x)dx = \int_a^b p_n(x)dx \qquad \ldots (9.2)$$

where n is the degree of the polynomial $p_n(x)$.

Since Eq. (9.2) will not be exact as the polynomial is not identical to $f(x)$, so we can find the expression for the error by integrating the error term of $p_n(x)$.

There are various ways in which we can employ Eq. (9.2), and as a result we get various Newton-Cotes formulae. These are set of integration rules corresponding to the varying degree of the interpolating polynomial. The first three, with a degree of polynomial equal to 1, 2, or 3 known as Trapezoidal rule, Simpson's 1/3rd rule, and Simpson's 3/8th rule respectively, are particularly important.

If the degree of the polynomial is too high, errors due to rounding-off and local irregularities can cause a problem. That is why only the lower degree Newton-Cotes formulae are used.

9.2.1 Trapezoidal Rule

The Trapezoidal rule approximates the area under a curve by connecting successive points on the curve to form trapezoids of uniform width, and then summing the area under these trapezoids to obtain the approximate area under the curve.

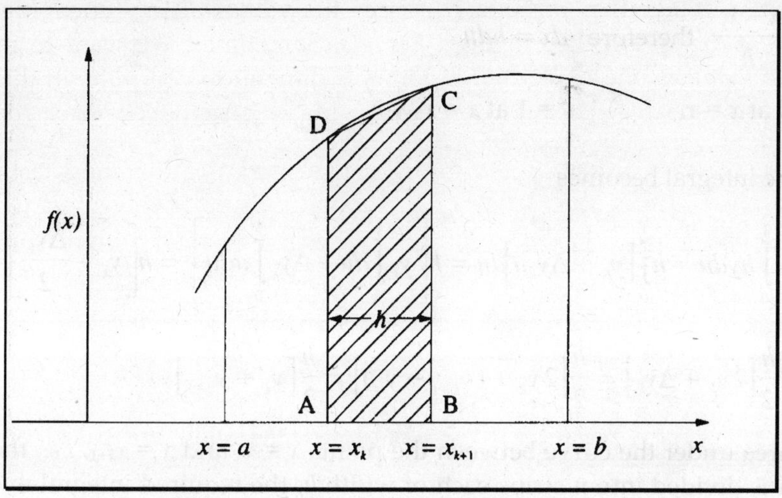

Figure 9.3: Approximation of area by Trapezoidal rule

Consider the function $f(x)$, whose graph between $x = a$ and $x = b$ is shown in Figure 9.2. An approximation to the area under the curve is obtained by dividing the interval $[a, b]$ into n strips of width h each, and approximating the area of each strip by that of a trapezoid as shown by the shaded area.

For the derivation of the formula for the trapezoid rule, we assume that the function $f(x)$ is given in the following form

x	x_1	x_1+h	x_1+2h	...	x_1+nh
$y = f(x)$	y_1	y_2	y_3	...	y_{n+1}

We consider the trapezoid formed by connecting the points (x_k, y_k) and (x_{k+1}, y_{k+1}). Further, consider the first two terms of the Newton's forward difference interpolating polynomial to represent the straight line function $f(x)$ as shown below:

$$y = f(x) = y_k + \Delta y_k u \qquad \text{... (9.3)}$$

where $\Delta y_k = y_{k+1} - y_k \qquad u = \dfrac{x - x_k}{h} \qquad h = x_{k+1} - x_k$

Then the area of the kth strip is given by evaluating the following integral

$$I_k = \int_{x_k}^{x_{k+1}} y dx$$

Since $u = \dfrac{x - x_k}{h}$, therefore $dx = hdu$

And $u = 0$ at $x = x_k$ $\quad u = 1$ at $x = x_{k+1}$

With this, the integral becomes

$$I_k = \int_0^1 hydu = h\int_0^1 [y_k + \Delta y_k u]du = h\left[y_k \int_0^1 du + \Delta y_k \int_0^1 udu\right] = h\left[y_k + \frac{\Delta y_k}{2}\right]$$

$$= \frac{h}{2}[2y_k + \Delta y_k] = \frac{h}{2}[2y_k + (y_{k+1} - y_k)] = \frac{h}{2}[y_k + y_{k+1}]$$

This is the area under the curve between the points $x = x_k$ and $x = x_{k+1}$, *i.e.*, the kth strip. If the function is divided into n strips each of width h, the required integral over a range $x = a$ and $x = b$ is given by

$$I = \sum_{k=1}^{n} I_k$$

$$= I_1 + I_2 + I_3 + \ldots + I_n$$

$$= \frac{h}{2}[y_1 + y_2] + \frac{h}{2}[y_2 + y_3] + \frac{h}{2}[y_3 + y_4] + \ldots + \frac{h}{2}[y_n + y_{n+1}]$$

$$I = \frac{h}{2}[y_1 + 2y_2 + 2y_3 + 2y_4 + ..+ 2y_n + y_{n+1}] \qquad \ldots (9.4)$$

This is the formula for the trapezoidal rule. It gives the correct value of the integral only if $f(x)$ is a linear function.

Example 9.1: The function $f(x)$ is given as follows

x	0	0.1	0.2	0.3	0.4	0.5	0.6	0.7	0.8	0.9	1.0
y	1	1.2	1.4	1.6	1.8	2.0	2.2	2.4	2.6	2.8	3.0

Compute the integral of $f(x)$ between $x = 0$ and $x = 1.0$.

Solution: Given $h = 0.1$ and $n = 10$

Therefore formula of Eq. (9.4) becomes

$$I = \frac{h}{2}\left[y_1 + 2(y_2 + y_3 + y_4 + y_5 + y_6 + y_7 + y_8 + y_9 + y_{10}) + y_{11}\right]$$

Substituting the values of y_i's and h, we get

$$I = \frac{0.1}{2}\left[1 + 2 \times (1.2 + 1.4 + 1.6 + 1.8 + 2.0 + 2.2 + 2.4 + 2.6 + 2.8) + 3.0\right]$$

$$= 2.00$$

Example 9.2: Evaluate $\int_1^2 e^{-\frac{1}{2}x}\,dx$ using four intervals.

Solution: With four intervals, the interval size $h = 0.25$, the function is tabulated as (rounded to three decimal digits)

x	1	1.25	1.50	1.75	2.0
$f(x)$	0.607	0.535	0.472	0.417	0.368

Substituting these values, we get

$$I = \frac{h}{2}\left[y_1 + 2(y_2 + y_3 + y_4) + y_5\right]$$

$$= \frac{0.25}{2}\left[0.607 + 2(0535 + 0.472 + 0.417) + 0.368\right]$$

$$= 0.478$$ (rounded to three decimal places)

Algorithm 9.1: Trapezoidal rule for tabulated functions

One dimensional arrays named x and y are used to hold the table of $(n+1)$ values for tabulated function. The interval size is h.

```
Begin
    read: n, h
    for i = 1 to (n+1) by 1 do
        read: xi, yi
    endfor
    set sum = (y1 + yn+1)/2
    for i = 2 to n by 1 do
        set sum = sum + yi
    endfor
```

```
    set integral = h × sum
    write: "Value of the integral = ", integral
End.
```

Algorithm 9.2: Trapezoidal rule for known functions

The function $f(x)$ is to be integrated from a to b by dividing the integration range in to n intervals..

```
Begin
    read: a, b, n
    set h = (b-a)/n
    set sum = (f(a)+f(b))/2
    for i = 1 to (n-1) by 1 do
        set sum = sum + f(a+i×h)
    endfor
    set integral = h × sum
    write: "Value of the integral = ", integral
End.
```

9.2.1.1 Estimation of Truncation Error in Trapezoidal Rule

To analyze the error involved in using Trapezoidal rule, we suppose that the integration of the function $f(x)$ is desired in the interval $[0, h]$.

If we take the interval size equal to h, the integration by the Trapezoidal rule is given by

$$I_t = \frac{h}{2}\left[f(0) + f(h)\right]$$

Expanding $f(h)$ by Taylor's series, we get

$$I_t = \frac{h}{2}\left[f(0) + \left[f(0) + hf^1(0) + \frac{h^2}{2!}f^2(0) + \frac{h^3}{3!}f^3(0) + \ldots\right]\right]$$

$$= hf(0) + \frac{h^2}{2}f^1(0) + \frac{h^3}{4}f^2(0) + \frac{h^4}{12}f^3(0) + \ldots$$

We will now find the integration of $f(x)$. Expanding $f(x)$ by Taylor series, we get

$$f(x) = f(0) + xf^1(0) + \frac{x^2}{2!}f^2(0) + \frac{x^3}{3!}f^3(0) + \ldots$$

Integrating it between limits $x = 0$ and $x = h$, we get

$$I_c = \int_0^h f(x)dx = \int_0^h \left[f(0) + xf^1(0) + \frac{x^2}{2!}f^2(0) + \frac{x^3}{3!}f^3(0) + \ldots\right]dx$$

Simplifying, we get

$$I_c = hf(0) + \frac{h^2}{2} f^1(0) + \frac{h^3}{6} f^2(0) + \frac{h^4}{24} f^3(0) + \ldots$$

The difference of these two integrals will give the amount of truncation error involved per step in applying the trapezoidal rule, *i.e.*,

$$E = I_t - I_c = \left[hf(0) + \frac{h^2}{2} f^1(0) + \frac{h^3}{4} f^2(0) + \frac{h^4}{12} f^3(0) + \ldots \right]$$

$$- \left[hf(0) + \frac{h^2}{2} f^1(0) + \frac{h^3}{6} f^2(0) + \frac{h^4}{24} f^3(0) + \ldots \right]$$

$$= -\frac{h^3}{12} f^2(0) + \text{ higher order terms in } h \qquad \ldots (9.5)$$

Since h will always be less than 1, the major contribution will be from the term involving h^3. This may be shown equivalent to

$$E = -\frac{h^3}{12} f^2(\mu_k) \qquad \text{where } x_k < \mu_k < x_{k+1}$$

If there are n strips of same width, then total error in integrating a function is given by

$$E_T = -\frac{h^3}{12} \left[f^2(\mu_1) + f^2(\mu_2) + f^2(\mu_3) + \ldots + f^2(\mu_n) \right] \qquad \ldots (9.6)$$

where μ_1 is some x value in the first strip, μ_2 is some x value in the second strip, μ_3 is some x value in the third strip, and so on.

In order to simplify the calculations, let us consider $f^2(\mu)_{av}$ to be the arithmetic mean of the $f^2(\mu)$ values, *i.e.*,

$$f^2(\mu)_{av} = \frac{f^2(\mu_1) + f^2(\mu_2) + f^2(\mu_3) + \ldots + f^2(\mu_n)}{n}$$

$$\Rightarrow \quad f^2(\mu_1) + f^2(\mu_2) + f^2(\mu_3) + \ldots + f^2(\mu_n) = nf^2(\mu)_{av}$$

Therefore, substituting this value in Eq. (9.6), we get

$$E_T = -\frac{h^3}{12} \left(nf^2(\mu)_{av} \right)$$

But nh equals the total range of integration. Therefore, if we are integrating from $x = a$ to $x = b$, then

$$nh = b - a$$

and the above equation becomes

$$E_T = -\frac{h^2}{12}(b-a)f^2(\mu)_{av} \qquad \ldots (9.7)$$

Now $(b - a)$ is a constant, and will not vary with the number of strips. The average second order derivative $f^2(\mu)_{av}$ will also be constant as long as the number of strips does not become too small. Thus, we can write

$$E_T \cong Ch^2$$

where C is some constant.

Another error which is introduces in obtaining the approximate area of each strip is the round-off error. This error arises because of the arithmetic operations with the numerical values having a limited number of significant digits. Round-off errors can be minimised by the use of double-precision arithmetic.

Thus, the total error over the desired interval of integration is the sum of the truncation and round-off errors. If the total error is due to truncation error, it could be made as small as desired by reducing the strip width. But decreasing the strip width also affects the total error as it increases the total round-off error due to large number of calculations. Thus in decreasing the strip width to decrease the total error, there is an optimum point at which further decrease in strip width will cause the total error to increase as the round-off error becomes dominant.

9.2.2 Simpson's 1/3rd Rule

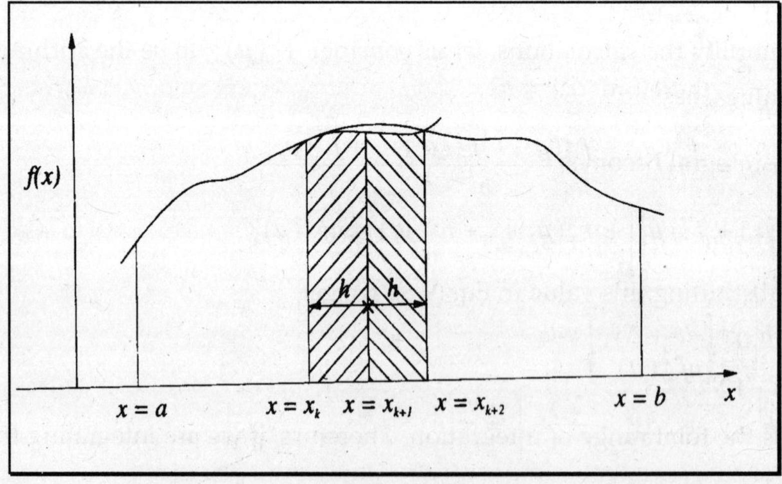

Figure 9.4: Approximation of area by Simpson's 1/3rd rule

Simpson's 1/3rd rule gives more accurate approximation of the integral value since it connects three points on the curve by second order parabolas, and then sums the areas under the parabolas to obtain the approximate area under the curve.

Consider the function $f(x)$, whose graph between $x = a$ and $x = b$ is shown in Figure 9.4. An approximation to the area under the curve is obtained by dividing it into n strips of width h each, and approximating the area of two strips by that of the area under the parabolas as shown by the shaded area in Figure 9.4.

For the derivation of the formula for the Simpson's 1/3rd rule, we assume that the function is given in the following form:

x	x_1	x_1+h	x_1+2h	...	x_1+nh
$y = f(x)$	y_1	y_2	y_3	...	y_{n+1}

We consider that the parabola passes through the points (x_k, y_k), (x_{k+1}, y_{k+1}) and (x_{k+2}, y_{k+2}). Further consider the first two terms of the Newton's forward difference interpolating polynomial to represent parabolic function $f(x)$ as shown:

$$y = f(x) = y_k + \Delta y_k u + \frac{\Delta^2 y_k}{2!} u(u-1) \qquad \ldots (9.8)$$

where $\Delta y_k = y_{k+1} - y_k$, $\Delta^2 y_k = y_{k+2} - 2y_{k+1} + y_k$, $u = \dfrac{x - x_k}{h}$, $h = x_{k+1} - x_k$

The area under the parabola is given by evaluating the following integral

$$I_k = \int_{x_k}^{x_{k+2}} y dx$$

Since $u = \dfrac{x - x_k}{h}$, therefore $dx = hdu$ and $u = 0$ at $x = x_k$, $u = 2$ at $x = x_{k+2}$

With this, the integral becomes

$$I_k = \int_0^2 hydu = h\int_0^2 \left[y_k + \Delta y_k u + \frac{\Delta^2 y_k}{2!} u(u-1) \right] du$$

$$= h\left[y_k \int_0^2 du + \Delta y_k \int_0^2 udu + \frac{\Delta^2 y_k}{2!} \int_0^2 u(u-1)du \right]$$

$$= h\left[2y_k + 2\Delta y_k + \frac{\Delta^2 y_k}{3} \right] = \frac{h}{3}\left[6y_k + 6\Delta y_k + \Delta^2 y_k \right]$$

$$= \frac{h}{3}\left[6y_k + 6(y_{k+1} - y_k) + (y_{k+2} - 2y_{k+1} + y_k) \right] = \frac{h}{3}\left[y_k + 4y_{k+1} + y_{k+2} \right]$$

This is the area under the curve between the points $x = x_k$ and $x = x_{k+2}$. This covers two intervals of width h each. If we use this process repetitively $n/2$ times, we can get the area under the curve for n intervals. Thus, if the range of integration is divided into n intervals, each of width h, the required integral over a range $x = a$ and $x = b$ is given by

$$I = \sum_{k=1}^{n/2} I_{2k-1} = I_1 + I_3 + I_5 + \dots + I_{n-1}$$

$$= \frac{h}{3}\left[y_1 + 4y_2 + y_3\right] + \frac{h}{3}\left[y_3 + 4y_4 + y_5\right] + \frac{h}{3}\left[y_5 + 4y_6 + y_7\right] +$$

$$\dots + \frac{h}{3}\left[y_{n-1} + 4y_n + y_{n+1}\right]$$

$$I = \frac{h}{3}\left[y_1 + 4y_2 + 2y_3 + 4y_4 + \dots + 2y_{n-1} + 4y_n + y_{n+1}\right] \quad \dots (9.9)$$

This is the formula for the Simpson's 1/3rd rule. It gives the correct value of the integral only if $f(x)$ is a second order (quadratic) function.

Example 9.3: The function $f(x)$ is given as follows

x	0	0.1	0.2	0.3	0.4	0.5	0.6	0.7	0.8	0.9	1.0
y	1	1.01	1.04	1.09	1.16	1.25	1.36	1.49	1.64	1.81	2.0

Compute the integral of $f(x)$ between $x = 0$ and $x = 1.0$.

Solution: Given $h = 0.1$ and $n = 10$. Therefore formula of Eq. (9.9) becomes

$$I = \frac{h}{3}\left[y_1 + 4y_2 + 2y_3 + 4y_4 + 2y_5 + 4y_6 + 2y_7 + 4y_8 + 2y_9 + 4y_{10} + y_{11}\right]$$

Substituting the values of y_i's and h, we get

$$I = \frac{0.1}{3}\begin{bmatrix} 1 + 4\times(1.01 + 1.09 + 1.25 + 1.49 + 1.81) \\ +2\times(1.04 + 1.16 + 1.36 + 1.64) + 2.0 \end{bmatrix}$$

$= 1.333$ (rounded to three decimal places)

Example 9.4: Evaluate $\int_1^2 e^{-\frac{1}{2}x}\,dx$ using four intervals.

Solution: With four intervals, the interval size $h = 0.25$, the function is evaluated as (rounded to three decimal digits)

x	1	1.25	1.50	1.75	2.0
$f(x)$	0.607	0.535	0.472	0.417	0.368

Substituting these values, we get

$$I = \frac{h}{3}\left[y_1 + 4y_2 + 2y_3 + 4y_4 + y_5\right]$$

$$= \frac{0.25}{3}[0.607 + 4 \times 0.535 + 2 \times 0.472 + 4 \times 0.417 + 0.368]$$

$= 0.477$ (rounded to three decimal places)

Algorithm 9.3: Simpson's 1/3rd rule for tabulated functions

One dimensional arrays named x and y are used to hold the table of $(n+1)$ values for tabulated function. The interval size is h.

```
Begin
    read: n, h
    for i = 1 to (n+1) by 1 do
        read: xi, yi
    endfor
    set s = y1 + yn+1
    set s2 = 0
    set s4 = 0
    for i = 2 to n by 2 do
        set s4 = s4 + yi
    endfor
    for i = 3 to (n-1) by 2 do
        set s2 = s2 + yi
    endfor
    set integral = (h/3) × (s + 2×s2 + 4×s4)
    write: "Value of the integral = ", integral
End.
```

Algorithm 9.4: Simpson's 1/3rd rule for known functions

The function $f(x)$ is to be integrated from a to b by dividing the integration range in to n intervals.

```
Begin
    read: a, b, n
    set h = (b - a)/n
    set s = f(a)+f(b)
    set s2 = 0
    set s4 = 0
    for i = 1 to n by 2 do
```

```
        set s4 = s4 + f(a+i×h)
    endfor
    for i = 2 to (n-1) by 2 do
        set s2 = s3 + f(a+i×h)
    endfor
    set integral = (h/3) × (s + 2×s2 + 4×s4)
    write: "Value of the integral = ", integral
End.
```

9.2.2.1 Estimation of Truncation Error in Simpson's 1/3rd Rule

To analyze the error involved in using Simpson's 1/3rd rule, we suppose that the integration of the function $f(x)$ is desired in the interval $[0, 2h]$.

If we take the interval size equal to h, the integration by the Simpson's 1/3rd rule is given by

$$I_s = \frac{h}{3}\left[f(0) + 4f(h) + f(2h)\right]$$

Expanding $f(h)$ and $f(2h)$ by Taylor's series, we get

$$I_s = \frac{h}{2}\left[f(0) + \left\{f(0) + hf^1(0) + \frac{h^2}{2!}f^2(0) + \frac{h^3}{3!}f^3(0) + \ldots\right\}\right.$$

$$\left. + \left\{f(0) + 2hf^1(0) + \frac{4h^2}{2!}f^2(0) + \frac{8h^3}{3!}f^3(0) + \ldots\right\}\right]$$

$$= 2hf(0) + 2h^2 f^1(0) + \frac{8h^3}{6}f^2(0) + \frac{12h^4}{18}f^3(0) + \frac{20h^5}{72} + \ldots$$

We will now find the integration of $f(x)$. Expanding $f(x)$ by Taylor series, we get

$$f(x) = f(0) + xf^1(0) + \frac{x^2}{2!}f^2(0) + \frac{x^3}{3!}f^3(0) + \frac{x^4}{4!}f^4(0) + \ldots$$

Integrating it between limits $x = 0$ and $x = 2h$, we get

$$I_c = \int_0^{2h} f(x)dx = \int_0^{2h}\left[f(0) + xf^1(0) + \frac{x^2}{2!}f^2(0) + \frac{x^3}{3!}f^3(0) + \frac{x^3}{3!}f^3(0) + \ldots\right]dx$$

Simplifying, we get

$$I_c = 2hf(0) + 2h^2 f^1(0) + \frac{8h^3}{6}f^2(0) + \frac{16h^4}{24}f^3(0) + \frac{32h^5}{120}f^4(0) + \ldots$$

The difference of these two integrals will give the amount of truncation error involved per step in applying the Simpson's 1/3rd rule, *i.e.*,

$$E = I_s - I_c = \left[2hf(0) + 2h^2 f^1(0) + \frac{8h^3}{6} f^2(0) + \frac{12h^4}{18} f^3(0) + \frac{20h^4}{72} f^4(0) + \ldots\right]$$

$$- \left[2hf(0) + 2h^2 f^1(0) + \frac{8h^3}{6} f^2(0) + \frac{16h^4}{24} f^3(0) + \frac{32h^5}{120} f^3(0) + \ldots\right]$$

$$= -\frac{h^5}{90} f^4(0) + \text{ higher order terms in } h \qquad \ldots (9.10)$$

Since h will always be less than 1, major contribution will be from the **term** involving h^5.

This may be shown equivalent to

$$E = -\frac{h^5}{90} f^4(\mu_k) \qquad \text{where } x_{k-1} < \mu_k < x_{k+1}$$

If there are n strips of same width, then total error in integrating a function is given by

$$E_T = -\frac{h^5}{90}\left[f^4(\mu_1) + f^4(\mu_2) + f^4(\mu_3) + \ldots + f^4(\mu_n)\right] \qquad \ldots (9.11)$$

where μ_1 is some x value in the first strip, μ_2 is some x value in the second strip, μ_3 is some x value in the third strip, and so on.

In order to simplify the calculations, let us consider $f^4(\mu)_{av}$ to be the arithmetic mean of the $f^4(\mu)$ values, *i.e.*,

$$f^4(\mu)_{av} = \frac{f^4(\mu_1) + f^4(\mu_2) + f^4(\mu_3) + \ldots + f^4(\mu_{n/2})}{n/2}$$

$$\Rightarrow \quad f^4(\mu_1) + f^4(\mu_2) + f^4(\mu_3) + \ldots + f^4(\mu_{n/2}) = \frac{n}{2} f^4(\mu)_{av}$$

Therefore, substituting this value in Eq. (9.11), we get

$$E_T = -\frac{h^5}{90}\left(\frac{n}{2} f^4(\mu)_{av}\right)$$

But nh equals the total range of integration. Therefore, if we are integrating from $x = a$ to $x = b$, then

$$nh = b - a$$

and the above equation becomes

$$E_T = -\frac{h^4}{180}(b - a) f^4(\mu)_{av} \qquad \ldots (9.12)$$

Now $(b - a)$ is a constant, and will not vary with the number of strips. The average fourth order derivative $f^4(\mu)_{av}$ will also be constant as long as the number of strips does not become too small. Thus, we can write

$$E_T \cong Ch^4$$

where C is some constant.

9.2.3 Simpson's 3/8th Rule

Assume that the tabulated function can be approximated by a third order polynomial. In this case, we can use the first four terms of the Newton's forward difference interpolation polynomial to represent $f(x)$ as shown below:

$$y = f(x) = y_k + \Delta y_k u + \frac{\Delta^2 y_k}{2!} u(u-1) + \frac{\Delta^3 y_k}{3!} u(u-1)(u-2) \qquad \ldots (9.13)$$

where Δy_k, $\Delta^2 y_k$, $\Delta^3 y_k$ are forward differences at $x = x_k$, and

$$u = \frac{x - x_k}{h}, \text{ and } h = x_{k+1} - x_k$$

Since $f(x)$ is a third order polynomial, we assume it passes through the points (x_k, y_k), (x_{k+1}, y_{k+1}), (x_{k+2}, y_{k+2}) and (x_{k+3}, y_{k+3}), then

$$I_k = \int_{x_k}^{x_{k+3}} y dx$$

Since $u = \dfrac{x - x_k}{h}$, therefore $dx = hdu$ and $u = 0$ at $x = x_k$, $u = 3$ at $x = x_{k+3}$

With this, the integral becomes

$$I_k = \int_0^3 hy du$$

After simplification, which is left as an exercise for the readers, we get

$$I_k = \frac{3h}{8}\left[y_k + 3y_{k+1} + 3y_{k+2} + y_{k+3}\right]$$

This is the area under the curve between the points $x = x_k$ and $x = x_{k+3}$. This covers three intervals of width h each. If we use this process repetitively $n/3$ times, we can get the area under the curve for n intervals.

Thus, if the range of integration is divided into n intervals, each of width h, the required integral over a range $x = a$ and $x = b$ is given by

$$I = \sum_{k=1}^{n/3} I_{3k-2} = I_1 + I_4 + I_7 + \ldots + I_{n-2}$$

$$= \frac{3h}{8}\left[y_1 + 3y_2 + 3y_3 + y_4\right] + \frac{3h}{8}\left[y_4 + 3y_5 + 3y_6 + y_7\right]$$

$$+ \frac{3h}{8}\left[y_7 + 3y_8 + 3y_9 + y_{10}\right] + \ldots + \frac{3h}{8}\left[y_{n-2} + 3y_{n-1} + 3y_n + y_{n+1}\right]$$

$$I = \frac{3h}{8}\begin{bmatrix} y_1 + 3y_2 + 3y_3 + 2y_4 + 3y_5 + 3y_6 + 2y_7 + \ldots \\ +2y_{n-2} + 3y_{n-1} + 3y_n + y_{n+1} \end{bmatrix} \qquad \ldots (9.14)$$

This is the formula for Simpson's 3/8th rule. It gives the correct value of the integral only if $f(x)$ is a cubic. The truncation error per step, *i.e.*, local error, in using Simpson's 3/8th rule is $O(h^5)$, whereas the total truncation error, *i.e.*, global error, is $O(h^4)$. It is left as an exercise for the readers to prove.

Example 9.5: The function $f(x)$ is given as follows

x	0.1	0.2	0.3	0.4	0.5	0.6	0.7	0.8	0.9	1.0
y	1.001	1.008	1.027	1.064	1.125	1.216	1.343	1.512	1.729	2.0

Compute the integral of $f(x)$ between $x = 0.1$ and $x = 1.0$.

Solution: Given $h = 0.1$ and $n = 9$

Therefore formula of Eq. (9.14) becomes

$$I = \frac{3h}{8}\left[y_1 + 3y_2 + 3y_3 + 2y_4 + 3y_5 + 3y_6 + 2y_7 + 3y_8 + 3y_9 + y_{10}\right]$$

Substituting the values of y_i's and h, we get

$$I = \frac{3 \times 0.1}{8}[1.001 + 3 \times 1.008 + 3 \times 1.027 + 2 \times 1.064 + 3 \times 1.125$$

$$+ 3 \times 1.216 + 2 \times 1.343 + 3 \times 1.512 + 3 \times 1.729 + 2.0]$$

$$= 1.150 \qquad \text{(rounded to three decimal places)}$$

Algorithm 9.5: Simpson's 3/8th rule for tabulated functions

One dimensional arrays named x and y are used to hold the table of $(n+1)$ values for tabulated function. The interval size is h.

```
Begin
    read: n, h
    for i = 1 to (n+1) by 1 do
        read: xi, yi
    endfor
    set s1 = y1 + yn+1
    set s2 = 0
    set s3 = 0
    for i = 2 to (n-1) by 3 do
        set s3 = s3 + yi + yi+1
    endfor
    for i = 4 to (n-2) by 3 do
        set s2 = s2 + yi
    endfor
    set integral = (3×h/8) × (s1 + 2×s2 + 3×s3)
    write: "Value of the integral = ", integral
End.
```

Algorithm 9.6: Simpson's 3/8th rule for known functions

The function $f(x)$ is to be integrated from a to b by dividing the integration range into n intervals.

```
Begin
    read: a, b, n
    set h = (b-a)/n
    set s1 = f(a)+f(b)
    set s2 = 0
    set s3 = 0
    for i = 1 to n by 3 do
        set x = a+i×h
        set s3 = s3 + f(x) + f(x+h)
    endfor
    for i = 3 to n by 3 do
        set s2 = s2 + f(a+i×h)
    endfor
    set integral = (3×h/8) × (s1 + 2×s2 + 3×s3)
    write: "Value of the integral = ", integral
End.
```

9.3 ADAPTIVE INTEGRATION

As we have seen in the preceding sections, we use Trapezoidal rule, Simpson's 1/3rd rule, and Simpson's 3/8th rule to find the integral of $f(x)$ over a fixed interval $[a, b]$ using a uniform strip of size h. When $f(x)$ is a known function, we can choose the value of h arbitrarily. The problem with this approach is that we do not know in advance what value of h should be chosen to attain the desired accuracy. Practically, when the function is known, the procedure of integration starts with one strip for the Trapezoid rule, with two strips for Simpson's 1/3rd rule, and with three strips for Simpson's 3/8th rule. Then the strip size is halved and the formula is applied again, and then the results are compared. If the new value of the integral is sufficiently close to its predecessor, then the new value of the integral is taken as the approximate value of the integral. If the new result is not close enough to its predecessor, the value of h is again halved, and the procedure is repeated.

Therefore in the adaptive scheme, the user needs to specify only the integral limits and the desired accuracy, the procedure will take care of the value of h to be used to give the desired accuracy.

Another point, which is worth mentioning, is that as the value of h is halved in each iteration, there will be repeated evaluations of the function at some of the points. Therefore, the procedure should be modified in such a way that repeated evaluations of the function are avoided.

This approach will make the procedure more effective in the following ways:

(i) Computational time will be reduced considerably, and
(ii) As the number of computations increases, the round-off errors will also increase considerable.

9.3.1 Modified Trapezoidal Rule for Known Functions

In a modified approach of the Trapezoidal rule to integrate a known function $f(x)$ from $x = a$ to $x = b$, the procedure starts with strip size $h = b - a$, and then successively halves the strip size to re-compute the integral. Such a halving is continued till the difference between two successive integrations become insignificant.

We reduce the number of computations, because h is halved, all of the old points at which the function was evaluated appear in the new computations, and thus we can avoid repeating the evaluations. This is illustrated overleaf:

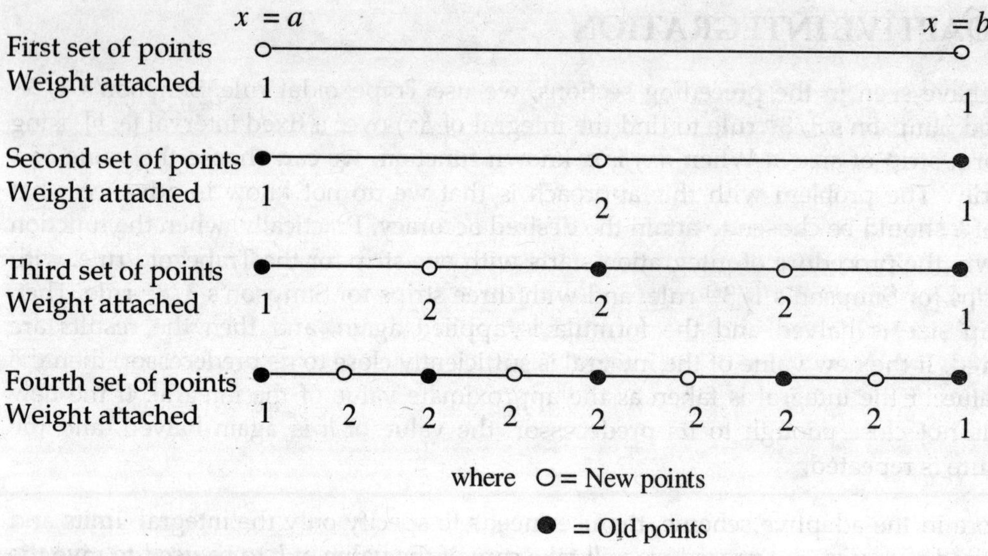

The successive steps of computations of the function values are summed up in the following table:

Table 9.1

a $f(a)$	$sum = \dfrac{f(a) + f(b)}{2}$	b $f(a)$
	$a + \dfrac{h}{2}$ $sum = sum + f\left(a + \dfrac{h}{2}\right)$	
$a + \dfrac{h}{4}$	$sum = sum + f\left(a + \dfrac{h}{4}\right)$ $+ f\left(a + \dfrac{3h}{4}\right)$	$a + \dfrac{3h}{4}$
	⋮	

Algorithm 9.7: Trapezoidal rule with adaptive scheme for known functions

The function $f(x)$ is to be integrated from a to b, and the prescribed tolerance in the final integrated value is *epsilon*.

```
Begin
    read: a, b, epsilon
    set h = b - a
    set sum = [f(a)+f(b)]/2
    set i_old = 0
    set i_new = h × sum
    set k = 1
    while ( | (i_new  - i_old)/i_new |  > epsilon ) do
        set k = k × 2
        set h = h / 2
        for i = 1 to k by 2 do
            set x = a + i × h
            set sum = sum + f(x)
        endfor
        set i_old =  i_new
        set i_new = h × sum
    endwhile
    write: "Value of the integral = ", i_new
End.
```

9.3.2 Modified Simpson's1/3rd Rule for Known Functions

In a modified approach of the Simpson's 1/3rd rule to integrate a known function $f(x)$ from $x = a$ to $x = b$, the procedure starts with strip size $h = (b - a)/2$, and then successively halves the strip size to re-compute the integral. Such a halving is continued till the difference between two successive integrations become insignificant.

Here also, we reduce the number of computations, because h is halved, all of the old points at which the function was evaluated appear in the new computations, and thus we can avoid repeating the evaluations. This is illustrated below.

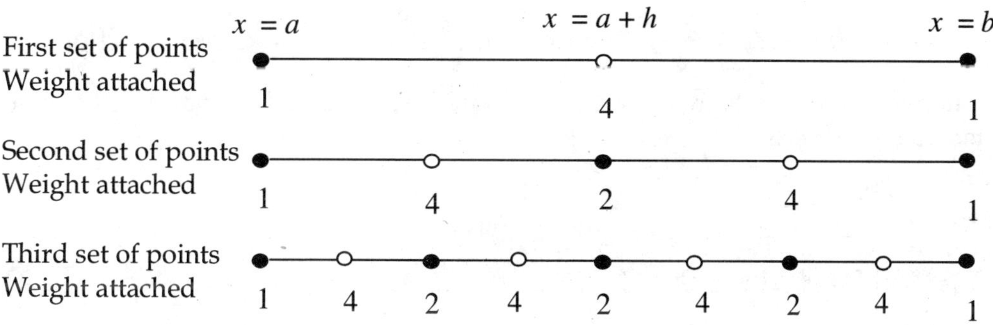

The successive steps of computations of the function values are summed up in the following table:

Table 9.2

a $f(a)$	$a+\frac{h}{2}$ $s_1 = f(a) + f(b)$ $s_2 = 0$ $s_4 = f\left(a+\frac{h}{2}\right)$	b $f(a)$
$f\left(a+\frac{h}{4}\right)$	$s_2 = s_2 + s_4$ $s_4 =$ $f\left(a+\frac{h}{4}\right)+f\left(a+\frac{h}{4}\right)$	$f\left(a+\frac{3h}{4}\right)$
$f\left(a+\frac{h}{8}\right)$	$s_2 = s_2 + s_4$ $s_4 = f\left(a+\frac{h}{8}\right)$ $+f\left(a+\frac{3h}{8}\right)+$ $f\left(a+\frac{5h}{8}\right)$	$f\left(a+\frac{5h}{8}\right)$
	. . .	

Algorithm 9.8: Simpson's 1/3rd rule with adaptive scheme for known functions

The function $f(x)$ is to be integrated from a to b, and the prescribed tolerance in the final integrated value is *epsilon*.

```
Begin
    read: a, b, epsilon
    set h = (b - a)/2
    set s1 = [f(a)+f(b)]/2
```

```
    set s2 = 0
    set s4 = f(a+h)
    set i_old = 0
    set i_new = (h/3) × (s1 + 4 × s4)
    set k = 2
    while ( | (i_new - i_old)/i_new | > epsilon ) do
        set k = k × 2
        set h = h / 2
        set s2 = s2 + s4
        set s4 = 0
        for i = 1 to k by 2 do
            set x = a + i × h
            set s4 = s4 + f(x)
        endfor
        set i_old = i_new
        set i_new = (h/3) × (s1 + 4 × s4 + 2 × s2)
    endwhile
    write: "Value of the integral = ", i_new
End.
```

The development of an algorithm for Simpson's 3/8th rule is left as an exercise for the readers.

9.4 ROMBERG INTEGRATION

The Romberg method of integration is basically a recursive extrapolation procedure that refines the values of integrals obtained from the trapezoidal rule. To describe the procedure, recall that the error for trapezoidal rule can be expressed as

$$E_T \cong Ch^2$$

where C is a constant and is numerically equal to $-\dfrac{b-a}{12} f^2(\mu)_{av}$, and h is the strip width and takes on variable values.

For strip width h_1 $\quad E_T \cong Ch_1^2$

For different strip width h_2 $\quad E_T \cong Ch_2^2$

Thus, if I is the exact value of the integral, then we can write

$$I = I_1 + Ch_1^2 \qquad I = I_1 + Ch_2^2$$

where I_1 is the value of the integral obtained with strip width h_1, and I_2 is the value of the integral obtained with strip width h_2.

Solving for C, we get

$$C = \frac{I_1 - I_2}{h_2^2 - h_1^2}$$

Substituting the value of C, we get

$$I = I_2 + \left[\frac{I_1 - I_2}{h_2^2 - h_1^2}\right] h_2^2$$

$$I = I_2 + \frac{I_1 - I_2}{(h_1 / h_2)^2 - 1} \qquad \ldots (9.15)$$

The second term on the right hand side of Eq. (9.15) is the truncation error estimate for the integration with strip width h_2. Adding it to I_2 will not give the exact integral value because the value of C is only approximate. However, it does give a much-improved value of the integral.

With the inclusion of the higher order terms, the exact error can be expressed as

$$E_T = C_1h + C_2h^4 + C_3h^6 + C_4h^8 \qquad \ldots (9.16)$$

But for small value of h, only the first term dominates. Therefore, the above equation can be written as

$$E_T = Ch^2 + O(h^4)$$

If we obtain the integral value I_1 with a particular strip width h_1, and then half the strip width to h_2 to get second integral value I_2, and then apply the Eq. (9.15), then the term C_1h^2 is removed from the error. This leaves an error $O(h^4)$, which might be expressed as $C_2h^4 + O(h^6)$.

Now if we determine the third integral value again by halving the strip width, we can combine it with the second integral value to obtain another estimate containing an error of $C_2h^4 + O(h^6)$. If we combine these two estimates containing an error of $C_2h^4 + O(h^6)$, we obtain new integral value in which the C_2h^4 is removed.

This process is continued, next removing the error term C_3h^6 and higher order terms from the error until the desired accuracy is obtained. This procedure will be clearer after doing a numerical example. But before that we will generalise this procedure so that it can be implemented on computer.

Rewriting Eq. (9.15) as

$$I = \frac{I_2\left[(h_1/h_2)^2 - 1\right] + I_2 - I_1}{(h_1/h_2)^2 - 1} \quad \Rightarrow \quad I = \frac{I_2(h_1/h_2)^2 - I_1}{(h_1/h_2)^2 - 1} \qquad \ldots (9.17)$$

Since the second integral is obtained by halving the strip width, therefore

$$h_1 / h_2 = 2$$

Thus Eq. (9.17) becomes

$$I = \frac{4I_2 - I_1}{4-1} \quad \Rightarrow \quad I_{Improved} = \frac{4I_{More\_accurate} - I_{Less\_accuate}}{4-1} \qquad \ldots (9.18)$$

Eq. (9.18) is used for first order extrapolation. For the second order extrapolation, the equation becomes

$$I_{Improved} = \frac{4^2 I_{More\_accurate} - I_{Less\_accuate}}{4^2 - 1}$$

In general, for nth order extrapolation, the equation becomes

$$I_{Improved} = \frac{4^n I_{More\_accurate} - I_{Less\_accuate}}{4^n - 1} \qquad \text{for } n = 1, 2, 3, \ldots \qquad \ldots (9.19)$$

First we find the two successive integrals of the function using one and two strips, and then we extrapolate these integrals using the first order extrapolation. Now we find the integral using three strips, and we again extrapolate the integral using the first order extrapolation. And the using these two first order extrapolations, we extrapolate using the second order extrapolation.

As a result of these successive extrapolations, we build the following Romberg table:

R_{11}			
R_{21}	R_{22}		
R_{31}	R_{32}	R_{33}	
R_{41}	R_{42}	R_{43}	R_{44}

where R_{44} is the final answer, *i.e.*, the last element in the last row of the Romberg table.

In the above table, the first column represents the integral values for a single strip, two strips, four strips, and so on. The second column represents the first order extrapolated values, the third column represent the second order extrapolated value, and so on.

Example 9.6: Evaluate $\int_0^{\frac{\pi}{2}} \sin x dx$ significant to five digits, *i.e.*, the error must be less than or equal to 0.00001.

Solution: The procedure begins by calculating the integral value for a single strip, for two strips, and then finding the first order extrapolated value to form the following table:

Number of strips	1	2
Integral value from Trapezoidal rule	0.7853982	0.9480594
First order extrapolated value	1.0022798	

By comparing the extrapolated value, *i.e.*, 1.0022798 and the more accurate integral value, *i.e.*, 0.9480594, the absolute is greater than the prescribed error.

Now we find the integral value for four strips:

Number of strips	1	2	3
Integral value from Trapezoidal rule	0.7853982	0.9480594	0.9871158
First order extrapolated value	1.0022798	1.0001346	
Second order extrapolated value	0.9999917		

By comparing the second order extrapolated value, *i.e.*, 0.9999917 and the more accurate first order extrapolated value, *i.e.*, 1.0001346, the error is still greater than the prescribed error.

Now we find the integral for eight strips:

Number of strips	1	2	3	4
Integral value from Trapezoidal rule	0.7853982	0.9480594	0.9871158	0.9967852
First order extrapolated value	1.0022798	1.0001346	1.0000083	
Second order extrapolated value	0.9999917	0.9999999		
Third order extrapolated value	1.0000000			

By comparing the third order extrapolated value, *i.e.*, 1.0000000 and the more accurate second order extrapolated value, *i.e.*, 0.9999999, the error becomes less than the prescribed error.

Thus $\int_0^{\frac{\pi}{2}} \sin x dx = 1.0$

Algorithm 9.9: Romberg Integration

The function *f*(*x*) is to be integrated from *a* to *b*, and the prescribed tolerance in the final integrated value is *epsilon*.

```
Begin
    read: a, b, epsilon
    set i = 1, n = 1, h = b - a
    set sum = [ f(a) + f(b) ]/2
    set r_i1 = h × sum
    set flag = false
    while ( flag = false) do
        set n = 2 × n, h = h/2, i = i + 1
        for j = 1 to n by 2 do
            set x = a + j × h
            set sum = sum + f(x)
        endfor
        set r_i1 = h × sum
        for k = 2 to i by 1 do
            set r_ik = (4^(k-1) × r_i(k-1) - r_(i-1)(k-1))/(4^(k-1) - 1)
        endfor
        if ( | r_ii - r_i(i-1) | ≤ epsilon ) then
            write: "Element of the Romberg table are"
            for k = 1 to i by 1 do
                for j = 1 to k by 1 do
                    write: r_kj
                endfor
            endfor
            set flag = true
        endif
    endwhile
End.
```

9.5 GAUSS'S QUADRATURE FORMULA

The most accurate of the quadrature formulae is the Gauss's formula. In the Trapezoidal rule and the Simpson's rules, the range of integration is divided into intervals of equal size. But Gauss found that some other spacing might give a better result, so he developed this formula.

Consider the integral $I = \int_a^b f(x)dx$

The basic requirement of the Gauss's formula is that the integral should be in the following form $I = \int_{-1}^{1} f(x)dx$

This can be accomplished by a change of variable that will also transform the integration limits to this required form. We will now see how this can be done.

Substitute $x = py + q$

where p and q are unknown and chosen in such a way that the values of p and q satisfy the following conditions:

At $x = a,\ y = -1$ and $x = b, y = 1$

As $y = \dfrac{x-q}{p}$

Therefore, at $x = a$, we have $y = \dfrac{a-q}{p} = -1$

And at $x = b$, we have $y = \dfrac{b-q}{p} = -1$

These conditions yields the following equations

$$-p + q = a$$

$$p + q = b$$

Solving these equations for p and q, we get

$$p = \frac{b-a}{2} \qquad q = \frac{a+b}{2}$$

Thus $x = \dfrac{b-a}{2}y + \dfrac{a+b}{2}$ and $dx = \dfrac{b-a}{2}dy$

Therefore, the given integral becomes

$$I = \frac{b-a}{2}\int_{-1}^{1} g(y)dy$$

This integral is then solved as $I = \dfrac{b-a}{2}\displaystyle\int_{-1}^{1} g(y)dy = \dfrac{b-a}{2}\sum_{i=1}^{n} R_i g(y_i)$ ... (9.20)

where R_i's are constants called weights, and y_i's are the values of y where the transformed function $g(y)$ is to be computed, and n is the number of points for computing the function. These parameters are standard and some of them are listed in Table 9.3.

Table 9.3: Weights and Y Co-ordinates

n	i	R_i	y_i	Truncation Error
2	1	1.0000000	–0.5773505	$O(f^{4}(y_0))$
	2	1.0000000	0.5773505	
3	1	0.5555556	–0.7745967	$O(f^{6}(y_0))$
	2	0.8888889	0.0000000	
	3	0.5555556	0.7745967	
4	1	0.3478549	–0.8611363	$O(f^{8}(y_0))$
	2	0.6521452	–0.3399810	
	3	0.6521452	0.3399810	
	4	0.3478549	0.8611363	
5	1	0.2369269	–0.9061799	$O(f^{10}(y_0))$
	2	0.4786287	–0.5384693	
	3	0.5688889	0.0000000	
	4	0.4786287	0.5384693	
	5	0.2369269	0.9061799	
6	1	0.1713245	–0.9324695	$O(f^{12}(y_0))$
	2	0.3607616	–0.6612094	
	3	0.4679139	–0.2386192	
	4	0.4679139	0.2386192	
	5	0.3607616	0.6612094	
	6	0.1713245	0.9324695	
7	1	0.1294850	–0.9491079	$O(f^{14}(y_0))$
	2	0.2797054	–0.7415312	
	3	0.3818301	–0.4058452	
	4	0.4179591	0.0000000	
	5	0.3818301	0.4058452	
	6	0.2797054	0.7415312	
	7	0.1294850	0.9491079	
8	1	0.1012285	–0.9602899	$O(f^{16}(y_0))$
	2	0.2223810	–0.7966665	
	3	0.3137066	–0.5255324	
	4	0.3626838	–0.1834346	
	5	0.3626838	0.1834346	
	6	0.3137066	0.5255324	
	7	0.2223810	0.7966665	
	8	0.1012285	0.9602899	
9	1	0.0812744	–0.9681602	$O(f^{18}(y_0))$
	2	0.1806482	–0.8360311	
	3	0.2606107	–0.6133714	
	4	0.3123471	–0.3242534	
	5	0.3302394	0.0000000	
	6	0.3123471	0.3242534	
	7	0.2606107	0.6133714	
	8	0.1806482	0.8360311	
	9	0.0812744	0.9681602	
10	1	0.0666713	–0.9379065	$O(f^{20}(y_0))$
	2	0.1494513	–0.8650633	
	3	0.2190864	–0.6794096	
	4	0.2692667	–0.4333954	
	5	0.2955242	–0.1488743	
	6	0.2955242	0.1488743	
	7	0.2692667	0.4333954	
	8	0.2190864	0.6794096	
	9	0.1494513	0.8650633	
	10	0.0666713	0.9379065	

where $-1 \le y_0 \le 1$, and values of R_i's and y_i's are rounded-off to seven decimal digits.

To use the Gauss's formula, perform the following steps:

(i) Transform the independent variable,
(ii) Choose the appropriate value of n, and
(iii) Substitute the values of R_i's and y_i's, and solve the formula.

Example 9.7: Evaluate $\int_2^4 (x^2 + 2x)dx$

Solution: Substitute $x = \frac{b-a}{2}y + \frac{a+b}{2} = \frac{4-2}{2}y + \frac{2+4}{2} = y + 3$

We get $g(y) = (y+3)^2 + 2(y+3) = y^2 + 8y + 15$

By taking $n = 2$, the Gauss's formula becomes

$$I = \frac{b-a}{2}\left[R_1 g(y_1) + R_2 g(y_2)\right]$$

$$= \frac{b-a}{2}\left[R_1 \times \left\{y_1^2 + 8y_1 + 15\right\} + R_2 \times \left\{y_2^2 + 8y_2 + 15\right\}\right]$$

$$= \frac{4-2}{2}\left[1.0 \times \left\{(-0.5773505)^2 + 8 \times (-0.5773505) + 15\right\}\right.$$

$$\left. + 1.0 \times \left\{(0.5773505)^2 + 8 \times (0.5773505) + 15\right\}\right]$$

$= 30.667$ (rounded-off to three decimal places)

Algorithm 9.10: Gauss's Quadrature formula

Two dimensional arrays named r and y holds the weights and the y co-ordinates. Integration limits are from a to b, and n is the number of points.

```
Begin
    read: a, b, n
    for i = 1 to n by 1 do
        read: ri, yi
    endfor
    set sum = 0
    for i = 1 to n by 1 do
        set sum = sum + ri, × g(yi)
    endfor
    set intgeral = (b - a)/2 × sum
    write: "Value of the integral = ", intgeral
End.
```

9.6 BOOLE'S RULE

The *Boole's rule* approximates the area under the curve by using fourth-order interpolating polynomial that gives

$$I_k = \int_{x_k}^{x_{k+4}} ydx = \frac{2h}{45}[7y_k + 32y_{k+1} + 12y_{k+2} + 32y_{k+3} + 7y_{k+4}]$$

This is the area under the curve between the points $x = x_k$ and $x = x_{k+4}$. This covers four intervals of width h each. If we use this process repetitively $n/4$ times, we can get the area under the curve for n intervals. Thus, if the range of integration is divided into n intervals, each of width h, the required integral over a range $x = a$ and $x = b$ is given by

$$I = \sum_{k=1}^{n/4} I_{4k-3} = I_1 + I_5 + I_9 + \ldots + I_{n-3}$$

$$I = \frac{2h}{45}\begin{bmatrix} 7y_1 + 32y_2 + 12y_3 + 32y_4 + 14y_5 + 32y_6 + 12y_7 + 32y_8 \\ +14y_9 + \ldots + 14y_{n-3} + 32y_{n-2} + 12y_{n-1} + 32y_n + 7y_{n+1} \end{bmatrix} \quad \ldots (9.21)$$

This is the formula for the Boole rule. It gives the correct value of the integral only if $f(x)$ is a fourth order function.

9.7 WEDDLE'S RULE

The *Weddle's rule* approximates the area under the curve by using sixth-order interpolating polynomial that gives

$$I_k = \int_{x_k}^{x_{k+6}} ydx = \frac{3h}{10}[y_k + 5y_{k+1} + y_{k+2} + 6y_{k+3} + y_{k+4} + 5y_{k+5} + y_{k+6}]$$

This is the area under the curve between the points $x = x_k$ and $x = x_{k+6}$. This covers four intervals of width h each. If we use this process repetitively $n/6$ times, we can get the area under the curve for n intervals.

Thus, if the range of integration is divided into n intervals, each of width h, the required integral over a range $x = a$ and $x = b$ is given by

$$I = \sum_{k=1}^{n/6} I_{6k-5} = I_1 + I_7 + I_{13} + \ldots + I_{n-5}$$

$$I = \frac{3h}{10}\begin{bmatrix} y_1 + 5y_2 + y_3 + 6y_4 + y_5 + 5y_6 + 2y_7 + 5y_8 \\ +y_9 + 6y_{10} + y_{11} + 5y_{12} + 2y_{13} + \ldots + y_{n-5} + 5y_{n-4} \\ +y_{n-3} + 6y_{n-2} + y_{n-1} + 5y_n + y_{n+1} \end{bmatrix} \quad \ldots (9.22)$$

This is the formula for the Weddle rule. It gives the correct value of the integral only if $f(x)$ is a sixth order function.

9.8 DOUBLE INTEGRATION

The double integral of type

$$I = \int_c^d \int_a^b f(x, y) dx dy \qquad \ldots (9.23)$$

can be evaluated by two successive integration in x and y directions, considering one variable at a time. The details of this are described below in order to derive the formulae for trapezoidal rule and Simpson's 1/3rd rule for double integration.

9.8.1 Trapezoidal Rule

Consider the integral

$$I = \int_c^d \int_a^b f(x, y) dx dy$$

Divide the interval [a, b] into n equal sub-intervals each of length h, and the interval [c, d] into m equal sub-intervals each of length k, we have

$$x_i = x_0 + ih \qquad x_0 = a \qquad x_n = b$$

$$y_j = y_0 + jk \qquad y_0 = c \qquad y_m = d$$

Rewriting the double integral in the form

$$I = \int_c^d \left[\int_a^b f(x, y) dx \right] dy$$

Now applying trapezoidal rule in x direction, we have

$$I = \int_c^d \left[\frac{h}{2} \{ f(x_0, y) + 2f(x_1, y) + 2f(x_2, y) + \ldots + 2f(x_{n-1}, y) + f(x_n, y) \} \right] dy$$

$$I = \frac{h}{2} \left[\int_c^d f(x_0, y) dy + 2\int_c^d f(x_1, y) dy + 2\int_c^d f(x_2, y) dy + \ldots + 2\int_c^d f(x_{n-1}, y) dy + \int_c^d f(x_n, y) dy \right]$$

Now applying trapezoidal rule in y direction, we have

$$I = \frac{h}{2} \left[\frac{k}{2} \{ f(x_0, y_0) + 2f(x_0, y_1) + 2f(x_0, y_2) + \ldots + 2f(x_0, y_{m-1}) + f(x_0, y_m) \} \right.$$

$$+ \frac{k}{2} \{ f(x_1, y_0) + 2f(x_1, y_1) + 2f(x_1, y_2) + \ldots + 2f(x_1, y_{m-1}) + f(x_1, y_m) \}$$

$$+\frac{k}{2}\{f(x_2,y_0)+2f(x_2,y_1)+2f(x_2,y_2)+...+2f(x_2,y_{m-1})+f(x_2,y_m)\}+...$$

$$+\frac{k}{2}\{f(x_{n-1},y_0)+2f(x_{n-1},y_1)+2f(x_{n-1},y_2)+...+2f(x_{n-1},y_{m-1})+f(x_{n-1},y_m)\}$$

$$+\frac{k}{2}\{f(x_n,y_0)+2f(x_n,y_1)+2f(x_n,y_2)+...+2f(x_n,y_{m-1})+f(x_n,y_m)\}\Big]$$

$$I=\frac{hk}{4}\Big[\{f(x_0,y_0)+2f(x_0,y_1)+2f(x_0,y_2)+...+2f(x_0,y_{m-1})+f(x_0,y_m)\}$$

$$+\{f(x_1,y_0)+2f(x_1,y_1)+2f(x_1,y_2)+...+2f(x_1,y_{m-1})+f(x_1,y_m)\}$$

$$+\{f(x_2,y_0)+2f(x_2,y_1)+2f(x_2,y_2)+...+2f(x_2,y_{m-1})+f(x_2,y_m)\}+...$$

$$+\{f(x_{n-1},y_0)+2f(x_{n-1},y_1)+2f(x_{n-1},y_2)+...+2f(x_{n-1},y_{m-1})+f(x_{n-1},y_m)\}$$

$$+\{f(x_n,y_0)+2f(x_n,y_1)+2f(x_n,y_2)+...+2f(x_n,y_{m-1})+f(x_n,y_m)\}\Big]$$

$$I=\frac{hk}{4}\left[\left\{(f_{00}+f_{0m})+2\sum_{j=1}^{m-1}f_{0j}+(f_{n0}+f_{nm})+2\sum_{j=1}^{m-1}f_{nj}\right\}+2\sum_{i=1}^{n-1}\left\{(f_{i0}+f_{im})+2\sum_{j=1}^{m-1}f_{ij}\right\}\right] \quad ...(9.24)$$

where $f_{ij}=f(x_i, y_j)$. This is the formula for trapezoidal rule for double integration.

Example 9.8: Evaluate $\int_2^{2.6}\int_4^{4.4}\frac{1}{xy}dxdy$ by taking $h = 0.2$ and $k = 0.3$.

Solution: With $h = 0.2$ and $k = 0.3$, we have $m = n = 2$

$$I=\frac{hk}{4}\Big[\{f(4,2)+2f(4,2.3)+f(4,2.6)\}+2\{f(4.2,2)+2f(4.2,2.3)+f(4.2,2.6)\}$$

$$+\{f(4.4,2)+2f(4.4,2.3)+f(4.4,2.6)\}\Big]$$

$$=\frac{0.2\times0.3}{4}[\{0.125+2\times0.107+0.096\}+2\times\{0.119+2\times0.104+0.092\}$$

$$+\{0.114+2\times0.099+0.087\}]$$

$$=\frac{0.06}{4}[\{0.435\}+2\times\{0.419\}+\{0.399\}]=0.025$$

9.8.2 Simpson's 1/3rd Rule

Consider again the integral

$$I = \int_c^d \int_a^b f(x, y)\,dx\,dy$$

Divide the interval $[a, b]$ into n equal sub-intervals each of length h, and the interval $[c, d]$ into m equal sub-intervals each of length k, where n and m are even. Then we have

$$x_i = x_0 + ih \qquad x_0 = a \qquad x_n = b$$

$$y_j = y_0 + jk \qquad y_0 = c \qquad y_m = d$$

Rewriting the double integral in the form

$$I = \int_c^d \left[\int_a^b f(x, y)\,dx \right] dy$$

Now applying Simpson's 1/3rd rule in x direction, we have

$$I = \int_c^d \left[\frac{h}{3} \{ f(x_0, y) + 4f(x_1, y) + 2f(x_2, y) + \ldots + 4f(x_{n-1}, y) + f(x_n, y) \} \right] dy$$

$$I = \frac{h}{3}\left[\int_c^d f(x_0, y)dy + 4\int_c^d f(x_1, y)dy + 2\int_c^d f(x_2, y)dy + \ldots + 4\int_c^d f(x_{n-1}, y)dy + \int_c^d f(x_n, y)dy \right]$$

Now applying Simpson's 1/3rd rule in y direction, we have

$$I = \frac{h}{3}\left[\frac{k}{3}\{ f(x_0, y_0) + 4f(x_0, y_1) + 2f(x_0, y_2) + \ldots + 4f(x_0, y_{m-1}) + f(x_0, y_m) \} \right.$$

$$+ \frac{4k}{3}\{ f(x_1, y_0) + 4f(x_1, y_1) + 2f(x_1, y_2) + \ldots + 4f(x_1, y_{m-1}) + f(x_1, y_m) \}$$

$$+ \frac{2k}{3}\{ f(x_2, y_0) + 4f(x_2, y_1) + 2f(x_2, y_2) + \ldots + 4f(x_2, y_{m-1}) + f(x_2, y_m) \} + \ldots$$

$$+ \frac{4k}{3}\{ f(x_{n-1}, y_0) + 4f(x_{n-1}, y_1) + 2f(x_{n-1}, y_2) + \ldots + 4f(x_{n-1}, y_{m-1}) + f(x_{n-1}, y_m) \}$$

$$\left. + \frac{k}{3}\{ f(x_n, y_0) + 4f(x_n, y_1) + 2f(x_n, y_2) + \ldots + 4f(x_n, y_{m-1}) + f(x_n, y_m) \} \right]$$

$$
\begin{aligned}
I = \frac{hk}{9}\Big[&\{f(x_0,y_0)+2f(x_0,y_1)+2f(x_0,y_2)+...+2f(x_0,y_{m-1})+f(x_0,y_m)\}\\
&+4\{f(x_1,y_0)+2f(x_1,y_1)+2f(x_1,y_2)+...+2f(x_1,y_{m-1})+f(x_1,y_m)\}\\
&+2\{f(x_2,y_0)+2f(x_2,y_1)+2f(x_2,y_2)+...+2f(x_2,y_{m-1})+f(x_2,y_m)\}+...\\
&+4\{f(x_{n-1},y_0)+2f(x_{n-1},y_1)+2f(x_{n-1},y_2)+...+2f(x_{n-1},y_{m-1})+f(x_{n-1},y_m)\}\\
&+\{f(x_n,y_0)+2f(x_n,y_1)+2f(x_n,y_2)+...+2f(x_n,y_{m-1})+f(x_n,y_m)\}\Big]
\end{aligned}
\quad ...(9.25)
$$

This is the formula for Simpson's 1/3rd rule for double integration.

Example 9.9: Evaluate $\int_2^{2.6}\int_4^{4.4}\frac{1}{xy}dxdy$ by taking $h = 0.2$ and $k = 0.3$.

Solution: With $h = 0.2$ and $k = 0.3$, we have $m = n = 2$

$$
\begin{aligned}
I &= \frac{hk}{9}\Big[\{f(4,2)+4f(4,2.3)+f(4,2.6)\}+4\{f(4.2,2)+4f(4.2,2.3)+f(4.2,2.6)\}\\
&\qquad +\{f(4.4,2)+4f(4.4,2.3)+f(4.4,2.6)\}\Big]\\
&= \frac{0.2\times0.3}{9}[\{0.125+4\times0.107+0.096\}+4\times\{0.119+4\times0.104+0.092\}\\
&\qquad +\{0.114+4\times0.099+0.087\}]\\
&= \frac{0.06}{9}[\{0.649\}+4\times\{0.627\}+\{0.597\}] = 0.025
\end{aligned}
$$

9.9 IMPROPER INTEGRALS

In our discussion to evaluate the integral of type

$$I = \int_a^b f(x)dx$$

We have assume that

(i) the function $f(x)$ is continuous for all values of x from a to b, and

(ii) a and b are finite.

If any of the above or both the stated assumptions are not true, the integral is said to be an improper integral. We will consider these cases separately.

Case I: Function is discontinuous

Consider the integral

$$I = \int_0^{\frac{\pi}{4}} \frac{1}{\cos 2x} dx$$

At $x = \frac{\pi}{4}$, the function becomes infinite. However, this integral can be expressed as

$$I = \int_0^{\vartheta_1} \frac{1}{\cos 2x} dx + \int_{\vartheta_1}^{\frac{\pi}{4}} \frac{1}{\cos 2x} dx$$

where $(\frac{\pi}{4} - \theta_1)$, *i.e.*, the range of the second integral, is very small. Now the first integral can be solved numerically, and we do hope that a reasonable estimate of the second integral can be obtained analytically.

Case II: Limits of Integration are not finite

Consider the integral

$$I = \int_1^{\infty} \frac{1}{1+x^2} dx$$

where the upper limit is infinite. This type of problem can be handled by a change of a variable that makes the infinite limit zero as shown below:

Substitute $x = \frac{1}{t}$, then $dx = -\frac{1}{t^2} dt$

At $x = 1$ $t = 1$

At $x = \infty$ $t = 0$

Therefore, the given integral becomes $I = -\int_1^0 \frac{1}{1+t^2} dt$

which is a proper integral. And if this type of transformation leads to Case I, then it can be solved as described under Case I.

EXERCISE . . .

1. Derive the formula for Trapezoidal rule. Also prove that error in the Trapezoidal rule is O(h^2), where h is the strip width.
2. Derive the formula for Simpson's 1/3rd. Also prove that error in the Simpson's 1/3rd rule is O(h^2), where h is the strip width.
3. Describe the Romberg method of integration.
4. Describe the Gaussian quadrature method of integration.
5. Evaluate $\int_0^2 \frac{1}{1+x^2}dx$ using Trapezoidal rule with h = 0.25. Also find the amount of error incurred in the approximation.
6. Evaluate $\int_{-2}^2 \frac{3x}{(4+x)^2}dx$ using Simpson's 1/3rd rule with six intervals.
7. Evaluate $\int_1^2 4e^{-x}dx$ using Simpson's 1/3rd rule choosing interval size of your choice. Also find the error incurred in the approximation.
8. It is required to evaluate $\int_1^2 3e^{-2x}dx$ with error not more than 0.0001. How many intervals must be used?
   (*a*) If Trapezoidal rule is used.
   (*b*) If Simpson's 1/3rd rule is used.
9. The table of values for a function $f(x)$ are

x :	1.8	2.0	2.4	2.6	3.0	3.8
$f(x)$:	6.050	7.389	11.023	12.464	20.086	44.701

   Can you apply Trapezoidal rule or Simpson's 1/3rd to obtain the integral of $f(x)$ in the interval from x = 1.8 to x = 3.8?
   (*a*) If not, why?
   (*b*) If yes, then find value of the integral.
10. Evaluate $\int_{0.2}^{1.5} e^{-x^2}dx$ correct to four significant digits using Romberg integration method.
11. The table of values for $f(x) = e^x$ is given as

x :	1.6	1.8	2.0	2.2	2.4	2.6	2.8
$f(x)$:	4.953	6.050	7.389	9.025	11.023	12.464	16.445

   Use Trapezoidal rule to obtain integral of $f(x)$ in interval from x = 1.8 to x = 2.4. Also find the amount of error incurred.

12. The table of values for function $f(x)$ is given as

x :	1.6	1.8	2.0	2.2	2.4	2.6	2.8	3.0
$f(x)$:	4.953	6.050	7.389	9.025	11.023	12.464	16.445	20.086

Evaluate $\int_{1.6}^{1.3.0} f(x)dx$ using Romberg integration method.

13. Evaluate $\int_{0.2}^{1.5} e^{-x^2} dx$ using Gaussian quadrature formula with $n = 3, 4, 5$. For which value of n the method gives better approximation.

14. Develop algorithm to solve a double integral of type

$$\int_c^d \int_a^b f(x,y)dxdy$$

using (*a*) Trapezoidal rule, (*b*) Simpson's 1/3rd rule

15. The function $f(x, y)$ is given in the following form

x \ y	0.1	0.2	0.3	0.4	0.5
0.5	0.165	0.428	0.687	0.942	1.190
1.0	0.271	0.640	1.003	1.359	1.703
1.5	0.447	0.990	1.524	2.045	2.549
2.0	0.738	1.568	2.384	3.177	3.943
2.5	1.216	2.520	3.800	5.044	6.241

Evaluate $\int_{0.1}^{0.5} \int_{0.5}^{2.5} f(x,y)dxdy$

Using (*a*) Trapezoidal rule, (*b*) Simpson's 1/3rd rule

10 Solution of Ordinary Differential Equations

Learning Outcomes

After reading this chapter, students will have the ability to

- define a differential equations and their usefulness
- explain the basic terms related to differential equations
- differentiate between initial value and boundary value problems
- describe the working of various methods to find the solution differential equations
- use these methods to find the solution of differential equations manually
- write computer programs for these methods

10.1 INTRODUCTION

The subject of ordinary differential equations (ODEs) is not only a fascinating part of mathematics but also an essential tool for modelling many physical processes. In addition, most of the scientific laws are expressed in terms of differential equations.

1. In thermodynamics, the differential equation

$$\frac{dT}{dt} = -0.27(u - 60)^{\frac{5}{4}}$$

describe the rate of change of temperature T of a body losing heat to its surroundings.

2. In probability theory, the differential equation

$$\frac{dp_r}{dt} = (r+1)(n-r)p_{r+1} - r(n-r+1)p_r$$

is used to study the spread of an epidemic. Here p_r is the probability that in a sample of size n, r people are still uninfected at a time t.

3. In mechanics, the differential equation

$$m\frac{dv}{dt} = mg - kv^2$$

describe the motion of a body falling freely under gravity through a medium which offers resistance proportional to the square of the velocity.

4. In physics, the solution of the differential equation

$$m\frac{d^2y}{dx^2} = a - y + by^2$$

was used by Einstein in his investigation of the orbital motion of planets under the assumption of general relativity.

5. Another very important differential equation encountered in the study of physics is Vander Pol's equation

$$m\frac{d^2y}{dt^2} - \varepsilon(1 - y^2)\frac{dy}{dt} + ay = 0$$

which has also been used to model the working of the heart.

6. In economics, the differential equation

$$\frac{dx}{dt} = sf(x) - g(x)$$

provides a simple model of economic growth, where x equals ratio of supply of labour, $f(x)$ is the productivity function, $g(x)$ represents the growth in the supply of labour, and s is a constant.

These examples emphasize the importance of differential equations in mathematical investigations.

10.2 SOLVING A DIFFERENTIAL EQUATION

Though it is relatively simple to formulate a differential equation, it is most oftenly difficult to solve it. If we are unable to solve it or its solution is awkward and cumbersome, and is almost unusable, then *what can we do*?

One answer is to compute the numerical solution i.e. instead of finding an algebraic (analytical) solution; we compute (approximately) the numerical values taken by the solution. Therefore, as a solution of a differential equation, instead of finishing up with an expression, we end up with the numerical values taken by that expression. This is known as the numerical solution of a differential equation.

10.3 BASIC TERMINOLOGY OF DIFFERENTIAL EQUATIONS

Before we move to see how a numerical solution can be computed using different techniques, let us make ourselves familiar with the basic terms used related to the subject of differential equations.

10.3.1 Differential Equation

A *differential equation* is an equation that in addition to independent and dependent variables also contains one or more derivatives of the dependent variables. These derivatives can be ordinary or partial.

10.3.2 Ordinary Derivative

If y is a function of x, i.e. $y = f(x)$, then $\frac{dy}{dx}$ is called the ordinary derivative. Physically it means the rate of change of the dependent variable with respect to the independent variable.

10.3.3 Partial Derivative

If u is a function of x and y, i.e. $u = f(x, y)$, then $\left.\frac{\partial u}{\partial x}\right|_y$ is called the partial derivative with respect to x keeping y constant, and $\left.\frac{\partial u}{\partial y}\right|_x$ is called the partial derivative with respect to y keeping x constant. Physically it means the rate of change of the dependent variable with respect to one of the independent variable keeping others fixed.

And the derivatives of higher order are interpreted as rates of rates, and hence differential equations are equations involving rates.

10.3.4 Ordinary Differential Equation

An *ordinary differential equation* is an equation involving only ordinary derivatives of one or more functions with respect to a single independent variable. Following are some examples of ordinary differential equations.

$$\frac{dy}{dx} = -\frac{x}{y} \qquad \frac{dy}{dx} = x^2 + y^2 \qquad \frac{d^2y}{dx^2} + y = 0$$

10.3.5 Partial Differential Equation

A *partial differential equation* is an equation involving partial derivatives of one or more functions with respect to a single independent variable. Following are some examples of partial differential equations.

$$\frac{\partial u}{dt} = k\frac{\partial^2 u}{\partial x^2} \qquad \frac{\partial^2 u}{dx^2} + k\frac{\partial^2 u}{\partial y^2} = 0$$

10.3.6 Order of a Differential Equation

The *order of differential equations* is the order of the highest order derivative in the differential equations. For example

$$m\frac{d^2y}{dx^2} + c\frac{dy}{dx} + ky = 0$$

is a second order ordinary differential equation.

10.3.7 Degree of a Differential Equation

The *degree of differential equations* is the power of the highest order derivative in the differential equations. For example

$$\left[\frac{d^2y}{dx^2}\right]^2 + c\frac{dy}{dx} + ky = 0$$

is a second degree ordinary differential equation.

10.3.8 Solution of a Differential Equation

Consider the first order ordinary differential equation of type

$$\frac{dy}{dx} = f(x,y) \qquad \ldots (10.1)$$

which can also be written as

$$y^1 = f(x, y)$$

where the function $f(x,y)$ may be a general non-linear function of x and y or in the form of a table of values.

The solution of such an ordinary differential equation is a two-dimensional curve $g(x,y)$ in the xy plane whose slope at every point (x, y) in the specified region is given by the Eq. (10.1).

10.4 INITIAL VALUE AND BOUNDARY VALUE PROBLEMS

An ordinary differential equation of nth order is of the form

$$f\left(x, y, \frac{dy}{dx}, \frac{d^2y}{dx^2}, \ldots, \frac{d^ny}{dx^n}\right) = 0 \qquad \ldots (10.2)$$

Its general solution contains n arbitrary constants and is of the form

$$\phi(x, y, c_1, c_2, c_3, \ldots, c_n) = 0 \qquad \ldots (10.3)$$

To obtain its solution, n conditions must be given so that the constants $c_1, c_2, c_3, \ldots, c_n$ can be determined. If these conditions are prescribed at one point only, say x_1, the differential equations together with the conditions constitute an *initial value problem* (IVP) of the nth order. However, if the conditions are prescribed at two or more points, then differential equations together with the conditions constitute a *boundary value problem* (BVP) of the nth order.

In this chapter, we will describe various methods for solving initial value problems.

10.5 PICARD'S METHOD

Rewriting $\frac{dy}{dx} = f(x,y)$ as $dy = f(x,y)dx$, and integrating this equation, we get

$$\int_{y_0}^{y} dy = \int_{x_0}^{x} f(x,y)dx$$

$$y = y_0 + \int_{x_0}^{x} f(x,y)dx \qquad \ldots (10.4)$$

This equation in which the unknown function y appears under the integral sign is called an *integral equation*. Such an equation can be solved by the method of successive approximations in which the first approximation to y is obtained by substituting y_0 for y on the right hand side of Eq. (10.4).

Thus, we write $$y_1 = y_0 + \int_{x_0}^{x} f(x, y_0)dx$$

The integral on the right hand side can now be solved and the resulting y_1 is substituted for y in the integrand of Eq. (10.4) to obtain the second approximation y_2 to y as

$$y_2 = y_0 + \int_{x_0}^{x} f(x, y_1)dx$$

Proceeding on the similar lines, we can obtain $(i+1)^{th}$ approximation from i^{th} approximations as

$$y_{i+1} = y_0 + \int_{x_0}^{x} f(x, y_i)dx \qquad \ldots (10.5)$$

Note that the Picard's method gives the solution of the ordinary differential equation at a point, say x, as a truncated power series in x as illustrated in Example 10.1.

Picard's method is of considerable theoretical value, but can be applied to a limited class of equations in which the successive integrations can be evaluated easily.

Example 10.1: Given $\frac{dy}{dx} = x + y$ with $y(0) = 1$. Find the solution of the given differential equation correct to fifth approximation.

Solution: Since $f(x, y) = x + y$, and $y_0 = 1$, the integral equation

$$y = y_0 + \int_{x_0}^{x} f(x, y)dx \qquad \ldots (10.6)$$

become

$$y = 1 + \int_{0}^{x} (x + y)dx$$

First Approximation: Substituting $y = 1$ on the right hand side of Eq. (10.6), we get

$$y_1 = 1 + \int_{0}^{x} (x + 1)dx = 1 + x + \frac{x^2}{2}$$

Second Approximation: Substituting $y = 1 + x + \frac{x^2}{2}$ on the right hand side of Eq. (10.6), we get

$$y_2 = 1 + \int_0^x \left(x + 1 + x + \frac{x^2}{2} \right) dx = 1 + \int_0^x \left(1 + 2x + \frac{x^2}{2} \right) dx = 1 + x + x^2 + \frac{x^3}{6}$$

Third Approximation: Substituting $y = 1 + x + x^2 + \frac{x^3}{6}$ on the right hand side of Eq. (10.6), we get

$$y_3 = 1 + \int_0^x \left(x + 1 + x + x^2 + \frac{x^3}{6} \right) dx = 1 + \int_0^x \left(1 + 2x + x^2 + \frac{x^3}{6} \right) dx$$

$$= 1 + x + x^2 + \frac{x^3}{3} + \frac{x^4}{24}$$

Fourth Approximation: Substituting y $= 1 + x + x^2 + \frac{x^3}{3} + \frac{x^4}{24}$ on the right hand side of Eq. (10.6), we get

$$y_4 = 1 + \int_0^x \left(x + 1 + x + x^2 + \frac{x^3}{3} + \frac{x^4}{24} \right) dx = 1 + \int_0^x \left(1 + 2x + x^2 + \frac{x^3}{3} + \frac{x^4}{24} \right) dx$$

$$= 1 + x + x^2 + \frac{x^3}{3} + \frac{x^4}{12} + \frac{x^5}{120}$$

Fifth Approximation: Substituting $y = 1 + x + x^2 + \frac{x^3}{3} + \frac{x^4}{12} + \frac{x^5}{120}$ on the right hand side of Eq. (10.6), we get

$$y_5 = 1 + \int_0^x \left(x + 1 + x + x^2 + \frac{x^3}{3} + \frac{x^4}{12} + \frac{x^5}{120} \right) dx$$

$$= 1 + \int_0^x \left(1 + 2x + x^2 + \frac{x^3}{3} + \frac{x^4}{12} + \frac{x^5}{120} \right) dx$$

$$= 1 + x + x^2 + \frac{x^3}{3} + \frac{x^4}{12} + \frac{x^5}{60} + \frac{x^6}{720}$$

Example 10.2: Find the solution, correct to three decimal positions, of the first order ordinary differential equation $\frac{dy}{dx} = x + y^2$ for $x = 0.1$ when $y(0) = 1$.

Solution: Since $f(x, y) = x + y^2$, $y_0 = 1$, $x_0 = 0$ and $x = 0.1$, the integral equation

$$y = y_0 + \int_{x_0}^{x} f(x, y)dx \quad \text{becomes} \quad y = 1 + \int_0^{0.1} (x + y^2)dx \qquad \ldots (10.7)$$

First Approximation: Substituting $y = 1$ on the right hand side of Eq. (10.7), we get

$$y_1 = 1 + \int_0^{0.1} (x + 1^2)dx = 1 + \left|\frac{x^2}{2} + x\right|_0^{0.1} = 1.105$$

Second Approximation: Substituting $y = 1.105$ on the right hand side of Eq. (10.7), we get

$$y_2 = 1 + \int_0^{0.1} (x + (1.105)^2)dx = 1 + \left|\frac{x^2}{2} + (1.105)^2 x\right|_0^{0.1} = 1.127$$

Third Approximation: Substituting $y = 1.127$ on the right hand side of Eq. (10.7), we get

$$y_3 = 1 + \int_0^{0.1} (x + (1.127)^2)dx = 1 + \left|\frac{x^2}{2} + (1.127)^2 x\right|_0^{0.1} = 1.132$$

Fourth Approximation: Substituting $y = 1.132$ on the right hand side of Eq. (10.7), we get

$$y_4 = 1 + \int_0^{0.1} (x + (1.132)^2)dx = 1 + \left|\frac{x^2}{2} + (1.132)^2 x\right|_0^{0.1} = 1.133$$

Fifth Approximation: Substituting $y = 1.133$ on the right hand side of Eq. (10.7), we get

$$y_5 = 1 + \int_0^{0.1} (x + (1.133)^2)dx = 1 + \left|\frac{x^2}{2} + (1.133)^2 x\right|_0^{0.1} = 1.1333$$

By looking at third, fourth, and fifth approximations, we conclude there is no change in the first three decimal places, hence the value of $y(0.1)$ correct to three decimal digits is 1.133 (rounded).

Example 10.3: Find the solution, correct to four decimal positions, of first order ordinary differential equation $\frac{dy}{dx} = x^2 + y^2$ for $x = 0.4$, given that when $y = 0$ when $x = 0$.

Solution: Since $f(x, y) = x^2 + y^2$, $y_0 = 0$, $x_0 = 0$ and $x = 0.4$, the integral equation

$$y = y_0 + \int_{x_0}^{x} f(x,y)dx \qquad \text{becomes} \qquad y = \int_{0}^{0.4} (x^2 + y^2)dx$$

First Approximation: Substituting $y = 0$ on the right hand side, we get

$$y_1 = \int_{0}^{0.4} (x^2 + 0^2)dx = \left|\frac{x^3}{3}\right|_0^{0.4} = 0.0213$$

Second Approximation: Substituting $y = 0.0213$ on the right hand side, we get

$$y_2 = \int_{0}^{0.4} (x^2 + 0.0213^2)dx = \left|\frac{x^3}{3} + 0.0213\,x\right|_0^{0.4} = 0.0214$$

Third Approximation: Substituting $y = 0.0214$ on the right hand side, we get

$$y_3 = \int_{0}^{0.4} (x^2 + 0.0214^2)dx = \left|\frac{x^3}{3} + 0.0214\,x\right|_0^{0.4} = 0.0214$$

By looking at second and third approximations, we conclude there is no change in the first four decimal places, hence the value of $y(0.4)$ correct to four decimal digits is 0.0214.

10.6 EULER'S METHOD

The Euler's method is one of the oldest and the simplest method. The Euler's method can be described as a technique of developing a piecewise linear approximation to the solution. In the initial value problem, the starting point of the solution curve and the slope of the curve at the starting point are given. With this information, the method extrapolates the solution curve using the specified step size.

Consider again the following first order ordinary differential equation

$$\frac{dy}{dx} = f(x,y)$$

with an initial condition: $y - y_1$ for $x = x_1$

Lets us know find the mathematical formula for the Euler's method.

Recall the mean value theorem states that

If a function is continuous and differentiable between two pints, say (x_1, y_1) and (x_2, y_2), then the slope of the line joining these points is equal to the derivative of the function at least at one other point, say (c, d), between these two points, i.e.

$$y^1(c) = \frac{y(x_2) - y(x_1)}{x_2 - x_1} \qquad \ldots (10.8)$$

If we substitute $c = x_1$ and $h = x_2 - x_1$, then Eq. (10.8) can be written as

$$y(x_2) - y(x_1) = hy^1(x_1)$$

Now from Eq. (10.1), we have

$$y^1(x_1) = f(x_1, y_1)$$

therefore

$$y(x_2) - y(x_1) = hf(x_1, y_1)$$

$$y(x_2) = y(x_1) + hf(x_1, y_1)$$

$$y_2 = y_1 + hf(x_1, y_1)$$

Using this equation, we can find the second point on the solution curve as (x_2, y_2).

Similarly, taking (x_2, y_2) as the starting point, we get

$$y_3 = y_2 + hf(x_2, y_2)$$

In general, the $(i+1)^{th}$ point of the solution curve is obtained from the ith point using the following formula

$$\boxed{y_{i+1} = y_i + hf(x_i, y_i)} \qquad \ldots (10.9)$$

Geometrical Interpretation

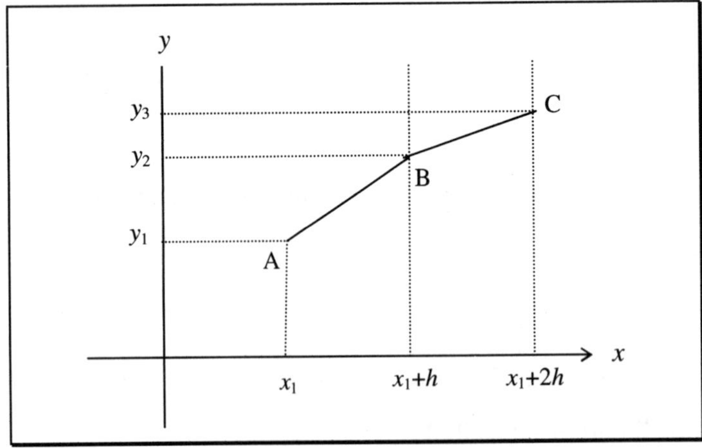

Figure 10.1: Illustration of Euler's method

We are given the initial point (x_1, y_1), say A, of the solution curve, and the slope $f(x_1, y_1)$, say s, of the curve at the initial point. Now from point A draw a line with slope s. The point where this line intersects the vertical line erected at $x_2 = x_1+h$ gives the next value of y. Let it be y_2. Now calculate the slope of the curve at this point, which is given by $f(x_1+h, y_2)$, and draw a line from (x_1+h, y_2) with slope $f(x_1+h, y_2)$; and where this line intersects the vertical line erected at $x_3 = x_1+2h = x_2+h$ gives the next value of y i.e. y_3, and so on. The error involved in each step is of the order of $\frac{h^2}{2} y^2(z)$, where z is a point in the region of computation, and $y^2(z)$ is a second order derivative. Therefore, in order to improve accuracy, the value of the step size h should be as small as possible.

Example 10.4: Given $\frac{dy}{dx} = xy$ with $y(1) = 5$. Find the solution correct to three decimal position in the interval $[1,1.5]$ using step size $h = 0.1$.

Solution: The formula for Euler's method is $y_{i+1} = y_i + hf(x_i, y_i)$

In our example

$$f(x, y) = xy \qquad x_1 = 1 \qquad y_1 = 5 \qquad h = 0.1$$

The Euler's formula can be written as

$$y_{i+1} = y_i + 0.1 \times xy$$

For $i = 1$

$$y_2 = y_1 + 0.1 \times x_1 \times y_1 = 5 + 0.1 \times 1 \times 5 = 5.5$$

Thus, the second solution point is obtained as $(x_2, y_2) = (1.1, 5.5)$.

For $i = 2$

$$y_3 = y_2 + 0.1 \times x_2 \times y_2 = 5.5 + 0.1 \times 1.1 \times 5.5 = 6.105$$

Thus, the third solution point is obtained as $(x_3, y_3) = (1.2, 6.105)$.

For $i = 3$

$$y_4 = y_3 + 0.1 \times x_3 \times y_3 = 6.105 + 0.1 \times 1.2 \times 6.105 = 6.838$$

Thus, the fourth solution point is obtained as $(x_4, y_4) = (1.3, 6.838)$.

For $i = 4$

$$y_5 = y_4 + 0.1 \times x_4 \times y_4 = 6.838 + 0.1 \times 1.3 \times 6.838 = 7.727$$

Thus, the fifth solution point is obtained as $(x_5, y_5) = (1.4, 7.727)$.

For $i = 5$

$$y_6 = y_5 + 0.1 \times x_5 \times y_5 = 7.727 + 0.1 \times 1.4 \times 7.727 = 8.809$$

Thus, the sixth and final solution point is obtained as $(x_6, y_6) = (1.5, 8.809)$.

The complete solution of the given differential equations is given as

i	1	2	3	4	5	6
x_i	1.0	1.1	1.2	1.3	1.4	1.5
y_i	5.0	5.5	6.105	6.838	7.727	8.809

Algorithm 10.1: Euler's Method

To find the solution of ODE $dy/dx = f(x,y)$ when the solution at the initial point, $x = x_1$, is given, and the solution is desired in the interval $[x_1, x_f]$.

```
Begin
    read: x1, y1, xf
    read: h
    set x = x1
    set y = y1
    set i = 1
    write: i, x, y
    while ( x < xf ) do
        set y = y  + hf (x, y)
        set x = x  + h
        set i = i  + 1
        write: i, x, y
    endwhile
End.
```

10.7 TAYLOR SERIES METHOD

Given $\frac{dy}{dx} = f(x,y)$ with an initial condition: $y = y_1$ at $x = x_1$

If the solution curve is expanded in Taylor series around $x = x_1$, we get

$$y(x) = y(x_1) + (x - x_1)y^1(x_1) + \frac{(x - x_1)^2}{2!}y^2(x_1) + ... \quad ...(10.10)$$

where $y^1(x_1)$, $y^2(x_1)$, ... are first and second order derivatives, and so on.

Then at $x = x_1+h$, the above equation becomes

$$y(x_1 + h) = y(x_1) + hy^1(x_1) + \frac{h^2}{2!}y^2(x_1) + ...$$

Now $y^1(x) = f(x, y)$ and y is a function of x, therefore

$$y^2(x) = \frac{d[y^1(x)]}{dx} = \frac{d}{dx}[f(x,y)] = \frac{\partial}{\partial x}f(x,y) + \frac{\partial}{\partial y}f(x,y)\frac{dy}{dx}$$

$$= \frac{\partial}{\partial x}f(x,y) + \frac{\partial}{\partial y}f(x,y) \times f(x,y)$$

If we take $f_x(x,y) = \frac{\partial}{\partial x}f(x,y)$ and $f_y(x,y) = \frac{\partial}{\partial y}f(x,y)$

Then $y^2(x) = f_x(x,y) + f_y(x,y) \times f(x,y)$

Thus $y(x_1 + h) = y(x_1) + hf(x_1, y_1) + \frac{h^2}{2!}\left[f_x(x_1, y_1) + f_y(x_1, y_1) \times f(x_1, y_1)\right] + ...$

$$\Rightarrow \quad y_2 = y(x_1 + h) = y_1 + hf(x_1, y_1) + \frac{h^2}{2!}\left[f_x(x_1, y_1) + f_y(x_1, y_1) \times f(x_1, y_1)\right] + ... \qquad ...(10.11)$$

Similarly, taking (x_2, y_2) as the starting point, we get

$$y_3 = y_2 + hf(x_2, y_2) + \frac{h^2}{2!}\left[f_x(x_2, y_2) + f_y(x_2, y_2) \times f(x_2, y_2)\right] + ...$$

In general

$$y_{i+1} = y_i + hf(x_i, y_i) + \frac{h^2}{2!}\left[f_x(x_i, y_i) + f_y(x_i, y_i) \times f(x_i, y_i)\right] + ... \qquad ...(10.12)$$

This formula is known as the *Taylor series formula* and can be sued repetitively to obtain $y(x)$ for successive values of x. If we reduce the above formula after the third term, the error per step will be $O(h^3)$.

Though the amount of error incurred per step is less that the Euler's method, this method is rarely used on computer as it is not always easy to obtain the partial derivatives of $f(x, y)$, and even if possible, considerable computational effort is required.

10.8 RUNGE-KUTTA METHODS

These methods were devised by *C. Runge* and extended by *W. Kutta* a few years later. The Runge-Kutta methods are actually a family of methods, of which the second order and fourth order methods are widely used. As we have seen in the preceding section, we can improve the accuracy of those methods by taking smaller step sizes. Much greater accuracy can be obtained by using Runge-Kutta methods. These methods are equivalent of approximating the exact solution by matching the first n terms of the Taylor series method. In this text, we will discuss only the second and fourth order methods. In these methods, first the slope at some of the intermediate points is computed, and then weighted average of slopes is used to extrapolate the next solution point.

10.8.1 Runge-Kutta Second Order Methods

The Runge-Kutta second order methods are actually a family of methods, each of that matches the Taylor series method up to the second-degree terms in h, where h is the step size. In these methods the interval $\left[x_1, x_f\right]$ is divided into subintervals and a weighted average of derivatives (slopes) at these intervals is used to determine the value of the dependent variable. One advantage of these methods is that they, like Euler's method, are single step methods i.e. in order to evaluate y_{i+1}, we need information only at the preceding point (x_i, y_i).

Consider the following differential equation

$$\frac{dy}{dx} = f(x,y)$$

with an initial condition:

$$y = y_1 \text{ at } x = x_1$$

At the starting point, compute the slope of the curve as $f(x_1, y_1)$. Let it be s_1. Now compute the slope of the curve at point (x_2, y_1+s_1h) as $f(x_2, y_1+s_1h)$, where $x_2 = x_1+h$. Let this new slope be s_2. Find the average of these two slopes, and then compute the value of the dependent variable y from the following equation

$$y_2 = y_1 + hs$$

where $s = \dfrac{s_1 + s_2}{2}$, $s_1 = f(x_1, y_1)$ and $s_2 = (x_1+h, y_1+s_1h)$

Hence, starting from point $f(x_1, y_1)$, we obtained the second point (x_2, y_2). Similarly, starting from second point, we can obtain the third point. And this process is repeated till we find the solution in the desired interval.

In general, the value of y for the $(i+1)^{th}$ point on the solution curve is obtained from the ith solution point using the formula

$$\boxed{y_{i+1} = y_i + hs} \qquad \ldots (10.13)$$

where $s = \dfrac{s_i + s_{i+1}}{2}$, $s_i = f(x_i, y_i)$, and $s_{i+1} = (x_i+h, y_i+s_ih)$

This formula for the Runge-Kutta second order method is also known as *Heun's method*.

Geometric Interpretation

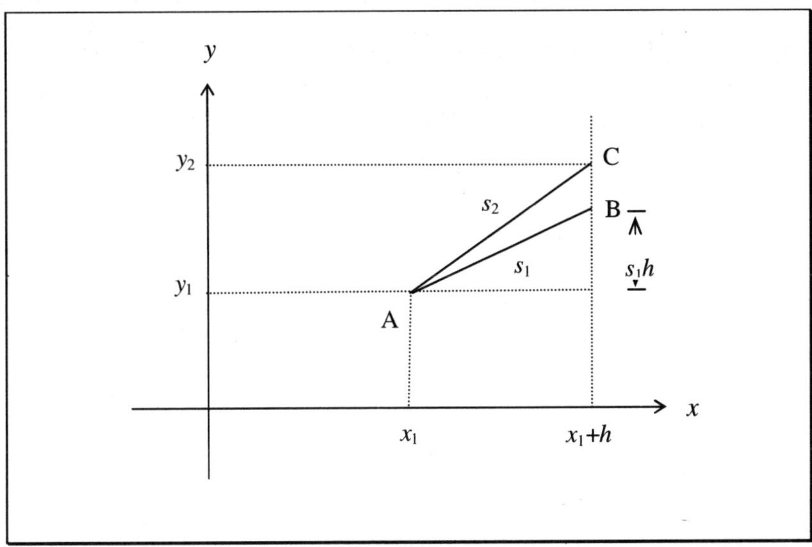

Figure 10.2: Illustration of second order Runge-Kutta method

From the starting point (x_1, y_1) labelled as A, draw a line with slope $s_1(=f(x_1,y_1))$. Let this line intersects the vertical line at x_1+h at a point labelled B, whose co-ordinates are (x_1+h, y_1+s_1h). Now compute the slope of the curve at point B. Let this slope be $s_2(=f(x_1+h, y_1+s_1h))$. Compute the average of these two slopes i.e. s_1 and s_2. Let this average slope be s. Now draw a line from point A with slope s. Let this new line intersects the vertical line at x_1+h at a point labelled C. The point C, whose co-ordinates are (x_1+h, y_1+sh), is taken as the second solution point. Then in the similar manner we proceed from the second solution point to obtain the third solution point, and so on.

We will now show that Runge-Kutta second order methods are identical to Taylor series method.

Since $y_{i+1} = y_i + \frac{h}{2}(s_i + s_{i+1}) \quad \Rightarrow \quad y_{i+1} = y_i + \frac{h}{2}\left[f(x_i, y_i) + f(x_i + h, y_i + s_i h)\right]$

Expanding $f(x_i+h, y_i+s_ih)$ in Taylor series, we get

$$y_{i+1} = y_i + \frac{h}{2}\left[f(x_i, y_i) + f(x_i, y_i) + hf_x(x_i, y_i) + s_i h f_y(x_i, y_i) + \ldots\right]$$

$$y_{i+1} = y_i + \frac{h}{2}\left[f(x_i, y_i) + f(x_i, y_i) + hf_x(x_i, y_i) + hf_y(x_i, y_i) \times f(x_i, y_i) + \ldots\right]$$

$$\Rightarrow \quad y_{i+1} = y_i + hf(x_i, y_i) + \frac{h^2}{2}\left[f_x(x_i, y_i) + f_y(x_i, y_i) \times f(x_i, y_i) + \ldots\right]$$

which is identical to the Taylor series method, and hence the proof.

Example 10.5: Given $\dfrac{dy}{dx} = xy$ with $y(1) = 5$. Find the solution correct to three decimal position in the interval $[1,1.5]$ using step size $h = 0.1$.

Solution: The formula for second order Runge-Kutta method (Heun's method) is

$$y_{i+1} = y_i + hs$$

where

$$s = \frac{s_1 + s_2}{2}, \qquad s_1 = f(x_1, y_1), \quad s_2 = f(x_1+h, y_1+hs_1)$$

In our example

$$f(x, y) = xy \qquad x_1 = 1 \qquad y_1 = 5 \qquad h = 0.1$$

Runge-Kutta second order formula can be written as

$$y_{i+1} = y_i + 0.1 \times s$$

For $i = 1$

$$y_2 = y_1 + 0.1 \times s$$
$$s_1 = f(x_1, y_1) = f(1, 5) = 1 \times 5 = 5$$
$$s_2 = f(x_1+0.1, y_1+0.1\times s_1) = f(1.1, 5+0.1\times 5) = f(1.1, 5.5) = 1.1 \times 5.5 = 6.05$$
$$s = (5+6.05)/2 = 5.525$$

Thus $y_2 = 5 + 0.1 \times 5.525 = 5.553$

Thus, the second solution point is obtained as $(x_2, y_2) = (1.1, 5.553)$.

For $i = 2$

$$y_3 = y_2 + 0.1 \times s$$
$$s_1 = f(x_2, y_2) = f(1.1, 5.553) = 1.1 \times 5.553 = 6.108$$
$$s_2 = f(x_2+0.1, y_2+0.1\times s_1) = f(1.2, 5.553+0.1\times 6.108) = f(1.2, 6.164)$$
$$= 1.2 \times 6.164 = 7.397$$
$$s = (6.108+7.397)/2 = 6.752$$

Therefore $y_3 = 5.553 + 0.1 \times 6.752 = 6.228$

Thus, the third solution point is obtained as $(x_3, y_3) = (1.2, 6.228)$.

For $i = 3$

$$y_4 = y_3 + 0.1 \times s$$
$$s_1 = f(x_3, y_3) = f(1.2, 6.228) = 1.2 \times 6.228 = 7.474$$

$s_2 = f(x_3+0.1, y_3+0.1 \times s_1) = f(1.3, 6.228+0.1 \times 7.474) = f(1.3, 6.975)$
$= 1.3 \times 6.975 = 9.068$

$s = (7.474+9.068)/2 = 8.271$

Therefore $y_4 = 6.228 + 0.1 \times 8.271 = 7.055$

Thus, the fourth solution point is obtained as $(x_4, y_4) = (1.3, 7.055)$.

For $i = 4$

$y_5 = y_4 + 0.1 \times s$

$s_1 = f(x_4, y_4) = f(1.3, 7.055) = 1.3 \times 7.055 = 9.172$

$s_2 = f(x_4+0.1, y_4+0.1 \times s_1) = f(1.4, 7.055+0.1 \times 9.172) = f(1.4, 7.972)$
$= 1.4 \times 7.972 = 11.161$

$s = (9.172+11.161)/2 = 10.166$

Therefore $y_5 = 7.055 + 0.1 \times 10.166 = 8.072$

Thus, the fifth solution point is obtained as $(x_5, y_5) = (1.4, 8.072)$.

For $i = 5$

$y_6 = y_5 + 0.1 \times s$

$s_1 = f(x_5, y_5) = f(1.4, 8.072) = 1.4 \times 8.072 = 11.301$

$s_2 = f(x_5+0.1, y_5+0.1 \times s_1) = f(1.5, 8.072+0.1 \times 11.301) = f(1.5, 9.202)$
$= 1.5 \times 9.202 = 13.803$

$s = (11.301+13.803)/2 = 12.552$

Therefore $y_6 = 8.072 + 0.1 \times 12.552 = 9.327$

Thus, the sixth and final solution point is obtained as $(x_6, y_6) = (1.5, 9.327)$.

The complete solution of the given differential equations is given as

i	1	2	3	4	5	6
x_i	1.0	1.1	1.2	1.3	1.4	1.5
y_i	5.0	5.525	6.228	7.055	8.072	9.327

Algorithm 10.2: Runge-Kutta Second Order Method

To find the solution of ODE of type $dy/dx = f(x,y)$ when the solution at the initial point, $x = x_1$, is given, and the solution is desired in the interval $[x_1, x_f]$.

```
Begin
    read: x1, y1, xf
    read: h
    set x = x1
```

```
    set y = y1
    set i = 1
    write: i, x, y
    while ( x < xf ) do
        set s1 = f (x, y)
        set x = x + h
        set s2 = f(x, y+hs1)
        set y = y + h(s1+s2)/2
        set i = i + 1
        write: i, x, y
    endwhile
End.
```

10.8.2 Runge-Kutta Fourth Order Methods

The error in the second order Runge-Kutta methods is O(h^3) per step. However, if more precision is required, then we can use the fourth order Runge-Kutta methods in which the error is O(h^5) per step.

In Runge-Kutta fourth order methods, the slope at four points including the starting point is computed, and then the weighted average of these slopes is computed as

$$s = \frac{1}{6}(s_1 + 2s_2 + 2s_3 + s_4)$$

where $s_1 = f(x_1, y_1)$, $s_2 = f(x_1 + \frac{h}{2}, y_1 + \frac{h}{2}s_1)$

$$s_3 = f(x_1 + \frac{h}{2}, y_1 + \frac{h}{2}s_2),\ s_4 = f(x_1 + h, y_1 + hs_3)$$

The value of the dependent variable y is computed as

$$y_2 = y_1 + hs$$

In the similar manner, starting from the second solution point we compute the third point. This process is repeated till we find the solution in the desired interval.

In general, the $(i+1)^{th}$ point of the solution curve is obtained from the ith point using the following equation

$$y_{i+1} = y_i + hs \qquad \text{. . . (10.14)}$$

where $s = \frac{1}{6}(s_1 + 2s_2 + 2s_3 + s_4)$, $s_1 = f(x_i, y_i)$, $s_2 = f(x_i + \frac{h}{2}, y_i + \frac{h}{2}s_1)$,

$$s_3 = f(x_i + \frac{h}{2}, y_i + \frac{h}{2}s_2),\ s_4 = f(x_i + h, y_i + hs_3)$$

Geometric Interpretation

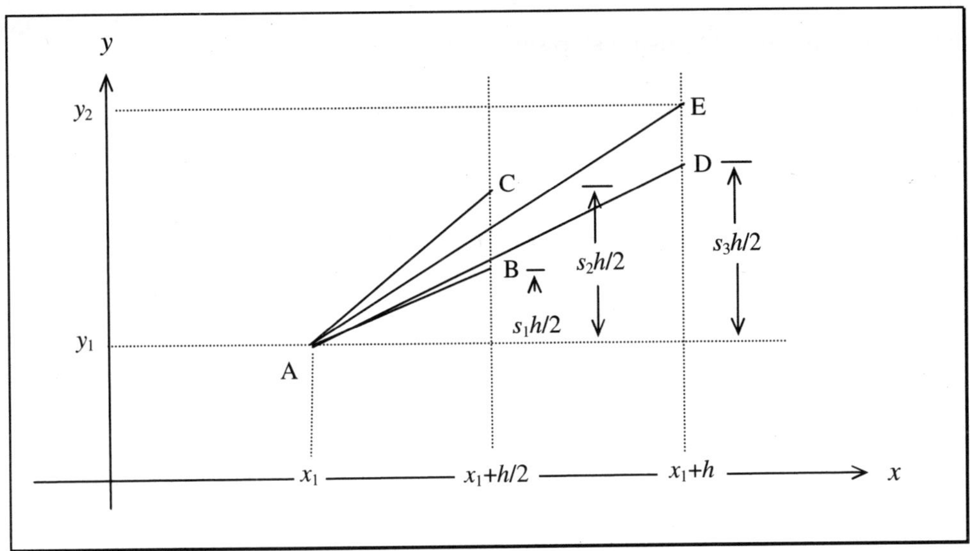

Figure 10.3: Illustration of fourth order Runge-Kutta method

From the starting point (x_1, y_1) labelled as A, draw a line with slope $s_1(=f(x_1,y_1))$. Let this line intersects the vertical line at $x_1+h/2$ at a point labelled B, whose co-ordinates are $(x_1+h/2,\ y_1+s_1h/2)$. Now compute the slope of the curve at point B. Let this slope be $s_2(=f(x_1+h/2,\ y_1+s_1h/2))$. Now draw a line from point A with slope s_2. Let this new line intersects the vertical line at $x_1+h/2$ at a point labelled C, whose co-ordinates are $(x_1+h/2,\ y_1+s_2h/2)$. Now compute the slope of the curve at point C. Let this slope be $s_3(=f(x_1+h/2,\ y_1+s_2h/2))$. Now draw a line from point A with slope s_3. Let this new line intersects the vertical line at x_1+h at a point labelled D, whose co-ordinates are $(x_1+h,\ y_1+s_3h)$. Now compute the slope of the curve at point D. Let this slope be $s_4(=f(x_1+h,\ y_1+s_3h))$. Now compute the weighted slope s as

$$s = \frac{1}{6}(s_1 + 2s_2 + 2s_3 + s_4)$$

Now draw a line from point A with slope s. Let this new line intersects the vertical line at x_1+h at a point labelled E, whose co-ordinates are $(x_1+h,\ y_1+sh)$. The point E is taken as the second solution point.

In the similar manner, starting from the second solution point, we obtain the third solution point and so on. This process goes on till we obtain the solution in the desired interval.

The formula for the fourth order Runge-Kutta method is identical to the Taylor series truncated after the term containing h^4. The proof is left as an exercise for the students.

Example 10.6: Given $\frac{dy}{dx} = xy$ with $y(1) = 5$. Find the solution correct to three decimal position in the interval $[1,1.5]$ using step size $h = 0.1$.

Solution: The formula for fourth order Runge-Kutta method is

$$y_{i+1} = y_i + hs$$

where $s = \frac{1}{6}(s_1 + 2s_2 + 2s_3 + s_4)$, $s_1 = f(x_i, y_i)$, $s_2 = f(x_i + \frac{h}{2}, y_i + \frac{h}{2}s_1)$

$$s_3 = f(x_i + \frac{h}{2}, y_i + \frac{h}{2}s_2), \; s_4 = f(x_i + h, y_i + hs_3)$$

In our example

$$f(x, y) = xy$$

$$x_1 = 1 \quad y_1 = 5 \quad h = 0.1$$

The Runge-Kutta fourth order formula can be written as

$$y_{i+1} = y_i + 0.1 \times s$$

For $i = 1$

$$y_2 = y_1 + 0.1 \times s$$

$$s_1 = f(x_1, y_1) = f(1, 5) = 1 \times 5 = 5$$

$$s_2 = f(x_1+0.05, y_1+0.05\times s_1) = f(1.05, 5+0.05\times5) = f(1.05, 5.25)$$
$$= 1.05 \times 5.25 = 5.513$$

$$s_3 = f(x_1+0.05, y_1+0.05\times s_2) = f(1.05, 5+0.05\times5.513) = f(1.05, 5.276)$$
$$= 1.05 \times 5.276 = 5.540$$

$$s_4 = f(x_1+0.1, y_1+0.1\times s_3) = f(1.1, 5+0.1\times5.540) = f(1.1, 5.554)$$
$$= 1.1 \times 5.554 = 6.109$$

$$s = \frac{1}{6}[5 + 2 \times 5.513 + 2 \times 5.540 + 6.109] = 5.536$$

Therefore $y_2 = 5 + 0.1 \times 5.536 = 5.554$

Thus, the second solution point is obtained as $(x_2, y_2) = (1.1, 5.554)$.

For $i = 2$

$$y_3 = y_2 + 0.1 \times s$$

$$s_1 = f(x_2, y_2) = f(1.1, 5.554) = 1.1 \times 5.554 = 6.109$$

$s_2 = f(x_2+0.05,\ y_2+0.05\times s_1) = f(1.15,\ 5.554+0.05\times 6.109) = f(1.15,\ 5.859)$
$= 1.15 \times 5.859 = 6.738$

$s_3 = f(x_2+0.05,\ y_2+0.05\times s_2) = f(1.15,\ 5.554+0.05\times 6.738) = f(1.15,\ 5.891)$
$= 1.15 \times 5.891 = 6.775$

$s_4 = f(x_2+0.1,\ y_2+0.1\times s_3) = f(1.2,\ 5.554+0.1\times 6.775) = f(1.2,\ 6.232)$
$= 1.2 \times 6.232 = 7.478$

$$s = \frac{1}{6}[6.109 + 2\times 6.738 + 2\times 6.775 + 7.478] = 6.769$$

Therefore $y_3 = 5.554 + 0.1 \times 6.769 = 6.231$

Thus, the third solution point is obtained as $(x_3, y_3) = (1.2, 6.231)$.

For $i = 3$

$y_4 = y_3 + 0.1 \times s$

$s_1 = f(x_3, y_3) = f(1.2, 6.231) = 1.2 \times 6.231 = 7.477$

$s_2 = f(x_3+0.05,\ y_3+0.05\times s_1) = f(1.25,\ 6.231+0.05\times 7.477) = f(1.25,\ 6.604)$
$= 1.25 \times 6.604 = 8.256$

$s_3 = f(x_3+0.05,\ y_3+0.05\times s_2) = f(1.25,\ 6.231+0.05\times 8.256) = f(1.25,\ 6.644)$
$= 1.25 \times 6.644 = 8.305$

$s_4 = f(x_3+0.1,\ y_3+0.1\times s_3) = f(1.3,\ 6.231+0.1\times 8.305) = f(1.3,\ 7.062)$
$= 1.3 \times 7.062 = 9.181$

$$s = \frac{1}{6}[7.477 + 2\times 8.256 + 2\times 8.305 + 9.181] = 8.297$$

Therefore $y_4 = 6.231 + 0.1 \times 8.297 = 7.061$

Thus, the fourth solution point is obtained as $(x_4, y_4) = (1.3, 7.061)$.

For $i = 4$

$y_5 = y_4 + 0.1 \times s$

$s_1 = f(x_4, y_4) = f(1.3, 7.061) = 1.3 \times 7.061 = 9.179$

$s_2 = f(x_4+0.05,\ y_4+0.05\times s_1) = f(1.35,\ 7.061+0.05\times 9.179) = f(1.35,\ 7.520)$
$= 1.35 \times 7.520 = 10.152$

$s_3 = f(x_4+0.05,\ y_4+0.05\times s_2) = f(1.35,\ 7.061+0.05\times 10.152) = f(1.35,\ 7.569)$
$= 1.35 \times 7.569 = 10.218$

$s_4 = f(x_4+0.1,\ y_4+0.1\times s_3) = f(1.4,\ 7.061+0.1\times 10.218) = f(1.4,\ 9.402)$
$= 1.4 \times 9.402 = 13.163$

$$s = \frac{1}{6}[9.179 + 2 \times 10.152 + 2 \times 10.218 + 14.305] = 10.514$$

Therefore $y_5 = 7.061 + 0.1 \times 10.514 = 8.112$

Thus, the fifth solution point is obtained as $(x_5, y_5) = (1.4, 8.112)$.

For $i = 5$

$y_6 = y_5 + 0.1 \times s$

$s_1 = f(x_5, y_5) = f(1.4, 8.112) = 1.4 \times 8.112 = 11.357$

$s_2 = f(x_5+0.05, y_5+0.05 \times s_1) = f(1.45, 8.112+0.05 \times 11.357) = f(1.45, 8.680)$
$= 1.45 \times 8.680 = 12.586$

$s_3 = f(x_5+0.05, y_5+0.05 \times s_2) = f(1.45, 8.112+0.05 \times 12.586) = f(1.45, 8.741)$
$= 1.45 \times 8.741 = 12.675$

$s_4 = f(x_5+0.1, y_5+0.1 \times s_3) = f(1.5, 8.112+0.1 \times 12.675) = f(1.5, 9.380)$
$= 1.5 \times 9.380 = 14.070$

$$s = \frac{1}{6}[11.357 + 2 \times 12.586 + 2 \times 12.675 + 14.070] = 12.658$$

Therefore $y_6 = 8.112 + 0.1 \times 12.658 = 9.378$

Thus, the sixth and final solution point is obtained as $(x_6, y_6) = (1.5, 9.378)$.

The complete solution of the given differential equations is given as

i	1	2	3	4	5	6
x_i	1.0	1.1	1.2	1.3	1.4	1.5
y_i	5.0	5.554	6.231	7.061	8.112	9.378

Algorithm 10.3: Runge-Kutta Fourth Order Method

To find the solution of ODE of type $dy/dx = f(x,y)$ when the solution at the initial point, $x = x_1$, is given, and the solution is desired in the interval $[x_1, x_f]$.

```
Begin
    read: x1, y1, xf
    read: h
    set x = x1
    set y = y1
    set i = 1
    write: i, x, y
    while ( x < xf ) do
       set s1 = f (x, y)
       set s2 = f (x +h/2, y +hs1/2)
```

```
        set s3 = f (x +h/2, y +hs2/2)
        set s4 = f (x +h, y +hs3)
        set s = (s1+2s2+2s3+s4)/6
        set y = y  + hs
        set x = x  + h
        set i = i  + 1
        write: i, x, y
    endwhile
End.
```

10.9 PREDICTOR CORRECTOR METHODS

The methods discussed so far are *single step methods, i.e.,* in order to extrapolate the solution curve they use only information available at the previous point. In contrast, the *multiple step methods* use the past information of the curve to extrapolate the solution curve. This class of methods is grouped under the name *predictor-corrector methods.* Some of the predictor corrector methods use the information about the solution curve at two previous points, some use at three points, still others use at more. But the only problem with predictor corrector methods, excluding *Modified Euler's method,* is that they are not self starting, *i.e.,* the values at first few points are computed using some other methods and then from there on these methods can take on.

10.9.1 Modified Euler's Method

This method is also known as *modified Euler's method.* As we have seen in the basic Euler's method, we use the slope at the starting point of the solution curve to determine the next point of the solution curve. This technique will work correctly only if the function is linear. The alternate approach is to use the average slope within the interval. This can be approximated by the mean of the slopes at both end points of the interval.

Suppose we use interval bounded by the point $x = x_i$ and $x = x_{i+1}$, then average slope is

$$\frac{y^1(x_i) + y^1(x_{i+1})}{2}$$

Then we have

$$y_{i+1} = y_i + \frac{h}{2}\left[y^1(x_i) + y^1(x_{i+1})\right]$$

$$\Rightarrow \quad y_{i+1} = y_i + \frac{h}{2}\left[f(x_i, y_i) + f(x_{i+1}, y_{i+1})\right] \qquad \ldots (10.15)$$

This is an improved estimate for y_{i+1} at x_{i+1}. But we are unable to apply Eq. (10.15) directly since we cannot evaluate $f(x_{i+1}, y_{i+1})$ as the value of y_{i+1} is unknown. This method works by estimating or *predicting* the value of y_{i+1} by basic **Euler's** method, Eq. (10.9).

Then it uses this value to compute $f(x_{i+1}, y_{i+1})$ giving an improved estimate (a corrected value) for y_{i+1}. Thus the value of y_{i+1} is predicted using the equation

$$y_{i+1}^{p} = y_i + hf(x_i, y_i) \qquad \ldots (10.16)$$

This equation is known as the *predictor formula.* The superscript p indicates the predicted value of y_{i+1}.

Using this predicted value of y i.e. y_{i+1}^{p}, a more accurate value of y_{i+1} is computed using the equation

$$y_{i+1}^{c} = y_i + \frac{h}{2}\left[f(x_i, y_i) + f(x_{i+1}, y_{i+1}^{p})\right] \qquad \ldots (10.17)$$

This equation is known as the *corrector formula.* The superscript c indicates the corrected value of y_{i+1}. The corrector formula can be used repeatedly to improve the accuracy of the corrected value of y_{i+1}.

These two equations i.e. Eq. (10.16) and Eq. (10.17), constitute the *Euler's predictor corrector method,* and is a two step method comprising of following steps:

Step 1. Predict y_{i+1} using the predictor formula.

Step 2. Correct y_{i+1} using the corrector formula.

If we need a higher precision, the step 2 can be executed repeatedly till the difference between two successive values of y_{i+1} agrees with some prescribed precision.

However, if more than two re-corrections are required, it is more efficient to reduce the step (interval) size or use a different method.

Example 10.7: Given $\dfrac{dy}{dx} = xy$ with $y(1) = 5$. Find the solution in the interval $[1, 1.5]$ using step size $h = 0.1$.

Solution: The formula for predictor corrector method is

$$y_{i+1}^{p} = y_i + hf(x_i, y_i)$$

$$y_{i+1}^{c} = y_i + \frac{h}{2}\left[f(x_i, y_i) + f(x_{i+1}, y_{i+1}^{p})\right]$$

For $i = 1$

Starting with initial conditions, we first predict value of y_2 using the predictor formula

$$y_2^{p} = y_1 + hf(x_1, y_1) = 5 + 0.1 \times f(1, 5) = 5 + 0.1 \times 1 \times 5 = 5.5$$

Using the predicted value of y_2, we make the correction on it using the corrector formula

$$y_2^c = y_1 + \frac{h}{2}\left[f(x_1, y_1) + f(x_2, y_2^p)\right]$$

$$= 5 + \frac{0.1}{2}\left[f(1,5) + f(1.1,5.5)\right] = 5 + \frac{0.1}{2}[1 \times 5 + 1.1 \times 5.5] = 5.553$$

Thus, the second solution point is obtained as $(x_2, y_2) = (1.1, 5.553)$.

For $i = 2$

Starting with second solution point $(x_2, y_2) = (1.1, 5.553)$, we first predict the value of y_3 using the predictor formula

$$y_3^p = y_2 + hf(x_2, y_2) = 5.553 + 0.1 \times f(1.1, 5.553)$$

$$= 5.553 + 0.1 \times 1.1 \times 5.553 = 6.164$$

Using the predicted value of y_3, we make the correction on it using the corrector formula

$$y_3^c = y_2 + \frac{h}{2}\left[f(x_2, y_2) + f(x_3, y_3^p)\right] = 5.553 + \frac{0.1}{2}\left[f(1.1,5.553) + f(1.2,6.164)\right]$$

$$= 5.553 + \frac{0.1}{2}[1.1 \times 5.553 + 1.2 \times 6.164] = 6.231$$

Thus, the third solution point is obtained as $(x_3, y_3) = (1.2, 6.231)$.

For $i = 3$

Starting with third solution point $(x_3, y_3) = (1.2, 6.231)$, we first predict the value of y_4 using the predictor formula

$$y_4^p = y_3 + hf(x_3, y_3) = 6.231 + 0.1 \times f(1.2, 6.231)$$

$$= 6.231 + 0.1 \times 1.2 \times 6.231 = 6.979$$

Using the predicted value of y_4, we make the correction on it using the corrector formula

$$y_4^c = y_3 + \frac{h}{2}\left[f(x_3, y_3) + f(x_4, y_4^p)\right] = 6.231 + \frac{0.1}{2}\left[f(1.2,6.231) + f(1.3,6.979)\right]$$

$$= 6.231 + \frac{0.1}{2}[1.2 \times 6.231 + 1.3 \times 6.979] = 7.059$$

Thus, the fourth solution point is obtained as $(x_4, y_4) = (1.3, 7.059)$.

For $i = 4$

Starting with fourth solution point $(x_4, y_4) = (1.3, 7.059)$, we first predict the value of y_5 using the predictor formula

$$y_5^p = y_4 + hf(x_4, y_4) = 7.059 + 0.1 \times f(1.3, 7.059)$$

$$= 7.059 + 0.1 \times 1.3 \times 7.059 = 7.977$$

Using the predicted value of y_5, we make the correction on it using the corrector formula

$$y_5^c = y_4 + \frac{h}{2}\left[f(x_4, y_4) + f(x_5, y_5^p)\right]$$
$$= 7.059 + \frac{0.1}{2}\left[f(1.3, 7.059) + f(1.4, 7.977)\right]$$
$$= 7.059 + \frac{0.1}{2}[1.3 \times 7.059 + 1.4 \times 7.977] = 8.076$$

Thus, the fifth solution point is obtained as $(x_5, y_5) = (1.4, 8.076)$.

For $i = 5$

Starting with fifth solution point $(x_5, y_5) = (1.4, 8.076)$, we first predict the value of y_6 using the predictor formula

$$y_6^p = y_5 + hf(x_5, y_5) = 8.076 + 0.1 \times f(1.4, 8.076)$$
$$= 8.076 + 0.1 \times 1.4 \times 8.076 = 9.207$$

Using the predicted value of y_6, we make the correction on it using the corrector formula

$$y_6^c = y_5 + \frac{h}{2}\left[f(x_5, y_5) + f(x_6, y_6^p)\right]$$
$$= 8.076 + \frac{0.1}{2}\left[f(1.4, 8.076) + f(1.5, 9.207)\right]$$
$$= 8.076 + \frac{0.1}{2}[1.4 \times 8.076 + 1.5 \times 9.207] = 9.332$$

Thus, the sixth and final solution point is obtained as $(x_6, y_6) = (1.5, 9.332)$.

The complete solution of the given differential equations is given as

i	1	2	3	4	5	6
x_i	1.0	1.1	1.2	1.3	1.4	1.5
y_i	5.0	5.553	6.231	7.059	8.076	9.332

Algorithm 10.3: Modified Euler's Method

To find the solution of ODE of type $dy/dx = f(x,y)$ when the solution at the initial point, $x = x_1$, is given, and the solution is desired in the interval $[x_1, x_f]$.

```
Begin
    read: x1, y1, xf
    read: h, epsilon
    set x = x1
    set y = y1
    set i = 1
    write: i, x, y
```

```
    while ( x < xf ) do
        set y2 = y + hf (x, y)
        do
            set y1 = y2
            set y2 = y + (h/2)[f (x, y) + f (x +h, y2)]
        while ( |(y2 - y1)/y2) | > epsilon )
        set y = y2
        set x = x +h
        set i = i  +1
        write: i, x, y
    endwhile
End.
```

10.9.2 Adams-Bashforth Method

The Adams-Bashforth method uses the information at past four solution points to extrapolate the solution at the next point. Therefore, in order to apply Adams-Bashforth method three more solution points, in addition to the starting solution point, are computed using any of the methods discussed in the preceding sections.

In order to derive the mathematical expression for Adams-Bashforth method, we rewrite the first order ordinary differential equation

$$\frac{dy}{dx} = f(x,y) \quad \text{as} \quad dy = f(x,y)dx$$

Integrating this equation between $x = x_k$ and $x = x_{k+1}$, we get

$$\int_{x_k}^{x_{k+1}} dy = \int_{x_k}^{x_{k+1}} f(x,y)dx \quad \Rightarrow \quad y_{k+1} = y_k + \int_{x_k}^{x_{k+1}} f(x,y)dx \qquad \ldots(10.18)$$

To solve the integral on the right, we approximate $f(x, y)$ as a polynomial in x. We derive this by making it a fit through the past points (x_{k-3}, y_{k-3}), (x_{k-2}, y_{k-2}), (x_{k-1}, y_{k-1}) and (x_k, y_k). Suppose we approximate it using Newton's backward difference interpolating polynomial as

$$f(x,y) = f_k + u\nabla f_k + \frac{u(u+1)}{2}\nabla^2 f_k + \frac{u(u+1)(u+2)}{6}\nabla^3 f_k \qquad \ldots(10.19)$$

where $u = \dfrac{x - x_k}{h}$, and ∇f_k, $\nabla^2 f_k$, $\nabla^3 f_k$ are backward differences at point $x = x_k$.

Therefore $y_{k+1} = y_k + \displaystyle\int_{x_k}^{x_{k+1}} \left[f_k + u\nabla f_k + \frac{u(u+1)}{2}\nabla^2 f_k + \frac{u(u+1)(u+2)}{6}\nabla^3 f_k \right] dx$

Since $u = \dfrac{x - x_k}{h}$, therefore $dx = hdu$

And at $x = x_k$ $u = 0$

at $x = x_{k+1}$ $u = 1$

Substituting these values, we get

$$y_{k+1} = y_k + h\int_0^1 \left[f_k + u\nabla f_k + \frac{u(u+1)}{2}\nabla^2 f_k + \frac{u(u+1)(u+2)}{6}\nabla^3 f_k \right] du$$

$$= y_k + h\left[f_k + \frac{1}{2}\nabla f_k + \frac{5}{12}\nabla^2 f_k + \frac{3}{8}\nabla^3 f_k \right]$$

Expressing the backward differences in terms of function values, and substituting in the above expression, we get

$$y_{k+1}^p = y_k + \frac{h}{24}[-9f_{k-3} + 37f_{k-2} - 59f_{k-1} + 55f_k] \qquad \text{... (10.20)}$$

This is called *Adams-Bashforth formula* and is used as a *predictor formula*. The superscript p indicates the predicted value of y_{k+1}.

The corrector formula is developed similarly. Now we construct a Newton's backward difference interpolating polynomial passing through the points (x_{k-2}, y_{k-2}), (x_{k-1}, y_{k-1}), (x_k, y_k) & (x_{k+1}, y_{k+1}^p) as

$$f(x, y) = f_{k+1} + u\nabla f_{k+1} + \frac{u(u+1)}{2}\nabla^2 f_{k+1} + \frac{u(u+1)(u+2)}{6}\nabla^3 f_{k+1} \qquad \text{... (10.21)}$$

where $u = \dfrac{x - x_{k+1}}{h}$, and ∇f_{k+1}, $\nabla^2 f_{k+1}$, $\nabla^3 f_{k+1}$ are backward differences at point $x = x_{k+1}$.

Now substituting $f(x, y)$ in Eq. (10.18) and integrating between $x = x_k$ and $x = x_{k+1}$, we get

$$y_{k+1} = y_k + \int_{x_k}^{x_{k+1}} \left[f_{k+1} + u\nabla f_{k+1} + \frac{u(u+1)}{2}\nabla^2 f_{k+1} + \frac{u(u+1)(u+2)}{6}\nabla^3 f_{k+1} \right] dx$$

Since $u = \dfrac{x - x_{k+1}}{h}$, therefore $dx = h du$

And

at $x = x_k$ $u = -1$

at $x = x_{k+1}$ $u = 0$

Substituting these values, we get

$$y_{k+1} = y_k + h\int_{-1}^{0}\left[f_{k+1} + u\nabla f_{k+1} + \frac{u(u+1)}{2}\nabla^2 f_{k+1} + \frac{u(u+1)(u+2)}{6}\nabla^3 f_{k+1}\right]du$$

$$= y_k + h\left[f_{k+1} - \frac{1}{2}\nabla f_{k+1} - \frac{1}{12}\nabla^2 f_{k+1} - \frac{1}{24}\nabla^3 f_{k+1}\right]$$

Expressing the backward differences in terms of function values, and substituting in the above expression, we get

$$y_{k+1}^c = y_k + \frac{h}{24}\left[f_{k-2} - 5f_{k-1} + 19f_k + 9f_{k+1}^p\right] \qquad \ldots (10.22)$$

This is called *Adams-Moulten formula* and is used as a *corrector formula.* The superscript c indicates the predicted value of y_{k+1}. This formula can be used repeatedly to improve the accuracy of the computed value of y_{k+1}.

The Eqs. (10.20) and (10.22) constitute the *Adams-Bashforth-Moulten predictor corrector method.* This method is popularly known as *Adams-Bashforth method.*

Example 10.8: Given $\frac{dy}{dx} = 1 + y^2$ with $y(0) = 0$, $y(0.2) = 0.2027$, $y(0.4) = 0.4228$, and $y(0.6) = 0.6841$. Compute $y(0.8)$.

Solution: Given $h = 0.2$ and $f(x, y) = 1 + y^2$. The function values are computed at the starting points. These values are tabulated along with the starting points in the following table.

i	x	y	$f(x, y) = 1 + y^2$
0	0.0	0.0000	1.0000
1	0.2	0.2027	1.0411
2	0.4	0.4228	1.1787
3	0.6	0.6841	1.4681

To obtain $y(0.8)$, we use the predictor formula (at $k = 3$) as

$$y_{0.8}^p = y_3 + \frac{h}{24}[-9f_0 + 37f_1 - 59f_2 + 55f_3]$$

$$= 0.6841 + \frac{0.2}{24}[-9\times 1.0 + 37\times 1.0411 - 59\times 1.1787 + 55\times 1.4681]$$

$$= 1.0233 \qquad \text{(Correct to four decimal places)}$$

Therefore

$$f_{0.8}^p = f(0.8, 1.0233) = 1 + (1.0233)^2 = 2.0471$$

To correct the predicted value of $y(0.8)$, we use the corrector formula as

$$y_{0.8}^c = y_3 + \frac{h}{24}\left[f_1 - 5f_2 + 19f_3 + 9f_{0.8}^p\right]$$

$$= 0.6841 + \frac{0.2}{24}[1.0411 - 5 \times 1.1787 + 19 \times 1.4681 + 9 \times 2.0471]$$

$$= 1.0296 \qquad \text{(Correct to four decimal places)}$$

The accuracy in the value of $y(0.8) = 1.0296$ can be improved by repeatedly using the corrector formula.

Algorithm 10.4: Adams-Bashforth method

To find the solution of ODE of type $dy/dx = f(x,y)$ when the solution at the initial point, $x = x_0$, is given, and the solution is desired in the interval $[x_0, x_f]$.

```
Begin
   read: x0, xf, y0
   read: n
   set h = (xf  - x0)/n
   set f0 = f (x0, y0)
   for k = 1 to 3 by 1 do  //compute three additional solution points
      set yk = yk-1 + hf (xk-1, yk-1)  // using Modified Euler's method
      set xk = xk-1 + h
      set yk = yk-1 + (h/2)[f (xk-1, yk-1)+ hf (xk, yk)]
      set fk = f (xk, yk)
   endfor
   for k = 3 to (n-1) by 1 do    //Adams-Bashforth method starts from here
      set y^p  = yk+(h/24)×[-9fk-3+37fk-2-59fk-1+55fk]
      set xk+1 = xk + h
      set f^p_k+1 = f (xk+1, y^p)
      set yk+1 = yk+(h/24)×[fk-2-5fk-1+19fk+9 f^p_k+1]
      set fk-3 = fk-2              // save previous function values
      set fk-2 = fk-1
      set fk-1 = fk
      set fk+1 = f (xk+1, yk+1)
   endfor
   write: "Solution of the differential equation is"
   for k = 1 to n by 1 do
      write: k, xk, yk
   endfor
End.
```

10.9.3 Milne's Method

The Milne's method, like Adams-Bashforth method, uses the information at past four solution points to extrapolate the solution at the next point. Therefore, in order to apply Milne's method three more solution points, in addition to the starting solution point, are computed using any of the methods discussed in the preceding sections.

To derive the expression for Milne's method, we rewrite the first order ordinary differential equation

$$\frac{dy}{dx} = f(x,y) \quad \text{as} \quad dy = f(x,y)dx$$

Integrating this equation between $x = x_{k-3}$ and $x = x_{k+1}$, we get

$$\int_{x_{k-3}}^{x_{k+1}} dy = \int_{x_{k-3}}^{x_{k+1}} f(x,y)dx \quad \Rightarrow \quad y_{k+1} = y_{k-3} + \int_{x_{k-3}}^{x_{k+1}} f(x,y)dx \qquad \text{. . . (10.23)}$$

To solve the integral on the right, we approximate $f(x, y)$ as a polynomial in x. We derive this by making it a fit through the past points (x_{k-3}, y_{k-3}), (x_{k-2}, y_{k-2}), (x_{k-1}, y_{k-1}) and (x_k, y_k). Suppose we approximate it using Newton's forward difference interpolating polynomial as

$$f(x,y) = f_{k-3} + u\Delta f_{k-3} + \frac{u(u-1)}{2}\Delta^2 f_{k-3} + \frac{u(u-1)(u-2)}{6}\Delta^3 f_{k-3} \qquad \text{. . . (10.24)}$$

where $u = \dfrac{x - x_{k-3}}{h}$, and Δf_{k-3}, $\Delta^2 f_{k-3}$, $\Delta^3 f_{k-3}$ are forward differences at point $x = x_{k-3}$.

Therefore

$$y_{k+1} = y_{k-3} + \int_{x_{k-3}}^{x_{k+1}} \left[f_{k-3} + u\nabla f_{k-3} + \frac{u(u-1)}{2}\Delta^2 f_{k-3} + \frac{u(u-1)(u-2)}{6}\Delta^3 f_{k-3} \right] dx$$

Since $u = \dfrac{x - x_{k-3}}{h}$, therefore $dx = hdu$

And at $x = x_{k-3}$ $\quad u = 0$ $\quad$ and $\quad$ at $x = x_{k+1}$ $\quad u = 4$.

Substituting these values, we get

$$y_{k+1} = y_{k-3} + h\int_0^4 \left[f_{k-3} + u\Delta f_{k-3} + \frac{u(u-1)}{2}\Delta^2 f_{k-3} + \frac{u(u-1)(u-2)}{6}\Delta^3 f_{k-3} \right] du$$

$$= y_{k-3} + h\left[4f_{k-3} + 8\Delta f_{k-3} + \frac{20}{3}\Delta^2 f_{k-3} + \frac{8}{3}\Delta^3 f_{k-3} \right]$$

Expressing the forward differences in terms of function values, and substituting in the above expression, we get

$$y_{k+1}^{p} = y_{k-3} + \frac{4h}{3}\left[2f_{k-2} - f_{k-1} + 2f_{k}\right] \qquad \ldots (10.25)$$

This is called *Milne's formula* and is used as a *predictor formula.* The superscript p indicates the predicted value of y_{k+1}.

Now to derive the corrector formula, we construct a Newton's forward difference interpolating polynomial passing through the points (x_{k-1}, y_{k-1}), (x_k, y_k) and (x_{k+1}, y_{k+1}^{p}) as

$$f(x,y) = f_{k-1} + u\Delta f_{k-1} + \frac{u(u-1)}{2}\Delta^2 f_{k-1} \qquad \ldots (10.26)$$

to evaluate the integral $\int_{x_{k-1}}^{x_{k+1}} f(x,y)dx$ in order to obtain the value of y_{k+1} as

$$y_{k+1} = y_{k-1} + \int_{x_{k-1}}^{x_{k+1}} f(x,y)dx$$

According to Simpson's 1/3rd rule for numerical integration

$$\int_{x_{k-1}}^{x_{k+1}} f(x,y)dx = \frac{h}{3}\left[f_{k-1} + 4f_k + f_{k+1}^{p}\right]$$

Therefore

$$y_{k+1}^{c} = y_{k-1} + \frac{h}{3}\left[f_{k-1} + 4f_k + f_{k+1}^{p}\right] \qquad \ldots (10.27)$$

This is called *Simpson's formula* and is used as a *corrector formula.* The superscript 'c' indicates the predicted value of y_{k+1}. This formula can be used repeatedly to improve the accuracy of the computed value of y_{k+1}.

The Eqs. (10.25) and (10.27) constitute the *Milne-Simpson predictor corrector method.* This method is popularly known as *Milne's method.*

Example 10.9: Given $\frac{dy}{dx} = 1 + y^2$ with $y(0) = 0$, $y(0.2) = 0.2027$, $y(0.4) = 0.4228$, and $y(0.6) = 0.6841$. Compute $y(0.8)$.

Solution: Given h = 0.2 and $f(x, y) = 1 + y^2$. The function values are computed at the starting points. These values are tabulated along with the starting points in the following table.

i	x	y	$f(x, y) = 1 + y^2$
0	0.0	0.0000	1.0000
1	0.2	0.2027	1.0411
2	0.4	0.4228	1.1787
3	0.6	0.6841	1.4681

To obtain $y(0.8)$, we use the predictor formula as

$$y^{P}_{0.8} = y_0 + \frac{4h}{3}[2f_1 - f_2 + 2f_3]$$

$$= 0.0 + \frac{4 \times 0.2}{3}[2 \times 1.0411 - 1.1787 + 2 \times 1.4681]$$

$$= 1.0239 \qquad \text{(Correct to four decimal places)}$$

Therefore

$$f^{P}_{0.8} = f(0.8, 1.0239) = 1 + (1.0239)^2 = 2.0484$$

To correct the predicted value of $y(0.8)$, we use the corrector formula as

$$y^{c}_{0.8} = y_2 + \frac{h}{3}[f_2 + 4f_3 + f^{P}_{0.8}]$$

$$= 0.4228 + \frac{0.2}{3}[1.1787 + 4 \times 1.4681 + 2.0484]$$

$$= 1.0294 \qquad \text{(Correct to four decimal places)}$$

The accuracy in the value of $y(0.8)$ = 1.0296 can be improved by repeatedly using the corrector formula.

Algorithm 10.6: Milne's method

To find the solution of ODE of type $dy/dx = f(x,y)$ when the solution at the initial point, $x = x_0$, is given, and the solution is desired in the interval $[x_0, x_f]$.

```
Begin
    read: x0, xf, y0
    read: n
    set h = (xf - x0)/n
    set f0 = f (x0, y0)
    for k = 1 to 3 by 1 do   //compute three additional solution points
        set yk = yk-1 + hf (xk-1, yk-1)   // using Modified Euler's method
        set xk = xk-1 + h
        set yk = yk-1 + (h/2)[f (xk-1, yk-1)+ hf (xk, yk)]
```

```
        set fk = f (xk, yk)
    endfor
    for k = 3 to (n -1) by 1 do    // Milne's method starts from here
        set y^p = yk-3+(4h/3)×[2fk-2 - fk-1 + 2fk]
        set xk+1 = xk + h
        set f^p_k+1 = f (xk+1, y^p)
        set yk+1 = yk-1+(h/3)×[fk-1 + 4fk + f^p_k+1]
        set fk-2 = fk-1                 // save previous function values
        set fk-1 = fk
        set fk+1 = f (xk+1, yk+1)
    endfor
    write: "Solution of the differential equation is"
    for k = 1 to n by 1 do
        write: k, xk, yk
    endfor
End.
```

10.10 SYSTEM OF DIFFERENTIAL EQUATIONS

Any of the previously discussed methods can be used to solve a system of differential equations. To illustrate this, let us consider a system of differential equations of type

$$\frac{dy}{dx} = f(x,y,z) \qquad \frac{dz}{dx} = g(x,y,z) \qquad \ldots (10.28)$$

with initial conditions

$$y(x_0) = y_0 \text{ and } z(x_0) = z_0.$$

The solution of this system is given by

(i) Euler's method as

$$y_{i+1} = y_i + hf(x_i, y_i, z_i) \qquad z_{i+1} = z_i + hg(x_i, y_i, z_i) \qquad \ldots (10.29)$$

(ii) Runge-Kutta second order (Heun's) method as

$$y_{i+1} = y_i + \frac{h}{2}(s_1 + s_2) \qquad z_{i+1} = z_i + \frac{h}{2}(p_1 + p_2) \qquad \ldots (10.30)$$

where

$$s_1 = f(x_i, y_i, z_i) \qquad s_2 = f(x_i + h, y_i + s_i h, z_i + p_i h)$$

$$p_1 = g(x_i, y_i, z_i) \qquad p_2 = g(x_i + h, y_i + s_i h, z_i + p_i h)$$

(iii) Runge-Kutta fourth order method as

$$\left.\begin{aligned} y_{i+1} &= y_i + \frac{h}{6}(s_1 + 2s_2 + 2s_3 + s_4) \\ z_{i+1} &= z_i + \frac{h}{6}(p_1 + 2p_2 + 2p_3 + p_4) \end{aligned}\right\} \quad \ldots (10.31)$$

where

$$s_1 = f(x_i, y_i, z_i) \qquad s_2 = f(x_i + \frac{h}{2}, y_i + \frac{h}{2}s_1, z_i + \frac{h}{2}p_1)$$

$$s_3 = f(x_i + \frac{h}{2}, y_i + \frac{h}{2}s_2, z_i + \frac{h}{2}p_2) \qquad s_4 = f(x_i + h, y_i + hs_3, z_i + hp_3)$$

$$p_1 = g(x_i, y_i, z_i) \qquad p_2 = g(x_i + \frac{h}{2}, y_i + \frac{h}{2}s_1, z_i + \frac{h}{2}p_1)$$

$$p_3 = g(x_i + \frac{h}{2}, y_i + \frac{h}{2}s_2, z_i + \frac{h}{2}p_2) \qquad p_4 = g(x_i + h, y_i + hs_3, z_i + hp_3)$$

Example 10.10: Given $\frac{dy}{dx} = x - z^2$ and $\frac{dz}{dx} = x + y$ with initial conditions as $y(0) = 1$ and $z(0) = 2$. Find the solution of the above system of equations in the interval [0,0.4] with interval size h=0.1 using second order Runge-Kutta (Heun's) method.

Solution: For second order Runge-Kutta (Heun's) method, we have

$$y_{i+1} = y_i + \frac{h}{2}(s_1 + s_2) \qquad z_{i+1} = z_i + \frac{h}{2}(p_1 + p_2)$$

where

$$s_1 = f(x_i, y_i, z_i) \qquad s_2 = f(x_i + h, y_i + s_i h, z_i + p_i h)$$

$$p_1 = g(x_i, y_i, z_i) \qquad p_2 = g(x_i + h, y_i + s_i h, z_i + p_i h)$$

For $i = 1$

$s_1 = f(x_1, y_1, z_1) = f(0, 1, 2) = 0 - 2^2 = -4$

$p_1 = g(0, 1, 2) = 0 + 1 = 1.0$

$s_2 = f(0.1, 1{+}0.1{\times}(-4), 2{+}0.1{\times}1) = f(0.1, 0.6, 2.1)$

$= 0.1 - (2.1)^2 = -4.31$

$p_2 = g(0.1, 0.6, 2.1) = 0.1 + 0.6 = 0.7$

Thus

$y_2 = 1 + 0.1 \times (-4 - 4.31) / 2 = 0.585$

$z_2 = 2 + 0.1 \times (1.0 + 0.7) / 2 = 2.085$

Thus, we get the second solution point as $(x_2, y_2, z_2) = (0.1, 0.585, 2.085)$.

For $i = 2$

Now, starting with the point $(x_2, y_2, z_2) = (0.1, 0.585, 2.085)$, we have

$$s_1 = f(x_2, y_2, z_2) = f(0.1, 0.585, 2.085) = 0.1 - (2.085)^2 = -4.247$$

$$p_1 = g(0.1, 0.585, 2.085) = 0.1 + 0.585 = 0.685$$

$$s_2 = f(0.2, 0.585+0.1\times(-4.247), 2.085+0.1\times0.685) = f(0.2, 0.160, 2.154)$$

$$= 0.2 - (2.154)^2 = -4.440$$

$$p_2 = g(0.2, 0.160, 2.154) = 0.2 + 0.160 = 0.360$$

Thus

$$y_3 = 0.585 + 0.1 \times (-4.247 - 4.440) / 2 = 0.151$$

$$z_3 = 2.085 + 0.1 \times (0.685 + 0.360) / 2 = 2.137$$

Thus, we get the third solution point as $(x_3, y_3, z_3) = (0.2, 0.151, 2.137)$.

For $i = 3$

Now, starting with the point $(x_3, y_3, z_3) = (0.2, 0.151, 2.137)$, we have

$$s_1 = f(x_3, y_3, z_3) = f(0.2, 0.151, 2.137) = 0.2 - (2.137)^2 = -4.367$$

$$p_1 = g(0.2, 0.151, 2.137) = 0.2 + 0.151 = 0.351$$

$$s_2 = f(0.3, 0.151+0.1\times(-4.367), 2.137+0.1\times0.351) = f(0.3, -0.286, 2.172)$$

$$= 0.1 - (2.172)^2 = -4.418$$

$$p_2 = g(0.3, -0.286, 2.172) = 0.3 - 0.286 = 0.140$$

Thus

$$y_4 = 0.151 + 0.1 \times (-4.367 - 4.418) / 2 = -0.288$$

$$z_4 = 2.137 + 0.1 \times (0.351 + 0.140) / 2 = 2.162$$

Thus, we get the fourth solution point as $(x_4, y_4, z_4) = (0.3, -0.288, 2.162)$.

For $i = 4$

Now, starting with the point $(x_4, y_4, z_4) = (0.3, -0.288, 2.162)$, we have

$$s_1 = f(x_4, y_4, z_4) = f(0.3, -0.288, 2.162) = 0.3 - (2.162)^2 = -4.374$$

$$p_1 = g(0.3, -0.288, 2.162) = 0.3 - 0.288 = 0.012$$

$$s_2 = f(0.4, -0.288+0.1\times(-4.374), 2.162+0.1\times0.012) = f(0.4, -0.725, 2.163)$$

$$= 0.4 - (2.163)^2 = -4.279$$

$$p_2 = g(0.4, -0.725, 2.163) = 0.4 - 0.725 = -0.325$$

Thus

$$y_5 = -0.288 + 0.1 \times (-4.374 - 4.279) / 2 = -0.721$$

$$z_5 = 2.162 + 0.1 \times (0.012 - 0.325) / 2 = 2.146$$

Thus, we get the fifth and final solution point as $(x_5, y_5, z_5) = (0.4, -0.721, 2.146)$.

The complete solution of the given differential equations is given as

i	1	2	3	4	5
x_i	0.0	0.1	0.2	0.3	0.4
y_i	1.0	0.585	0.151	–0.288	–0.721
z_i	2.0	2.085	2.137	2.162	2.146

10.11 HIGHER ORDER DIFFERENTIAL EQUATIONS

Higher order differential equations can be reduced to a system of first order differential equations. To illustrate this, consider the following second order differential equation

$$\frac{d^2y}{dx^2} = f\left(x, y, \frac{dy}{dx}\right) \qquad \ldots (10.32)$$

with initial conditions

$$y(x_1) = y_1 \text{ and } \left(\frac{dy}{dx}\right)_{x_1} = y_1^1$$

If we substitute $z = \frac{dy}{dx}$ then $\frac{dz}{dx} = \frac{d^2y}{dx^2}$

The Eq. (10.32) can be expressed as a following system of first order differential equations

$$\frac{dz}{dx} = g(x, y, z) \qquad \ldots (10.33)$$

with initial conditions

$$y(x_1) = y_1 \text{ and } z(x_1) = y_1^1.$$

In order to solve it, any of the methods described in the preceding section can be used.

Example 10.11: Given $\frac{d^2y}{dx^2} + 2\frac{dy}{dx} + y = 0$ with initial conditions $y(0) = 0$, and $\left.\frac{dy}{dx}\right|_{x=0} = 1$. Find the solution of the above system of equations in the interval [0,0.4] with interval size h=0.1 using second order Runge-Kutta (Heun's) method.

Solution: Substituting $\frac{dy}{dx} = z$, the problem of solving the given second order differential equation reduces to the problem of solving following simultaneous equations

$$\frac{dy}{dx} = z = f(x, y, z) \qquad \frac{dz}{dx} = -y - 2z = g(x, y, z)$$

with initial conditions

$y(0) = 0$ $\qquad z(0) = 1$

For second order Runge-Kutta (Heun's) method, we have

$$y_{i+1} = y_i + \frac{h}{2}(s_1 + s_2) \qquad z_{i+1} = z_i + \frac{h}{2}(p_1 + p_2)$$

where

$$s_1 = f(x_i, y_i, z_i) \qquad s_2 = f(x_i + h, y_i + s_i h, z_i + p_i h)$$

$$p_1 = g(x_i, y_i, z_i) \qquad p_2 = g(x_i + h, y_i + s_i h, z_i + p_i h)$$

In our problem

$s_1 = z_I$ $\qquad p_1 = -y_i - 2z_i$

$s_2 = z_i + hp_1$ $\qquad p_2 = -y_i - 2z_i + 2hy_i + 3hz_i$

For *i* = 1

Starting with $x_1 = 0$, $y_1 = 0$ and $z_1 = 1$ i.e. initial solution point of the second order differential equation as $(x_1, y_1) = (0, 0)$,, we get

$s_1 = 1$

$p_1 = -0 - 2 \times 1 = -2$

$s_2 = 1 + 0.1 \times (-2) = 0.8$

$p_2 = -0 - 2 \times 1 + 2 \times 0.1 \times 0 + 3 \times 0.1 \times 1 = -1.7$

Therefore

$y_2 = 0 + 0.1 \times (1.0 + 0.8)/2 = 0.09$

$z_2 = 1 + 0.1 \times (-2 - 1.7)/2 = 0.815$

Thus, we get the second solution point as $(x_2, y_2) = (0.1, 0.09)$.

For *i* = 2

Staring with the second solution point $(x_2, y_2) = (0.1, 0.09)$, we get

$s_1 = 0.815$

$p_1 = -0.09 - 2 \times 0.815 = -1.72$

$s_2 = 0.815 + 0.1 \times (-1.72) = 0.643$

$p_2 = -0.09 - 2 \times 0.815 + 2 \times 0.1 \times 0.09 + 3 \times 0.1 \times 0.815 = -1.458$

Therefore

$y_3 = 0.09 + 0.1 \times (0.815 + 0.643)/2 = 0.163$

$z_3 = 0.815 + 0.1 \times (-1.72 - 1.458)/2 = 0.656$

Thus, we get the third solution point as $(x_3, y_3) = (0.2, 0.163)$.

For $i = 3$

Staring with the second solution point $(x_3, y_3) = (0.2, 0.163)$, we get

$$s_1 = 0.656$$
$$p_1 = -0.163 - 2 \times 0.656 = -1.475$$
$$s_2 = 0.656 + 0.1 \times (-1.475) = 0.509$$
$$p_2 = -0.163 - 2 \times 0.656 + 2 \times 0.1 \times 0.169 + 3 \times 0.1 \times 0.656 = -1.244$$

Therefore

$$y_4 = 0.163 + 0.1 \times (0.656 + 0.509)/2 = 0.221$$
$$z_4 = 0.656 + 0.1 \times (-1.475 - 1.244)/2 = 0.520$$

Thus, we get the fourth solution point as $(x_4, y_4) = (0.3, 0.221)$.

For $i = 4$

Staring with the second solution point$(x_4, y_4) = (0.3, 0.221)$, we get

$$s_1 = 0.520$$
$$p_1 = -0.221 - 2 \times 0.520 = -1.261$$
$$s_2 = 0.520 + 0.1 \times (-1.261) = 0.394$$
$$p_2 = -0.221 - 2 \times 0.520 + 2 \times 0.1 \times 0.221 + 3 \times 0.1 \times 0.520 = -1.061$$

Therefore

$$y_5 = 0.221 + 0.1 \times (0.520 + 0.394)/2 = 0.267$$
$$z_5 = 0.520 + 0.1 \times (-1.261 - 1.061)/2 = 0.404$$

Thus, we get the fifth and final solution point as $(x_5, y_5) = (0.4, 0.267)$.

The complete solution of the given differential equations is as given in the following table.

i	1	2	3	4	5
x_i	0.0	0.1	0.2	0.3	0.4
y_i	0.0	0.09	0.163	0.221	0.267

EXERCISE . . .

1. Explain the following terms with suitable examples
   (*a*) Differential equation
   (*b*) Solution of a differential equation
   (*c*) Order of a differential equation
   (*d*) Degree of a differential equation
2. Differentiate between the following:
   (*a*) Single step and multiple step methods
   (*b*) Initial value and boundary value problems

(*c*) Ordinary and partial derivatives
(*d*) Ordinary and partial differential equations

3. Using Picard's method, obtain the solution, correct to four decimal digits, of $\frac{dy}{dx} = x - y^2$ for $x = 0.1$. Given that $y = 1$ when $x = 0$.
4. Find the solution of the following differential equation $\frac{dy}{dx} = 3x + y$ using Euler's method for $0.1 \le x \le 0.5$. Given that $y = 0$ when $x = 0$, and $h = 0.1$.
5. Find the solution of the following differential equation $\frac{dy}{dx} = x^2 + y$ using Runge-Kutta second order method for x = 0.1, 0.2, 03, and 0.4. Given that y=1 when x =0.
6. Using Runge-Kutta fourth order method, find the solution of the following differential equation $\frac{dy}{dx} = x + y$ for x = 0.1, 0.2, and 0.3. Given that $y = 0$ when $x = 0$.
7. Using Modified Euler's method, find the solution of the following differential equation $\frac{dy}{dx} = x + y^2$ for x = 1.1, 1.2, and 1.3. Given that $y = 1$ when $x = 1$.
8. Using Adams-Bashforth method, find the solution of the following differential equation $\frac{dy}{dx} = xy^2 + 1$ with $y(0) = 1$ for $x = 0.4$, when it is given that

x:	0.1	0.2	0.3
y:	1.105	1.223	1.355

9. Using Milne's method, find the solution of the following differential equation $\frac{dy}{dx} = x - y^2$ with $y(0) = 0$ for $x = 0.8$ when it is given that

x:	0.2	0.4	0.6
y:	0.0200	0.0795	0.1762

10. Solve the following simultaneous differential equations using the second order Runge-Kutta method $\frac{dy}{dx} = -x - yz$ and $\frac{dz}{dx} = -y - zx$ with initial conditions $y(0) = 0$, $z(0) = 1.0$ for x = 0.1, 0.2, 0.3
11. Find the solution of the following differential equation using the second order Runge-Kutta method $\frac{d^2y}{dx^2} - 2\frac{dy}{dx} - y = 2e^x$ with $y(3) = 1$ for x = 3.1, 3.2, 3.3, 3.4
13. Modify algorithms 10.5 and 10.6, so that these algorithms produce the numerical solution accurate to the prescribed accuracy.

❑ ❑ ❑

Appendix A

Listing of C Programs

Bisection Method . . .

```
/*  Program to implement Bisection method to find the
    root of equation x^3 - 4x - 9 = 0       */
#include<stdio.h>
#include<stdlib.h>
#include<math.h>
double f( double x ) {
   return pow((double)x,(double)3.0)-4*x-9;
}
void main() {
   float x1, x2, epsilon, x3;
   printf("Enter first point of the search interval : ");
   scanf("%f", &x1);
   printf("Enter second point of the search interval : ");
   scanf("%f", &x2);
   if ( ( f(x1) * f(x2) ) > 0 ) {
      printf("\nInitial approximations are unsuitable\n");
      exit(1);
   }
   printf("Enter prescribed tolerance : ");
   scanf("%f", &epsilon);
   do {
      x3 = (x1+x2)/2;
      if ( f(x1) * f(x3) < 0 )
         x2 = x3;
      else
         x1 = x3;
   }
   while( fabs((double)(x1-x2)) > epsilon);
   printf("\nApproximate root = %8.4f\n", x3);
}
```

False Position Method . . .

```
/*  Program to implement False position method to find root
    equation x^3 - 2x - 5 = 0         */
#include<stdio.h>
#include<stdlib.h>
#include<math.h>
double f( double x ) {
   return pow((double)x,(double)3.0) - 2 * x - 5.0;
}
void main() {
   float x1, x2, epsilon, x3 = 0, x3p, f1, f2, f3, delta;
   printf("Enter first point of the search interval : ");
   scanf("%f", &x1);
   printf("Enter second point of the search interval : ");
   scanf("%f", &x2);
   f1 = f(x1);
   f2 = f(x2);
   if ( f1 * f2 > 0 ) {
      printf("\nInitial approximations are unsuitable\n");
      exit (1);
   }
   printf("Enter prescribed tolerance : ");
   scanf("%f", &epsilon);
   printf("Enter permissible lower limit on slope : ");
   scanf("%f", &delta);
   do {
      if ( fabs ( f2 - f1 ) <= delta ) {
         printf("\nSlope of function becomes too small\n");
         exit (1);
      }
      x3p = x3;
      x3 = (x1*f2 - x2*f1) / (f2 - f1);
      f3 = f(x3);
      if ( f1 * f3 < 0 ) {
         x2 = x3;
         f2 = f3;
      } else {
         x1 = x3;
         f1 = f3;
      }
   }
   while ( fabs((double)((x3-x3p)/x3) ) > epsilon );
   printf("\nApproximate root = %.4f\n", x3);
}
```

Secant Method . . .

```
/* Program to implement Secant method to find root
   equation cos(x) - xe^x = 0        */
#include<stdio.h>
#include<stdlib.h>
#include<math.h>
double f( double x ) {
   return cos((double)x) - x * exp ((double)x);
}
void main() {
    int n, i;
    float x1, x2, x3 = 0, x3p, f1, f2, f3, epsilon, delta;
    printf("\nEnter first approximation : ");
    scanf("%f", &x1);
    printf("Enter second approximation : ");
    scanf("%f", &x2);
    printf("Enter prescribed tolerance : ");
    scanf("%f", &epsilon);
    printf("Enter lower limit on slope of function : ");
    scanf("%f", &delta);
    printf("Enter number of iterations permitted : ");
    scanf("%d", &n);
    f1 = f(x1);   f2 = f(x2);
    for ( i = 1; i <= n; i++ ) {
        if ( fabs(f2 - f1) <= delta ) {
            printf("Slope of the function becomes too small");
            exit(1);
        }
        x3p = x3;
        x3 = (x1*f2 - x2*f1) / (f2 - f1);
        f3 = f(x3);
        if ( fabs((double)((x3-x3p)/x3) ) <= epsilon ) {
            printf("\nApproximate root = %8.4f\n", x3);
            exit(1);
        }
        x1 = x2;  f1 = f2;
        x2 = x3;  f2 = f3;
    }
    printf("Solution does not converge in %d iterations", n);
}
```

Newton-Raphson Method . . .

```
/*  Program to implement Newton-Raphson method to find root
    of equation x^3- 4x - 9 = 0          */
#include<stdio.h>
#include<stdlib.h>
#include<math.h>
double f( double x ) {
```

```
   return x * x * x - 4.0 * x - 9.0;
}
double df( double x ) {
   return 3.0 * x * x - 4.0;
}
void main() {
   float x0, x1, epsilon, delta, relativeerror;
   int i, n;
   printf("Enter initial approximation : ");
   scanf("%f", &x0);
   printf("Enter number of iterations permitted : ");
   scanf("%d", &n);
   printf("Enter prescribed tolerance : ");
   scanf("%f", &epsilon);
   printf("Enter lower bound on slope : ");
   scanf("%f", &delta);
   for ( i = 1; i <= n; i++) {
      if ( fabs (df(x0)) < delta ) {
         printf("\nslope too small\n");
         exit(1);
      }
      x1 = x0 - f(x0) / df(x0);
      relativeerror = fabs ( ( x1 - x0 ) / x1 );
      x0 = x1;
      if ( relativeerror <= epsilon ) {
         printf("\nApproximate root = %.4f", x1 );
         exit(1);
      }
   }
   printf("\nDoes not converge in %d iterations\n", n);
}
```

Successive Approximation Method . . .

```
/*  Program to implement method of successive approximations
    to find root of equation 2^x - x - 3 = 0      */
#include<stdio.h>
#include<stdlib.h>
#include<math.h>
double g( double x ) {
   return pow((double)2.0,(double)x) - 3;
}
void main() {
   float x1, x2, epsilon, relativeerror;
   int i, n;
   printf("Enter intial approximation : ");
   scanf("%f", &x1);
   printf("Enter number of iterations permitted : ");
   scanf("%d", &n);
   printf("Enter prescribed tolerance : ");
```

```
    scanf("%f", &epsilon);
    for ( i = 1; i <= n; i++ ) {
        x2 = g(x1);
        relativeerror = fabs ( ( x2 - x1 ) / x2 );
        x1 = x2;
        if ( relativeerror <= epsilon ) {
            printf("\nRoot converges in %d iterations", i);
            printf("\nApproximate root is %.4f", x2);
            exit(1);
        }
    }
    printf("\nRoot does not converge in %d iterations\n", n);
}
```

Bairstow's Method . . .

```
/*  Program to implement Bairstow's method to find roots of
    polynomial of type x^n+a(1)*x^(n-1)+a(2)*x^(n-2)+...+a(n)      */
#include<stdio.h>
#include<stdlib.h>
#include<math.h>
#define MAXDEGREE  20
/*  function prototypes    */
void quad(float u,float v,float *r1,float *r2,char *type );
void printroots( int *it, float r1, float r2, char type );
void main() {
   float a[MAXDEGREE],b[MAXDEGREE],c[MAXDEGREE],r1,r2,u,v;
   float du,dv,epsilon,denom;
   int n, it, i, k;
   char type;
   printf("Enter order of polynomial : ");
   scanf("%d", &n);
   for ( i = 1; i <= n; i++ ) {
       printf("Enter coefficient of x^%d : ", n-i);
       scanf("%f", &a[i] );
   }
   printf("\nRequired precision : ");
   scanf("%f", &epsilon);
   printf("\n\nRoot#     Real part    Imaginary part");
   it = 0;
   while ( n > 2 ) {
      u = v = 0.0;
      do {
         b[1] = a[1] - u;
         b[2] = a[2] - b[1] * u - v;
         for ( k = 3; k <= n; k++ ) {
             b[k] = a[k] - b[k-1] * u - b[k-2] * v;
         }
         c[1] = b[1] - u;
         c[2] = b[2] - c[1] * u - v;
```

```
            for ( k = 3; k < n; k++ ) {
                c[k] = b[k] - c[k-1] * u - c[k-2] * v;
            }
            if ( n == 3 ) {
               denom = c[n-1] - c[n-2] * c[n-2];
               if ( denom == 0 ) {
                  printf("\nzero denominator\n");
                  exit(1);
               }
               du = ( b[n] - b[n-1] * c[n-2] ) / denom;
            } else {
               denom = c[n-1] * c[n-3] - c[n-2] * c[n-2];
               if ( denom == 0 ) {
                  printf("\nzero denominator\n");
                  exit(1);
               }
               du = ( b[n] * c[n-3] - b[n-1] * c[n-2] ) / denom;
            }
            dv = ( b[n-1] * c[n-1] - b[n] * c[n-2] ) / denom;
            u = u + du;
            v = v + dv;
         } while((fabs(du)+fabs(dv)) > epsilon);
         quad ( u, v, &r1, &r2, &type );
         it++;
         printroots ( &it, r1, r2, type );
         n = n - 2;
         for ( i = 1; i <= n; i++ ) {
             a[i] = b[i];
         }
      }
      if ( n == 1 ) {
         it++;
         r1 =  -a[1];
         printf("\n  %d\t%8.4f\n", it, r1);
      } else if ( n == 2 ) {
         u = a[1];
         v = a[2];
         it++;
         quad(u, v, &r1, &r2, &type);
         printroots(&it, r1, r2, type);
      }
   }
   /*  function to solve quadratic factor of type x^2+ux+v    */
   void quad(float u,float v,float *r1,float *r2,char *type ) {
      double d;
      d = u * u - 4.0 * v;
      if ( d > 0 ) {
         *type = 'r';
         d = sqrt ( (double) d );
         *r1 = ( - u + d ) / 2.0;
         *r2 = ( - u - d ) / 2.0;
```

```
    } else if ( d < 0 ) {
        *type = 'i';
        *r1 = -u / 2.0;
        *r2 = sqrt ( - (double) d ) / 2.0;
    } else {
        *type = 'r';
        *r1 = - u / 2.0;  *r2 = *r1;
    }
}
/*   function to print the roots    */
void printroots( int *it, float r1, float r2, char type )
{
    if ( type == 'r' ) {
        printf("\n  %d\t%8.4f", *it, r1);
        (*it)++;
        printf("\n  %d\t%8.4f", *it, r2);
    } else {
        printf("\n  %d\t%8.4f\t%8.4f", *it, r1, r2);
        (*it)++;
        printf("\n  %d\t%8.4f\t%8.4f", *it, r1, -r2);
    }
}
```

Muller's Method . . .

```
/* Program to implement Muller's method to find the
   root of equation 3x + sin(x) - e^x = 0          */
#include<stdio.h>
#include<stdlib.h>
#include<math.h>
double f ( double x ) {
    return 3*x+sin((double)x)-exp((double)x);
}
void main() {
    float a1, b1, x0, x1, x2, xrp, xrn, f0, f1, f2, epsilon;
    float h1, h2, rel_error, gama, a, b, c, disc;
    clrscr();
    printf("Enter first point of the search interval : ");
    scanf("%f", &a1);
    printf("Enter second point of the search interval : ");
    scanf("%f", &b1);
    printf("Enter prescribed tolerance : ");
    scanf("%f", &epsilon);
    x2 = a1;
    x0 = (a1+b1)/2;
    x1 = b1;
    xrp = 0.0;
    do {
        h1 = x1 - x0;
        h2 = x0 - x2;
```

```
        gama = h2/h1;
        f0 = f(x0);
        f1 = f(x1);
        f2 = f(x2);
        c = f0;
        a = (gama*f1-f0*(1.0+gama)+f2)/(gama*h1*h1*(1.0+gama));
        b = (f1-f0-a*h1*h1)/h1;
        disc = sqrt( (double)b*b-4*a*c );
        if ( b > 0 ) {
           xrn = x0-2.0*c/(b+disc);
        }
        else {
           xrn = x0-2.0*c/(b-disc);
        }
        rel_error = fabs((xrp-xrn)/xrn);
        xrp = xrn;
        if ( xrn > x0 ) {
           x2 = x0;
           x0 = xrn;
        }
        else {
           x1 = x0;
           x0 = xrn;
        }
     } while ( rel_error > epsilon );
     printf("\nApproximate root = %.4f\n", xrn);
}
```

Quotient-Difference Method . . .

```
/*  Program to implement Quotient-Difference method to find real
    roots of polynomial of type
       a(n)x^n+a(n-1)*x^(n-1)+a(n-2)*x^(n-2)+...+a(0)           */
#include<stdio.h>
#include<stdlib.h>
#include<math.h>
#define MAXDEGREE  20
void main() {
   float a[MAXDEGREE], e[MAXDEGREE], q[MAXDEGREE];
   float max, epsilon;
   int n, i;
   printf("Enter order of polynomial : ");
   scanf("%d", &n);
   for ( i = n; i >= 0; i-- ) {
      printf("Enter coefficient of x^%d : ", i);
      scanf("%f", &a[i] );
   }
   printf("\nRequired precision : ");
   scanf("%f", &epsilon);
   q[0] = -a[n-1]/a[n];
```

```
    for ( i = 1; i < n; i++ ) {
        q[i] = 0.0;
    }
    e[0] = e[n] = 0.0;
    for ( i = 1; i < n; i++ ) {
        e[i] = a[n-i-1]/a[n-i];
    }
    do {
        for ( i = 0; i < n; i++ ) {
            q[i] = e[i+1] - e[i] + q[i];
        }
        for ( i = 1; i < n; i++ ) {
            e[i] = ( q[i]/q[i-1] ) * e[i];
        }
        max = fabs(e[0]);
        for ( i = 1; i <= n; i++ ) {
            if ( fabs(e[i]) > max ) {
                max = fabs(e[i]);
            }
        }
    } while ( max > epsilon );
    printf("\n  Roots of polynomial are\n" );
    printf("\n  s#       root\n ---------------\n");
    for ( i = 0; i < n; i++ ) {
        printf( "%4d\t%.4f\n", i+1, q[i] );
    }
}
```

Inverse of a Matrix . . .

```
/* Program to find inverse of square matrix in place with pivoting */
#include<stdio.h>
#include<math.h>
#define MAX  20
void main() {
    float a[MAX][ MAX], max, temp;
    int inter[MAX][3];
    int i, j, k, m, n, p, q, r, s;
    printf("Enter size of matrix : ");
    scanf("%d", &n);
    printf("\nEnter elements of matrix row-wise\n\n");
    for (i = 1; i <= n; i++ ) {
        for (j = 1; j <= n; j++ ) {
            scanf("%f", &a[i][j]);
        }
    }
    for (k = 1; k <= n; k++ ) {
        max = fabs(a[k][k]);
        p = k;
        if ( k != n ) {
```

```
        for (m = 1; m <= n; m++ ) {
            if ( fabs(a[m][k]) > max ) {
                max = fabs(a[m][k]);
                p = m;
            }
        }
    }
    inter[k][1]= k;
    inter[k][2]= p;
    if ( p != k ) {
        for (q = 1; q <= n; q++) {
            temp = a[k][q];
            a[k][q] = a[p][q];
            a[p][q]= temp;
        }
    }
    for (j = 1; j <= n; j++) {
        if ( j != k ) {
            a[k][j] = a[k][j] / a[k][k];
        }
    }
    a[k][k] = 1.0 / a[k][k];
    for (i = 1; i <= n; i++ ) {
        if ( i != k ) {
            for (j = 1; j <= n; j++ ) {
                if ( j != k ) {
                    a[i][j] = a[i][j] - a[i][k]* a[k][j];
                }
            }
        }
    }
    for (i = 1; i <= n; i++ ) {
        if ( i != k ) {
            a[i][k] = - a[i][k]* a[k][k];
        }
    }
}
for (q = n; q >= 1; q--) {
    r = inter[q][1];
    s = inter[q][2];
    if ( r != s ) {
        for (i = 1; i <= n; i++) {
            temp = a[i][r];
            a[i][r] = a[i][s];
            a[i][s] = temp;
        }
    }
}
printf("\n\nElements of Inverse\n");
for (i = 1; i <= n; i++ ) {
```

```
        for (j = 1; j <= n; j++ ) {
            printf("%f\t", a[i][j]);
        }
        printf("\n");
    }
}
```

Expanding a Determinant . . .

```
/*  Program to expand a determinant using
    Gauss elimination technique with pivoting  */
#include<stdio.h>
#include<math.h>
#define MAX  20
void main() {
    float a[MAX][ MAX], max, temp, prod;
    int i, j, k, m, n, p, q, count;
    printf("Enter size of determinant : ");
    scanf("%d", &n);
    printf("\nEnter elements of determinant row-wise\n\n");
    for (i = 1; i <= n; i++ ) {
        for (j = 1; j <= n; j++ ) {
            scanf("%f", &a[i][j]);
        }
    }
    count = 0;
    for (k = 1; k <= n; k++ ) {
        max = fabs(a[k][k]);
        p = k;
        for (m = k+1; m <= n; m++ ) {
            if ( fabs(a[m][k]) > max ) {
                max = fabs(a[m][k]);
                p = m;
            }
        }
        if ( p != k ) {
            count++;
            for (q = 1; q <= n; q++) {
               temp = a[k][q];
               a[k][q] = a[p][q];
               a[p][q]= temp;
            }
        }
        for (i = k+1; i <= n; i++ ) {
            temp = a[i][k]/a[k][k];
            for (j = k; j <= n; j++ ) {
                a[i][j] = a[i][j] - temp * a[k][j];
            }
        }
    }
```

```
    prod = 1.0;
    for (i = 1; i <= n; i++) {
       prod = prod * a[i][i];
    }
    if ( count % 2 )
       prod = -prod;
    printf("\nValue of determinant = %.3f\n", prod);
}
```

Gauss Elimination Method . . .

```
/* Program to find solution of system of linear equation
   using Gauss elimination technique with pivoting        */
#include<stdio.h>
#include<math.h>
#define MAX  20
void main() {
    float a[MAX][MAX], x[MAX], max, temp, sum;
    int i, j, k, m, n, p, q;
    printf("Enter number of equations : ");
    scanf("%d", &n);
    printf("\nEnter coefficients equation wise\n");
    for (i = 1; i <= n; i++ ) {
        for (j = 1; j <= (n+1); j++ ) {
            scanf("%f", &a[i][j]);
        }
    }
    for (k = 1; k < n; k++ ) {
        max = fabs(a[k][k]);
        p = k;
        for (m = k+1; m <= n; m++ ) {
            if ( fabs(a[m][k]) > max ) {
                max = fabs(a[m][k]);  p = m;
            }
        }
        if ( p != k ) {
            for (q = k; q <= (n+1); q++) {
               temp = a[k][q];  a[k][q] = a[p][q];  a[p][q]= temp;
            }
        }
        for (i = k+1; i <= n; i++ ) {
            temp = a[i][k]/a[k][k];
            for (j = k; j <= (n+1); j++ )
                a[i][j] = a[i][j] - temp * a[k][j];
        }
    }
    x[n] = a[n][n+1]/a[n][n];
    for (i = (n-1); i >= 1; i--) {
        for (sum = 0.0, j = (i+1); j <= n; j++)
            sum = sum + a[i][j]*x[j];
```

```
        x[i] = (a[i][n+1]-sum)/a[i][i];
    }
    printf("\n\nRequired solution is\n");
    for (i = 1; i <= n; i++) {
        printf("\nx(%d) = %.3f", i, x[i]);
    }
}
```

Gauss-Jordon Method . . .

```
/*  Program to find solution of system of linear equation
    using Gauss-Jordon technique                  */
#include<stdio.h>
#include<math.h>
#define MAX  20
void main() {
    float a[MAX][MAX], temp, comm;
    int i, j, k, m, n, p, q;
    printf("Enter number of equations : ");
    scanf("%d", &n);
    printf("\nEnter coefficients equation wise\n");
    for (i = 1; i <= n; i++ ) {
        for (j = 1; j <= (n+1); j++ ) {
            scanf("%f", &a[i][j]);
        }
    }
    for (k = 1; k <= n; k++ ) {
        temp = a[k][k];
        for (j = k; j <= (n+1); j++ )
            a[k][j] = a[k][j]/temp;
        for (i = 1; i <= n; i++ ) {
            if ( i != k ) {
                comm = a[i][k];
                for (j = k; j <= (n+1); j++ )
                    a[i][j] = a[i][j] - comm * a[k][j];
            }
        }
    }
    printf("\n\nRequired solution is\n");
    for (i = 1; i <= n; i++) {
        printf("\nx(%d) = %.3f", i, a[i][n+1]);
    }
}
```

Matrix Inversion Method . . .

```
/* Program to find solution of system of linear equation
   using Matrix Inversion technique                    */
#include<stdio.h>
#include<math.h>
#define MAX  20
void main() {
    float a[MAX][MAX], b[MAX], x[MAX], temp, comm;
    int i, j, k, m, n, p, q, r, s;
    printf("Enter number of equations : ");
    scanf("%d", &n);
    printf("\nEnter coefficients equation wise\n");
    for (i = 1; i <= n; i++ ) {
        for (j = 1; j <= n; j++ ) {
            scanf("%f", &a[i][j]);
        }
        scanf("%f", &b[i]);
    }
    for (i = 1; i <= n; i++ ) {
        for (j = (n+1); j <= (n+n); j++ ) {
            if ((j - i) == n) {
                a[i][j]= 1.0;
            }
            else {
                a[i][j]= 0.0;
            }
        }
    }
    for (k = 1; k <= n; k++ ) {
        temp = a[k][k];
        for (j = k; j <= (n+n); j++ ) {
            a[k][j] = a[k][j]/temp;
        }
        for (i = 1; i <= n; i++ ) {
            if ( i != k ) {
                comm = a[i][k];
                for (j = k; j <= (n+n); j++ ) {
                    a[i][j] = a[i][j] - comm * a[k][j];
                }
            }
        }
    }
    for (i = 1; i <= n; i++ ) {
        for (j = (n+1); j <= (n+n); j++ ) {
            a[i][j-n] = a[i][j];
        }
    }
    for (i = 1; i <= n; i++ ) {
        x[i] = 0.0;
```

```
        for (j = 1; j <= n; j++ ) {
            x[i] = x[i] + a[i][j] * b[j];
        }
    }
    printf("\n\nRequired solution is\n");
    for (i = 1; i <= n; i++) {
       printf("\nx(%d) = %.3f", i, x[i]);
    }
}
```

Modified Crout's Method . . .

```
/* Program to solve a system of non-homogeneous linear equations using
   modified Crout's method    */
#include<stdio.h>
#include<math.h>
#define MAX  20
void main() {
    float a[MAX][MAX], x[MAX], temp, sum;
    int i, j, jj, k, n;
    printf("Enter number of equations : ");
    scanf("%d", &n);
    printf("\nEnter coefficients equation wise\n");
    for (i = 1; i <= n; i++ ) {
        for (j = 1; j <= (n+1); j++ ) {
            scanf("%f", &a[i][j]);
        }
    }
    temp = a[1][1];
    for ( j = 2; j <= n+1; j++ )
        a[1][j] = a[1][j] / temp;
    for ( j = 2; j <= n; j++ ) {
        for ( i = j; i <= n; i++ ) {
            for ( sum = 0.0, k = 1; k <= j-1; k++ ) {
                sum = sum + a[i][k] * a[k][j];
            }
            a[i][j] = a[i][j] - sum;
        }
        i = j;
        for ( jj = i+1; jj <= n+1; jj++ ) {
            for ( sum = 0.0, k = 1; k <= i-1; k++ )
                sum = sum + a[i][k] * a[k][jj];
            a[i][jj] = ( a[i][jj] - sum ) / a[i][i];
        }
    }
    x[n] = a[n][n+1];
    for (i = n-1; i >= 1; i-- ) {
        for ( sum = 0.0, j = i+1; j <= n; j++ ) {
            sum = sum + a[i][j] * x[j];
```

```
        x[i] = a[i][n+1] - sum;
    }
    printf("\n\nRequired solution is\n");
    for (i = 1; i <= n; i++) {
        printf("\nx(%d) = %.3f", i, x[i]);
    }
}
```

Jacobi's Method . . .

```
/*  Program to find solution of system of linear equation
    using Jacobi's method              */
#include<stdio.h>
#include<stdlib.h>
#include<math.h>
#define MAX  20
void main() {
    float a[MAX][MAX], old_x[MAX], new_x[MAX], sum, epsilon;
    float relerror, bigrelerror;
    int i, j, k, n, maxiter;
    printf("Enter number of equations : ");
    scanf("%d", &n);
    printf("\nEnter coefficients equation wise\n");
    for (i = 1; i <= n; i++ ) {
        for (j = 1; j <= (n+1); j++ ) {
            scanf("%f", &a[i][j]);
        }
    }
    printf("Enter maximum number of iterations permited : ");
    scanf("%d", &maxiter);
    printf("\nEnter required precision : ");
    scanf("%f", &epsilon);
    for (i = 1; i <= n; i++ ) {
        old_x[i] = 0.0;
    }
    for (k = 1; k <= maxiter; k++ ) {
        bigrelerror = 0.0;
        for (i = 1; i <= n; i++ ) {
            sum = 0.0;
            for (j = 1; j <= n; j++ ) {
                if ( i != j ) {
                    sum = sum + a[i][j] * old_x[j];
                }
            }
            new_x[i] = ( a[i][n+1]-sum)/a[i][i];
            relerror = fabs ( (new_x[i]-old_x[i])/new_x[i]);
            if ( relerror > bigrelerror ) {
                bigrelerror = relerror;
            }
        }
```

```
        if ( bigrelerror < epsilon ) {
            printf("\n\nSolution converges in %d iterations\n", k);
            for (i = 1; i <= n; i++) {
                printf("\nx(%d) = %.3f", i, new_x[i]);
            }
            exit(0);
        }
        for (i = 1; i <= n; i++) {
            old_x[i] = new_x[i];
        }
    }
    printf("\n\nSolution does not converges in %d iterations\n",
           maxiter);
}
```

Gauss-Seidel Method . . .

```
/*  Program to find solution of system of linear equation
    using Gauss-Seidel method         */
#include<stdio.h>
#include<stdlib.h>
#include<math.h>
#define MAX  20
void main() {
    float a[MAX][MAX], x[MAX], sum, epsilon;
    float relerror, bigrelerror, temp;
    int i, j, k, n, maxiter;
    printf("Enter number of equations : ");
    scanf("%d", &n);
    printf("\nEnter coefficients equation wise\n");
    for (i = 1; i <= n; i++ ) {
        for (j = 1; j <= (n+1); j++ ) {
            scanf("%f", &a[i][j]);
        }
    }
    printf("Enter maximum number of iterations permited : ");
    scanf("%d", &maxiter);
    printf("\nEnter required precision : ");
    scanf("%f", &epsilon);
    for (i = 1; i <= n; i++ )
        x[i] = 0.0;
    for (k = 1; k <= maxiter; k++ ) {
        bigrelerror = 0.0;
        for (i = 1; i <= n; i++ ) {
            for (sum = 0.0, j = 1; j <= n; j++ ) {
                if ( i != j )
                    sum = sum + a[i][j] * x[j];
            }
            temp = ( a[i][n+1]-sum)/a[i][i];
            relerror = fabs((temp-x[i])/temp);
```

```
            x[i] = temp;
            if ( relerror > bigrelerror )
                bigrelerror = relerror;
        }
        if ( big_error < epsilon ) {
            printf("\n\nSolution converges in %d iterations\n", k);
            for (i = 1; i <= n; i++)
                printf("\nx(%d) = %.3f", i, x[i]);
            return;
        }
    }
    printf("\n\nSolution does not converges in %d iterations\n",
            maxiter);
}
```

Power Method . . .

```
/*  Program to find the largest eigen value using Power method      */
#include<stdio.h>
#include<math.h>
#define MAX  20
void main() {
    float a[MAX][MAX], x[MAX], s[MAX], epsilon;
    float error, big_error, temp, lamda;
    int i, j, k, n;
    printf("Enter number of equations : ");
    scanf("%d", &n);
    printf("\nEnter coefficients equation wise\n");
    for (i = 1; i <= n; i++ ) {
        for (j = 1; j <= n; j++ ) {
            scanf("%f", &a[i][j]);
        }
    }
    printf("\nEnter required precision : ");
    scanf("%f", &epsilon);
    for (i = 1; i <= n; i++ ) {
        x[i] = 1.0;
    }
    do {
        big_error = 0.0;
        for (i = 1; i <= n; i++ ) {
            s[i] = 0.0;
            for (j = 1; j <= n; j++ ) {
                s[i] = s[i] + a[i][j] * x[j];
            }
        }
        lamda = fabs (s[1]);
        for (i = 1; i <= n; i++ ) {
            temp = s[i]/lamda;
            error = fabs((temp-x[i])/temp);
```

```
            x[i] = temp;
            if ( error > big_error ) {
                big_error = error;
            }
        }
    } while ( big_error > epsilon );
    printf("\n\nLargest eigen value = %.4f\n", lamda);
    printf("\nAssociated eigen vector is\n");
    for (i = 1; i <= n; i++) {
        printf("\nx(%d) = %.4f", i, x[i]);
    }
}
```

Lagrangian Interpolation . . .

```
/* Program to implement the Lagrangian method of interpolation to
   interpolate value of variable 'y' for given value of variable 'x' */
#include<stdio.h>
#include<stdlib.h>
void main()
{
   float x[20], y[20], sum, prod, a;
   int i, j, n;
   printf("Enter Number of table points : ");
   scanf("%d", &n);
   printf("Enter %d pair of values for (x,y)\n", n);
   for ( i = 0; i < n; ++i ) {
       printf("Enter value for x[%d] : ", i);
       scanf("%f", &x[i]);
       printf("Enter value for y[%d] : ", i);
       scanf("%f", &y[i]);
   }
   printf("Enter the value of x for interpolation : ");
   scanf("%f", &a);
   if ( ( a < x[1] ) && ( a > x[n] ) ) {
      printf("\nValue lies outside range\n");
      exit(1);
   }
   sum = 0;
   for ( i = 0; i < n; ++i ) {
       prod = 1.0;
       for ( j = 0; j < n; ++j ) {
           if ( i != j )
              prod *= (a-x[j])/(x[i]-x[j]);
       }
       sum += y[i] * prod;
   }
   printf("\nInterpolated value = %.4f\n", sum);
}
```

Newton's Forward Difference Interpolation . . .

```
/* Program to implement Newton's forward difference interpolation
formula  */
#include<stdio.h>
#include<stdlib.h>
void main() {
    int fact(int m);
    float x[10], y[10], d[10][10], a, u, sum, prod;
    int i, j, k, n;
    printf("Enter number of table points : ");
    scanf("%d", &n);
    printf("Enter %d values for (x,y)\n", n);
    for ( i = 1; i <= n; i++ ) {
      scanf("%f,%f", &x[i], &y[i]);
    }
    printf("Enter value for interpolation : ");
    scanf("%f", &a);
    if ( ( a < x[1] ) && ( a > x[n] ) ) {
      printf("\nValue lies outside range\n");
      exit(1);
    }
    i = 2;
    while ( a > x[i] ) {
       i++;
    }
    k = i-1;
    u = (a - x[k])/(x[k+1] - x[k]);
    for ( j = 1; j <= (n-1); j++ ) {
        for ( i = 1; i <= (n-j); i++ ) {
            if ( j == 1 ) {
               d[i][j] = y[i+1] - y[i];
            } else {
               d[i][j] = d[i+1][j-1] - d[i][j-1];
            }
        }
    }
    sum = y[k];
    for ( i = 1; i <= (n-k); i++ ) {
        prod = 1.0;
        for ( j = 0; j <= (i-1); j++ ) {
            prod = prod *(u-j);
        }
        sum = sum + d[k][i] * prod / fact(i);
    }
    printf("\nInterpolated value = %8.3f\n", sum);
}
/*  function to compute factorial of number 'm'   */
int fact(int m)
{
   int i, prod = 1;
   for ( i = 2; i <= m; i++ ) {
```

```
        prod = prod * i;
    }
    return prod;
}
```

Newton's Backward Difference Interpolation . . .

```
/*  Program to implement Newton's backward difference interpolation
formula  */
#include<stdio.h>
#include<stdlib.h>
void main() {
    int fact(int m);
    float x[10], y[10], d[10][10], a, u, sum, prod;
    int i, j, n, k;
    printf("Enter number of table points : ");
    scanf("%d", &n);
    printf("Enter %d values for (x,y)\n", n);
    for ( i = 1; i <= n; i++ ) {
        scanf("%f,%f", &x[i], &y[i]);
    }
    printf("Enter value for interpolation : ");
    scanf("%f", &a);
    if ( ( a < x[1] ) && ( a > x[n] ) ) {
        printf("\nValue lies outside range\n");
        exit(1);
    }
    i = 2;
    while ( a > x[i] ) {
        i++;
    }
    k = i;
    u = (a - x[k])/(x[k] - x[k-1]);
    for ( j = 1; j < n; j++ ) {
        for ( i = j+1; i <= n; i++ ) {
            if ( j == 1 ) {
                d[i][j] = y[i] - y[i-1];
            } else {
                d[i][j] = d[i][j-1] - d[i-1][j-1];
            }
        }
    }
    sum = y[k];
    for ( i = 1; i <= (k-1); i++ ) {
        prod = 1.0;
        for ( j = 0; j <= (i-1); j++ ) {
            prod = prod * (u+j);
        }
        sum = sum + d[k][i] * prod / fact(i);
    }
```

```
    printf("\nInterpolated value = %8.3f\n", sum);
}
/*   function to compute factorial of number 'm'     */
int fact(int m) {
   int i, prod = 1;
   for ( i = 2; i <= m; i++ ) {
       prod = prod * i;
   }
   return prod;
}
```

Newton's Divided Difference Interpolation . . .

```
/* Program to implement Newton's divided difference interpolation formula  */
#include<stdio.h>
#include<stdlib.h>
void main() {
    float x[10], y[10], d[10][10], a, sum, prod;
    int i, j, k, n;
    printf("Enter number of table points : ");
    scanf("%d", &n);
    printf("Enter %d values for (x,y)\n", n);
    for ( i = 1; i <= n; i++ ) {
        scanf("%f,%f", &x[i], &y[i]);
    }
    printf("Enter value for interpolation : ");
    scanf("%f", &a);
    if ( ( a < x[1] ) && ( a > x[n] ) ) {
       printf("\nValue lies outside range\n");
       exit(1);
    }
    i = 2;
    while ( a > x[i] ) {
      i++;
    }
    k = i-1;
    for ( j = 1; j < n; j++ ) {
        for ( i = 1; i <= (n-j); i++ ) {
            if ( j == 1 ) {
                d[i][j] = (y[i+1] - y[i])/(x[i+1] - x[i]);
            } else {
                d[i][j] = (d[i+1][j-1]-d[i][j-1])/(x[i+j]-x[i]);
            }
        }
    }
    sum = y[k];
    for ( i = 1; i <= (n-k); i++ ) {
        prod = 1.0;
        for ( j = 0; j <= (i-1); j++ ) {
            prod = prod * (a-x[k+j]);
```

```
        }
        sum = sum + d[k][i] * prod;
    }
    printf("\nInterpolated value = %8.3f\n", sum);
}
```

Fitting a Straight Line . . .

```
/* Program developed to fit a straight line of Y on X
  through a set of points using method of least squares     */
#include<stdio.h>
void main() {
   float x[20], y[20];
   int i, n;
   float sx, sx2, sy, sxy, denom, a1, a2;
   printf("Enter number of points : ");
   scanf("%d", &n);
   printf("Enter %d pair of values as x,y\n", n);
   for ( i = 1; i <= n; i++ ) {
       printf("Enter value for x[%d],y[%d] : ", i, i );
       scanf("%f,%f", &x[i], &y[i]);
   }
   sx = 0.0;        sx2 = 0.0;
   sy = 0.0;        sxy = 0.0;
   for ( i = 1; i <= n; i++ ) {
       sx  = sx + x[i];
       sx2 = sx2 + x[i] * x[i];
       sy  = sy + y[i];
       sxy = sxy + x[i] * y[i];
   }
   denom = n * sx2 - sx * sx;
   a2 = ( n * sxy - sx * sy ) / denom;
   a1 = ( sy * sx2 - sx * sxy ) / denom;
   printf("\nRegression coefficients are\n");
   printf("\nSlope of the regression line = %6.3f", a2);
   printf("\nY-intercept for the line = %6.3f\n", a1);
}
```

Fitting a Polynomial . . .

```
/*  Program to fit a polynomial of degree 'm' through a set
    of points using method of least squares              */
#include<stdio.h>
#include<math.h>
void main() {
   float x[20], y[20], a[20], c[20][20];
   int i, j, n, m, k, m1, m2, l;
   float p, q, max, temp, sum;
   printf("Enter degree of polynomial : ");
```

```
scanf("%d", &m);
printf("Enter number of points : ");
scanf("%d", &n);
printf("Enter %d pair of values as x,y\n", n);
for ( i = 1; i <= n; i++ ) {
    printf("Enter value for x[%d] : ");
    scanf("%f", &x[i]);
    printf("Enter value for y[%d] : ");
    scanf("%f", &y[i]);
}
m1 = m + 1;
m2 = m + 2;
for ( i = 1; i <= m1; i++ ) {
   for ( j = 1; j <= m1; j++ ) {
       l = i + j - 2;
       c[i][j] = 0.0;
       for ( k = 1; k <= n; k++ ) {
         c[i][j] = c[i][j] + pow((double)x[k], (double)l);
       }
   }
   l = i - 1;
   c[i][m2] = 0.0;
   for ( k = 1; k <= n; k++ ) {
       c[i][m2] = c[i][m2]+ y[k] * pow((double)x[k], (double)l);
   }
}
for ( k = 1; k <= (m1-1); k++ ) {
    max = abs(c[k][k]);
    p = k;
    for ( q = k+1; q <= m1; q++ ) {
        if ( abs( c[q][k] ) > max ) {
            max = abs ( c[q][k] );
            p = q;
        }
    }
    if ( p != k ) {
        for ( q = k; q <= m2; q++ ) {
            temp = c[k][q];
            c[k][q] = c[p][q];
            c[p][q] = temp;
        }
    }
    for ( i = k+1; i <= m1; i++ ) {
        temp = c[i][k]/c[k][k];
        for ( j = k; j <= m2; j++ ) {
            c[i][j] = c[i][j] - temp * c[k][j];
        }
    }
}
a[m1] = c[m1][m2]/c[m1][m1];
```

```
    for ( i = (m1-1); i >= 1; i-- ) {
        sum = 0.0;
        for ( j = (i+1); j <= m1; j++ ) {
            sum = sum + c[i][j] * a[j];
        }
        a[i] = ( c[i][m2] - sum ) / c[i][i];
    }
    printf("\nRegression coefficients\n\n");
    for ( i = 1; i <= m1; i++ ) {
        printf("a[%d]= %8.4f\n", i, a[i]);
    }
}
```

Fitting an Exponential Curve . . .

```
/*  Program to fit an exponential curve of type  y = a*e^bx */
#include<stdio.h>
#include<math.h>
void main()
{
   float x[20], y[20];
   int i, n;
   float xx, yy, sx, sx2, slogy, sxlogy, denom, a, b, c;
   printf("Enter number of points : ");
   scanf("%d", &n);
   for ( i = 1; i <= n; i++ ) {
      printf("Enter x[%d] : ", i);
      scanf("%f", &x[i]);
      printf("Enter y[%d] : ", i);
      scanf("%f", &y[i]);
   }
   sx = 0.0;        sx2 = 0.0;
   slogy = 0.0;     sxlogy = 0.0;
   for ( i = 1; i <= n; i++ ) {
      xx = x[i];
      yy = log10((double)y[i]);
      sx   = sx + xx;
      sx2 = sx2 + xx * xx;
      slogy   = slogy + yy;
      sxlogy = sxlogy + xx * yy;
   }
   denom = n * sx2 - sx * sx;
   b = (n * sxlogy - sx * slogy) / denom;
   c = (slogy * sx2 - sx * sxlogy) / denom;
   a = exp((double)c);
   printf("\nRegression coefficients\n\n");
   printf("\na = %.5f", a);
   printf("\nb = %.5f", b);
}
```

Differentiating a Function . . .

```
/*  Program to compute first derivative at tabulated point
    when the function is tabulated at equal intervals        */
#include<stdio.h>
#include<stdlib.h>
void main() {
    float x[10], y[10], d[10][10], a, sum, derivative, h, term;
    int i, j, k, n, sign;
    printf("Enter number of table points : ");
    scanf("%d", &n);
    printf("Enter interval size : ");
    scanf("%f", &h);
    printf("Enter %d values for (x,y)\n", n);
    for ( i = 1; i <= n; i++ ) {
        scanf("%f,%f", &x[i], &y[i]);
    }
    printf("Enter tabulated point : ");
    scanf("%f", &a);
    if ((a < x[1]) || (a > x[n])) {
       printf("\nValue lies outside range\n");
       exit(1);
    }
    i = 1;
    while ( a != x[i] )
       i++;
    k = i;
    for ( j = 1; j <= (n-1); j++ ) {
        for ( i = 1; i <= (n-j); i++ ) {
            if ( j == 1 ) {
                d[i][j] = y[i+1] - y[i];
            }
            else {
                d[i][j] = d[i+1][j-1] - d[i][j-1];
            }
        }
    }
    sum = 0.0;
    sign = 1;
    for ( j = 1; j <= (n-k); j++ ) {
        term =  sign * d[k][j] / j;
        sum = sum + term;
        sign = - sign;
    }
    derivative = (1.0/h) * sum;
    printf("\nValue of derivative = %8.3f\n", derivative);
}

/* Program to compute first derivative at when the function
```

```
  is tabulated at unequal intervals                        */
#include<stdio.h>
#include<stdlib.h>
void main() {
    float x[10], y[10], d[10][10], a, sum1, sum2, prod;
    int i, j, k, n, m, p;
    printf("Enter number of table points : ");
    scanf("%d", &n);
    printf("Enter %d values for (x,y)\n", n);
    for ( i = 1; i <= n; i++ ) {
        scanf("%f,%f", &x[i], &y[i]);
    }
    printf("Enter tabulated point : ");
    scanf("%f", &a);
    if ((a < x[1]) || (a > x[n])) {
       printf("\nValue lies outside range\n");
       exit(1);
    }
    i = 1;
    while ( a > x[i])
       i++;
    k = i-1;
    for ( j = 1; j <= (n-1); j++ ) {
        for ( i = 1; i <= (n-j); i++ ) {
            if ( j == 1 ) {
               d[i][j] = (y[i+1]-y[i])/(x[i+1]-x[i]);
            }
            else {
               d[i][j] = (d[i+1][j-1]-d[i][j-1])/(x[i+j]-x[i]);
            }
        }
    }
    sum1 = d[k][1];
    for ( j = 2; j <= (n-k); j++ ) {
        sum2 = 0;
        for ( m = 0; m <= (j-1); m++ ) {
            prod = 1.0;
            for ( p = 0; p <= (j-1); p++ ) {
                if ( p != m ) {
                   prod = prod * ( a - x[k+p] );
                }
            }
            sum2 = sum2 + prod;
        }
       sum1 = sum1 + d[k][j] * sum2;
    }
    printf("\nValue of derivative = %8.3f\n", sum1);
}
```

Trapezoidal Rule . . .

```
/*  Program to implement Trapezoidal rule for tabulated function */
#include<stdio.h>
void main() {
    int i, n;
    float x[20], y[20], h, sum, integral;
    printf("Enter number of intervals : ");
    scanf("%d", &n);
    printf("Enter size of interval : ");
    scanf("%f", &h);
    printf("Enter %d pair of (x, y)\n", n+1);
    for ( i = 1; i <= (n+1); i++ ) {
        scanf("%f,%f", &x[i], &y[i]);
    }
    sum = ( y[1] + y[n+1] )/2;
    for ( i = 2; i <= n; i++ ) {
        sum = sum + y[i];
    }
    integral = h * sum;
    printf("\nValue of the integral = %9.4f\n", integral);
}

/*  Program to implement Trapezoidal rule for known function     */
#include<stdio.h>
#include<math.h>
float f(float x) {
    return exp(-0.5*x);
}
void main() {
    int i, n;
    float a, b, h, sum, integral;
    printf("Enter lower limit of integration : ");
    scanf("%f", &a);
    printf("Enter upper limit of integration : ");
    scanf("%f", &b);
    printf("Enter number of intervals : ");
    scanf("%d", &n);
    h = (b - a)/n;
    sum = (f(a)+f(b))/2.0;
    for ( i = 1; i <= (n-1); i++ ) {
        sum = sum + f(a+i*h);
    }
    integral = h * sum;
    printf("\nValue of the integral = %9.4f\n", integral);
}
```

Simpson 1/3rd Rule . . .

```
/* Program to implement Simpson's 1/3 rule for tabulated function   */
#include<stdio.h>
void main() {
```

```
    int i, n;
    float x[20], y[20], h, sum, integral, s2, s4;
    printf("Enter number of intervals : ");
    scanf("%d", &n);
    printf("Enter size of interval : ");
    scanf("%f", &h);
    printf("Enter %d pair of (x, y)\n", n+1);
    for ( i = 1; i <= (n+1); i++ ) {
        scanf("%f,%f", &x[i], &y[i]);
    }
    sum = y[1] + y[n+1];
    s2 = s4 = 0.0;
    for ( i = 2; i <= n; i += 2 ) {
        s4 = s4 + y[i];
    }
    for ( i = 3; i <= (n-1); i += 2 ) {
        s2 = s2 + y[i];
    }
    integral = (h/3.0) * ( sum + 2 * s2 + 4 * s4 );
    printf("\nValue of the integral = %9.4f\n", integral);
}
```

```
/*   Program to implement Simpson's 1/3rd rule for known function     */
#include<stdio.h>
#include<math.h>
float f(float x) {
    return exp(-0.5*x);
}
void main() {
    int i, n;
    float a, b, h, x, s2, s4, sum, integral;
    printf("Enter lower limit of integration : ");
    scanf("%f", &a);
    printf("Enter upper limit of integration : ");
    scanf("%f", &b);
    printf("Enter number of intervals : ");
    scanf("%d", &n);
    h = (b-a)/n;
    sum = f(a)+f(b);
    s2 = s4 = 0.0;
    for ( i = 1; i < n; i+=2 ) {
        x = a + i * h;
        s4 = s4 + f(x);
    }
    for ( i = 2; i < n; i+=2 ) {
        x = a + i * h;
        s2 = s2 + f(x);
    }
    integral = (h/3.0) * ( sum + 2 * s2 + 4 * s4 );
    printf("\nValue of the integral = %9.4f\n", integral);
```

```
}
```

Simpson 3/8th Rule . . .

```
/* Program to implement Simpson's 3/8th rule for tabulated function  */
#include<stdio.h>
void main() {
    int i, n;
    float x[20], y[20], h, sum, integral, s2, s3;
    printf("Enter number of intervals : ");
    scanf("%d", &n);
    printf("Enter size of interval : ");
    scanf("%f", &h);
    printf("Enter %d pair of (x, y)\n", n+1);
    for ( i = 1; i <= (n+1); i++ ) {
        scanf("%f,%f", &x[i], &y[i]);
    }
    sum = y[1] + y[n+1];
    s2 = s3 = 0.0;
    for ( i = 2; i <= (n-1); i += 3 ) {
        s3 = s3 + y[i] + y[i+1];
    }
    for ( i = 4; i <= (n-2); i += 3 ) {
        s2 = s2 + y[i];
    }
    integral = (3.0*h/8.0) * ( sum + 2 * s2 + 3 * s3 );
    printf("\nValue of the integral = %9.4f\n", integral);
}

/* Program to implement Simpson's 3/8th rule for known function  */
#include<stdio.h>
#include<math.h>
float f(float x)
{
   return  1.0+x*x*x;
}
void main()
{
    int i, n;
    float a, b, h, x, s2, s3, sum, integral;
    printf("Enter lower limit of integration : ");
    scanf("%f", &a);
    printf("Enter upper limit of integration : ");
    scanf("%f", &b);
    printf("Enter number of intervals : ");
    scanf("%d", &n);
    h = (b-a)/n;
    sum = f(a) + f(b);
    s2 = s3 = 0.0;
    for ( i = 1; i < n; i+=3 ) {
```

```
        x = a + i * h;
        s3 = s3 + f(x)+f(x+h);
    }
    for ( i = 3; i < n; i+=3 ) {
        x = a + i * h;
        s2 = s2 + f(x);
    }
    integral = (3.0*h/8.0) * ( sum + 2 * s2 + 3 * s3 );
    printf("\nValue of the integral = %9.4f\n", integral);
}
```

Romberg Integration . . .

```
/* Program to implement Romberg integration for known function    */
#include<stdio.h>
#include<math.h>
typedef enum { false, true } boolean;
float f ( float x ) {
  return (x*x+2.0*x);
}
void main() {
     float r[20][20], a, b, x, h, epsilon, sum, temp;
     int i, j, k, n;
     boolean flag;
     printf("Enter lower limit for integration : ");
     scanf("%f", &a);
     printf("Enter lower upper for integration : ");
     scanf("%f", &b);
     printf("Enter allowed precision : ");
     scanf("%f", &epsilon);
     i = 1;
     n = 1;
     h = b - a;
     sum = ( f(a) + f(b) )/2;
     r[i][1] = h * sum;
     flag = false;
     while ( flag == false ) {
      n *= 2;
      h /= 2;
      i++;
      for ( j = 1; j <=n; j +=2 ) {
         x = a + j * h;
         sum += f(x);
      }
      r[i][1] = h * sum;
      for ( k = 2; k <= i; k++ ) {
          temp = pow((double)4, (double)k-1);
          r[i][k] = (temp * r[i][k-1] - r[i-1][k-1])/(temp - 1);
      }
      if ( fabs(r[i][i] - r[i][i-1]) <= epsilon ) {
```

```
            printf("\nElements of Romberg table are\n\n");
            for ( k = 1; k <= i; k++ ) {
              for ( j = 1; j <= k; j++ )
                  printf("%12.4f", r[k][j]);
              printf("\n");
            }
            flag = true;
        }
      }
      printf("\nValue of integral = %10.4f\n", r[i][i]);

}
```

Gauss's Quadrature Formula . . .

```
/* Program to implement Gauss's Quadrature formula    */
#include<stdio.h>
float g(float y) {
    return y*y + 8.0*y + 15;
}
void main() {
    int i, n;
    float a, b, r[20], y[20], sum, integral;
    printf("Enter number of points : ");
    scanf("%d", &n);
    printf("Enter lower limit of integration : ");
    scanf("%f", &a);
    printf("Enter upper limit of integration : ");
    scanf("%f", &b);
    for ( i = 1; i <= n; i++ ) {
        printf("Enter value of r[%d],y[%d] : ", i, i);
        scanf("%f,%f", &r[i], &y[i]);
    }
    sum = 0;
    for ( i = 1; i <= n; i++ ) {
        sum = sum + r[i] * g(y[i]);
    }
    integral = ((b-a)/2.0) * sum;
    printf("\nValue of the integral = %9.4f\n", integral);
}
```

Euler's Method . . .

```
/*  Program to solve ODE of type dy/dx = xy using Euler's method  */
#include<stdio.h>
float f(float x, float y) {
    return (x*y);
}
void main() {
```

```
    int i;
    float x, y, x1, y1, xf, h, s, s1, s2, s3, s4;
    printf("Enter value of x1, y1, xf : ");
    scanf("%f,%f,%f", &x1, &y1, &xf);
    printf("Enter size of interval : ");
    scanf("%f", &h);
    x = x1;
    y = y1;
    i = 1;
    printf("\n\nPoints constituting solution curve are\n\n");
    printf("\n Point        x                y");
    printf("\n ================================");
    printf("\n %3d\t%8.3f\t%8.3f", i, x, y);
    while ( x < xf ) {
       y = y + h * f(x, y);
       x = x + h;
       i = i + 1;
       printf("\n %3d\t%8.3f\t%8.3f", i, x, y);
    }
    printf("\n ================================\n");
}
```

Runge-Kutta Second Order Method . . .

```
/* Program to solve ODE of type dy/dx = xy using
   second order Runge-Kutta method (Heun's method)    */
#include <stdio.h>
float f(float x, float y) {
    return (x*y);
}
void main() {
    int i;
    float x, y, x1, y1, xf, h, s, s1, s2, s3, s4;
    printf("Enter value of x1, y1, xf : ");
    scanf("%f,%f,%f", &x1, &y1, &xf);
    printf("Enter size of interval : ");
    scanf("%f", &h);
    x = x1;      y = y1;      i = 1;
    printf("\n\nPoints constituting solution curve are\n\n");
    printf("\n Point        x                y");
    printf("\n ================================");
    printf("\n %3d\t%8.3f\t%8.3f", i, x, y);
    while ( x < xf ) {
       s1 = f(x, y);
       x = x + h;
       s2 = f(x, y+h*s1);
       y = y + h * (s1+s2)/2;
       i = i + 1;
       printf("\n %3d\t%8.3f\t%8.3f", i, x, y);
    }
```

```
    printf("\n ================================\n");
}
```

Runge-Kutta Fourth Order Method . . .

```
/*  Program to solve ODE of type dy/dx = xy
    using fourth order Runge-Kutta method        */
#include <stdio.h>
float f(float x, float y) {
    return (x*y);
}
void main() {
    int i;
    float x, y, x1, y1, xf, h, s, s1, s2, s3, s4;
    printf("Enter value of x1, y1, xf : ");
    scanf("%f,%f,%f", &x1, &y1, &xf);
    printf("Enter size of interval : ");
    scanf("%f", &h);
    x = x1;
    y = y1;
    i = 1;
    printf("\n\nPoints constituting solution curve are\n\n");
    printf("\n Point       x              y");
    printf("\n ================================");
    printf("\n %3d\t%8.3f\t%8.3f", i, x, y);
    while ( x < xf ) {
       s1 = f(x, y);
       s2 = f(x+h/2, y+h*s1/2);
       s3 = f(x+h/2, y+h*s2/2);
       s4 = f(x+h, y+h*s3);
       s = (s1+2*s2+2*s3+s4)/6;
       y = y + h*s;
       x = x + h;
       i = i + 1;
       printf("\n %3d\t%8.3f\t%8.3f", i, x, y);
    }
    printf("\n ================================\n");
}
```

Euler's Predictor Corrector Method . . .

```
/* Program to solve ODE of type dy/dx = xy
   using Euler's predictor corrector method     */
#include <stdio.h>
typedef enum { false, true } boolean;
float f(float x, float y) {
    return (x*y);
}
void main() {
    int i;
    float x, y, x1, y1, y2, xf, h, epsilon;
    boolean flag;
    printf("Enter value of x1, y1, xf : ");
    scanf("%f,%f,%f", &x1, &y1, &xf);
    printf("Enter size of interval : ");
    scanf("%f", &h);
    printf("Enter allowed precision : ");
    scanf("%f", &epsilon);
    x = x1;
    y = y1;
    i = 1;
    printf("\n\nPoints constituting solution curve are\n\n");
    printf("\n Point       x              y");
    printf("\n =================================");
    printf("\n %3d\t%8.3f\t%8.3f", i, x, y);
    while ( x < xf ) {
       y2 = y + h * f(x, y);
       flag = true;
       while ( flag ) {
       y1 = y2;
       y2 = y + (h / 2) * (f(x, y) + f(x+h, y2));
       if ( fabs((y2-y1)/y2) <= epsilon )
          flag = false;
       }
       y = y2;
       x = x + h;
       i = i + 1;
       printf("\n %3d\t%8.3f\t%8.3f", i, x, y);
    }
    printf("\n ================================\n");
}
```

❑ ❑ ❑

Appendix B

Listing of C++ Programs

Bisection Method . . .

```
//  Program to implement Bisection method to find the
//  root of equation    2^x - x - 3 = 0
#include<iostream.h>
#include<iomanip.h>
#include<math.h>
inline float f( float x)  // inline function for expression 2^x-x-3
{
   return ( pow(double(2.0), double(x)) - x - 3 );
}
void main()
{
   float x1, x2, epsilon, x3;
   cout << "Enter first point of the search interval : ";
   cin >> x1;
   cout << "Enter second point of the search interval : ";
   cin >> x2;
   cout << "Enter prescribed tolerance : ";
   cin >> epsilon;
   do
   {
      x3 = ( x1+x2)/2;          // compute the mid point of the interval
      if ( f(x1)*f(x3) < 0 )  // select the appropriate subinterval
        x2 = x3;
      else
        x1 = x3;
   }
   while ( fabs(double(x1-x2)) > epsilon );
   cout << "Approximate root = " << setprecision(3) << x3 << endl;
}
```

False Position Method . . .

```
//  Program to implement False position method to find root
//  equation x^3 - 2x - 5 = 0
#include<iostream.h>
#include<iomanip.h>
#include<stdlib.h>
#include<math.h>
inline float f(float x) // inline function to compute x^3-2x-5
{
   return (x*x*x-2*x-5);
}
void main(void)
{
   float x1, x2, epsilon, x3, f1, f2, f3, delta;
   cout << "Enter first point of the search interval : ";
   cin >> x1;
   cout << "Enter second point of the search interval : ";
   cin >> x2;
   cout << "Enter prescribed tolerance : ";
   cin >> epsilon;
   cout << "Enter permissible lower limit on slope : ";
   cin >> delta;
   f1 = f(x1);
   f2 = f(x2);
   if ( f1 * f2 > 0 ) {
       cout << "\nInitial approximations are unsuitable\n";
       exit(1);
   }
   do {
       if ( fabs ( f2 - f1 ) <= delta ) {
          cout << "\nSlope of the function becomes too small\n";
          exit (1);
       }
       x3 = (x1*f2 - x2*f1) / (f2 - f1);
       f3 = f(x3);
       if ( f1 * f3 < 0 ) {
          x2 = x3;
          f2 = f3;
       } else {
          x1 = x3;
          f1 = f3;
       }
   }
   while ( fabs(double(x1-x2)) > epsilon );
   // output the computed root
   cout << "Approximate root = " << setprecision(4) << x3 << endl;
}
```

Secant Method . . .

```
// Program to implement Secant method to find root
// equation cos(x) - xe^x = 0
#include<iostream.h>
#include<iomanip.h>
#include<stdlib.h>
#include<math.h>
double f( double x ) {
   return cos(double(x)) - x * exp(double(x));
}
void main() {
    int n, i;
    float x1, x2, x3 = 0, x3p, f1, f2, f3, epsilon, delta;
    cout << "Enter first approximation : ";
    cin >> x1;
    cout << "Enter second approximation : ";
    cin >> x2;
    cout << "Enter prescribed tolerance : ";
    cin >> epsilon;
    cout << "Enter lower limit on slope of function : ";
    cin >> delta;
    cout << "Enter number of iterations permitted : ";
    cin >> n;
    f1 = f(x1);
    f2 = f(x2);
    for ( i = 1; i <= n; i++ ) {
        if ( fabs(f2 - f1) <= delta ) {
            cout << "\nSlope of the function becomes too small\n";
            exit(1);
        }
        x3p = x3;
        x3 = (x1*f2 - x2*f1) / (f2 - f1);
        f3 = f(x3);
        if ( fabs(double((x3-x3p)/x3) ) <= epsilon ) {
            cout << "\nApproximate root = " << setprecision(4)<< x3;
            exit(1);
        }
        x1 = x2;   f1 = f2;
        x2 = x3;   f2 = f3;
    }
    cout << "\nSolution does not converge in " << n
         << " iterations" << endl;
}
```

Newton-Raphson Method . . .

```
// Program to implement Newton-Raphson method to find root
// of equation x^3- 4x - 9 = 0
#include<iostream.h>
#include<iomanip.h>
#include<stdlib.h>
#include<math.h>
double f( double x ) {
   return x * x * x - 4.0 * x - 9.0;
}
double df( double x ) {
   return 3.0 * x * x - 4.0;
}

void main() {
   float x0, x1, epsilon, delta, relativeerror;
   int i, n;
   cout << "Enter initial approximation : ";
   cin >> x0;
   cout<< "Enter number of iterations permitted : ";
   cin >> n;
   cout << "Enter prescribed tolerance : ";
   cin >> epsilon;
   cout << "Enter lower bound on slope : ";
   cin >> delta;
   for ( i = 1; i <= n; i++) {
       if ( fabs (df(x0)) < delta ) {
          cout << "\nslope too small\n";
          exit(1);
       }
       x1 = x0 - f(x0) / df(x0);
       relativeerror = fabs ( ( x1 - x0 ) / x1 );
       x0 = x1;
       if ( relativeerror <= epsilon ) {
          cout << "\nApproximate root = " << setprecision(4)
               << x1;
          exit(1);
       }
   }
   cout << "\nSolution does not converge in " << n << " iterations" <<
endl;
}
```

Successive Approximation Method . . .

```
// Program to implement method of successive approximations
// to find root of equation 2^x - x - 3 = 0
#include<iostream.h>
#include<iomanip.h>
#include<stdlib.h>
#include<math.h>
double g( double x ) {
   return pow(double(2.0),double(x)) - 3;
}
void main() {
   float x1, x2, epsilon, relativeerror;
   int i, n;
   cout << "Enter intial approximation : ";
   cin >> x1;
   cout << "Enter number of iterations permitted : ";
   cin >> n;
   cout << "Enter prescribed tolerance : ";
   cin >> epsilon;
   for ( i = 1; i <= n; i++ ) {
       x2 = g(x1);
       relativeerror = fabs ( ( x2 - x1 ) / x2 );
       x1 = x2;
       if ( relativeerror <= epsilon ) {
           cout << setprecision(4);
           cout << "\nRoot converges in " << i << " iterations";
           cout << "\nApproximate root = " << x2;
           exit(1);
       }
   }
   cout << "\nSolution does not converge in " << n
        << " iterations" << endl;
}
```

Bairstow's Method . . .

```
//  Program to implement Bairstow's method to find roots of
//  polynomial of type x^n+a(1)*x^(n-1)+a(2)*x^(n-2)+...+a(n)
#include<iostream.h>
#include<iomanip.h>
#include<stdlib.h>
#include<math.h>
#define MAXDEGREE  20
//  function prototypes
void quad(float u,float v,float *r1,float *r2,char *type );
void printroots( int *it, float r1, float r2, char type );
void main() {
   float a[MAXDEGREE],b[MAXDEGREE],c[MAXDEGREE],r1,r2,u,v;
```

```
float du,dv,epsilon,denom;
int n, it, i, k;
char type;
cout << "Enter order of polynomial : ";
cin >> n;
for ( i = 1; i <= n; i++ ) {
     cout << "Enter coefficient of x^" << n-i << " : ";
     cin >> a[i];
}
cout << "\nRequired precision : ";
cin >> epsilon;
cout << "\n\nRoot#      Real part     Imaginary part";
it = 0;
while ( n > 2 ) {
    u = v = 0.0;
    do {
       b[1] = a[1] - u;
       b[2] = a[2] - b[1] * u - v;
       for ( k = 3; k <= n; k++ ) {
           b[k] = a[k] - b[k-1] * u - b[k-2] * v;
       }
       c[1] = b[1] - u;
       c[2] = b[2] - c[1] * u - v;
       for ( k = 3; k < n; k++ ) {
           c[k] = b[k] - c[k-1] * u - c[k-2] * v;
       }
       if ( n == 3 ) {
          denom = c[n-1] - c[n-2] * c[n-2];
          if ( denom == 0 ) {
           cout << "\nzero denominator\n";
           exit(1);
          }
          du = ( b[n] - b[n-1] * c[n-2] ) / denom;
       }
       else {
          denom = c[n-1] * c[n-3] - c[n-2] * c[n-2];
          if ( denom == 0 ) {
           cout << "\nzero denominator\n";
           exit(1);
          }
          du = ( b[n] * c[n-3] - b[n-1] * c[n-2] ) / denom;
       }
       dv = ( b[n-1] * c[n-1] - b[n] * c[n-2] ) / denom;
       u = u + du;
       v = v + dv;
    } while((fabs(du)+fabs(dv)) > epsilon);
    quad ( u, v, &r1, &r2, &type );
    it++;
    printroots ( &it, r1, r2, type );
    n = n - 2;
    for ( i = 1; i <= n; i++ ) {
```

```
            a[i] = b[i];
        }
    }
    if ( n == 1 ) {
        it++;
        r1 =  -a[1];
        cout << endl << setw(3) << it << "\t";
        cout << setprecision(4);
        cout.setf(ios::showpoint);
        cout << setw(8) << r1;
    }
    else if ( n == 2 ) {
        u = a[1];
        v = a[2];
        it++;
        quad(u, v, &r1, &r2, &type);
        printroots(&it, r1, r2, type);
    }
}
//  function to solve quadratic factor of type x^2+ux+v
void quad(float u,float v,float *r1,float *r2,char *type )
{
    double d;
    d = u * u - 4.0 * v;
    if ( d > 0 ) {
        *type = 'r';
        d = sqrt ( (double) d );
        *r1 = ( - u + d ) / 2.0;
        *r2 = ( - u - d ) / 2.0;
    } else if ( d < 0 ) {
        *type = 'i';
        *r1 = - u / 2.0;
        *r2 = sqrt ( - (double) d ) / 2.0;
    } else {
        *type = 'r';
        *r1 = - u / 2.0;
        *r2 = *r1;
    }
}
//   function to print the roots
void printroots( int *it, float r1, float r2, char type )
{
    cout << setprecision(4);
    cout.setf(ios::showpoint);
    if ( type == 'r' ) {
         cout << endl << setw(3) << *it;
         cout << "\t" << setw(8) << r1;
         (*it)++;
         cout << endl << setw(3) << *it;
         cout << "\t" << setw(8) << r2;
    } else {
```

```
        cout << endl << setw(3) << *it;
        cout << "\t" << setw(8) << r1;
        cout << "\t" << setw(8) << r2;
        (*it)++;
        cout << endl << setw(3) << *it;
        cout << "\t" << setw(8) << r1;
        cout << "\t" << setw(8) << -r2;
    }
}
```

Muller's Method . . .

```
//  Program to implement Muller's method to find the
//  root of equation 3x + sin(x) - e^x = 0
#include<iostream.h>
#include<iomanip.h>
#include<stdlib.h>
#include<math.h>
double f ( double x ) {
   return 3*x+sin(double(x))-exp(double(x));
}
void main() {
   float a1, b1, x0, x1, x2, xrp, xrn, f0, f1, f2, epsilon;
   float h1, h2, rel_error, gama, a, b, c, disc;
   cout << "Enter first point of the search interval : ";
   cin >> a1;
   cout << "Enter second point of the search interval : ";
   cin >> b1;
   cout << "Enter prescribed tolerance : ";
   cin >> epsilon;
   x2 = a1;
   x0 = (a1+b1)/2;
   x1 = b1;
   xrp = 0.0;
   do
   {
      h1 = x1 - x0;
      h2 = x0 - x2;
      gama = h2/h1;
      f0 = f(x0);
      f1 = f(x1);
      f2 = f(x2);
      c = f0;
      a = (gama*f1-f0*(1.0+gama)+f2)/(gama*h1*h1*(1.0+gama));
      b = (f1-f0-a*h1*h1)/h1;
      disc = sqrt( (double)b*b-4*a*c );
      if ( b > 0 ) {
         xrn = x0-2.0*c/(b+disc);
      } else {
         xrn = x0-2.0*c/(b-disc);
```

```
      }
      rel_error = fabs((xrp-xrn)/xrn);
      xrp = xrn;
      if ( xrn > x0 ) {
         x2 = x0;
         x0 = xrn;
      } else {
         x1 = x0;
         x0 = xrn;
      }
   } while ( rel_error > epsilon );
   cout << setw(8) << setprecision(4);
   cout << "\nApproximate root = " << xrn << endl;
}
```

Quotient-Difference Method . . .

```
//  Program to implement Quotient-Difference method to find real
//  roots of polynomial of type
//     a(n)x^n+a(n-1)*x^(n-1)+a(n-2)*x^(n-2)+...+a(0)
#include<iostream.h>
#include<iomanip.h>
#include<stdlib.h>
#include<math.h>
#define MAXDEGREE  20
void main() {
   float a[MAXDEGREE], e[MAXDEGREE], q[MAXDEGREE];
   float max, epsilon;
   int n, i;
   cout << "Enter order of polynomial : ";
   cin >> n;
   for ( i = n; i >= 0; i-- ) {
       cout << "Enter coefficient of x^" << i << " : ";
       cin >> a[i];
   }
   cout << "\nRequired precision : ";
   cin >> epsilon;
   q[0] = -a[n-1]/a[n];
   for ( i = 1; i < n; i++ ) {
       q[i] = 0.0;
   }
   e[0] = e[n] = 0.0;
   for ( i = 1; i < n; i++ ) {
       e[i] = a[n-i-1]/a[n-i];
   }
   do
   {
      for ( i = 0; i < n; i++ ) {
           q[i] = e[i+1] - e[i] + q[i];
      }
```

```
        for ( i = 1; i < n; i++ ) {
             e[i] = ( q[i]/q[i-1] ) * e[i];
        }
        max = fabs(e[0]);
        for ( i = 1; i <= n; i++ ) {
             if ( fabs(e[i]) > max ) {
                 max = fabs(e[i]);
             }
        }
    } while ( max > epsilon );
    cout << "\n  Roots of polynomial are\n";
    cout << "\n  s#        root\n ----------------\n';
    for ( i = 0; i < n; i++ ) {
        cout << setw(4) << i+1 << "\t";
        cout.setf(ios::showpoint);
        cout << setprecision(4) << setw(8) << q[i] << endl;
    }
}
```

Inverse of a Matrix . . .

```
//  Program to find the inverse of square matrix in place
//  with pivoting
#include<iostream.h>
#include<iomanip.h>
#include<math.h>
#define MAX  20
void main() {
    float a[MAX][ MAX], max, temp;
    int inter[MAX][3], i, j, k, m, n, p, q, r, s;
    cout << "Enter size of matrix : ";
    cin >> n;
    cout << "\nEnter elements of matrix row-wise\n\n";
    for (i = 1; i <= n; i++ ) {
        for (j = 1; j <= n; j++ ) {
            cin >> a[i][j];
        }
    }
    for (k = 1; k <= n; k++ ) {
        max = fabs(a[k][k]);
        p = k;
        if ( k != n ) {
            for (m = 1; m <= n; m++ ) {
                if ( fabs(a[m][k]) > max ) {
                   max = fabs(a[m][k]);
                   p = m;
                }
            }
        }
        inter[k][1]= k;
```

```
        inter[k][2]= p;
        if ( p != k ) {
             for (q = 1; q <= n; q++) {
                 temp = a[k][q];
                 a[k][q] = a[p][q];
                 a[p][q]= temp;
             }
        }
        for (j = 1; j <= n; j++) {
             if ( j != k ) {
                 a[k][j] = a[k][j] / a[k][k];
             }
        }
        a[k][k] = 1.0 / a[k][k];
        for (i = 1; i <= n; i++ ) {
             if ( i != k ) {
                 for (j = 1; j <= n; j++ ) {
                     if ( j != k ) {
                         a[i][j] = a[i][j] - a[i][k]* a[k][j];
                     }
                 }
             }
        }
        for (i = 1; i <= n; i++ ) {
           if ( i != k ) {
               a[i][k] = - a[i][k]* a[k][k];
           }
        }
    }
    for (q = n; q >= 1; q--) {
        r = inter[q][1];
        s = inter[q][2];
        if ( r != s ) {
           for (i = 1; i <= n; i++) {
                temp = a[i][r];
                a[i][r] = a[i][s];
                a[i][s] = temp;
           }
        }
    }
    cout << "\n\nElements of Inverse\n";
    cout.setf(ios::showpoint);
    cout << setw(8) << setprecision(4) << endl;
    for (i = 1; i <= n; i++ ) {
        for (j = 1; j <= n; j++ ) {
             cout << "\t" << a[i][j];
        }
        cout << endl;
    }
}
```

Expanding a Determinant . . .

```
//  Program to expand a determinant using
//  Gauss elimination technique with pivoting
#include<iostream.h>
#include<iomanip.h>
#include<math.h>
#define MAX  20
void main() {
    float a[MAX][ MAX], max, temp, prod;
    int i, j, k, m, n, p, q, count;
    cout << "Enter size of matrix : ";
    cin >> n;
    cout << "\nEnter elements of matrix row-wise\n\n";
    for (i = 1; i <= n; i++ ) {
        for (j = 1; j <= n; j++ ) {
            cin >> a[i][j];
        }
    }
    count = 0;
    for (k = 1; k <= n; k++ ) {
        max = fabs(a[k][k]);
        p = k;
        for (m = k+1; m <= n; m++ ) {
            if ( fabs(a[m][k]) > max ) {
                max = fabs(a[m][k]);
                p = m;
            }
        }
        if ( p != k ) {
            count++;
            for (q = 1; q <= n; q++) {
               temp = a[k][q];
               a[k][q] = a[p][q];
               a[p][q]= temp;
            }
        }
        for (i = k+1; i <= n; i++ ) {
            temp = a[i][k]/a[k][k];
            for (j = k; j <= n; j++ ) {
                a[i][j] = a[i][j] - temp * a[k][j];
            }
        }
    }
    prod = 1.0;
    for (i = 1; i <= n; i++) {
       prod = prod * a[i][i];
    }
    if ( count % 2 )
       prod = -prod;
```

```
    cout.setf(ios::showpoint);
    cout << setprecision(4) << endl;
    cout << "\nValue of determinant = " << prod << endl;
}
```

Gauss Elimination Method . . .

```
// Program to find solution of system of linear equation
// using Gauss elimination technique with pivoting
#include<iostream.h>
#include<iomanip.h>
#include<math.h>
#define MAX  20
void main() {
    float a[MAX][MAX], x[MAX], max, temp, sum;
    int i, j, k, m, n, p, q;
    cout << "Enter number of equations : ";
    cin >> n;
    cout << "\nEnter coefficients equation wise\n";
    for (i = 1; i <= n; i++ ) {
        for (j = 1; j <= (n+1); j++ ) {
            cin >> a[i][j];
        }
    }
    for (k = 1; k < n; k++ ) {
        max = fabs(a[k][k]);
        p = k;
        for (m = k+1; m <= n; m++ ) {
            if ( fabs(a[m][k]) > max ) {
                max = fabs(a[m][k]);
                p = m;
            }
        }
        if ( p != k ) {
            for (q = k; q <= (n+1); q++) {
                temp = a[k][q];
                a[k][q] = a[p][q];
                a[p][q]= temp;
            }
        }
        for (i = k+1; i <= n; i++ ) {
            temp = a[i][k]/a[k][k];
            for (j = k; j <= (n+1); j++ ) {
                a[i][j] = a[i][j] - temp * a[k][j];
            }
        }
    }
    x[n] = a[n][n+1]/a[n][n];
    for (i = (n-1); i >= 1; i--) {
        sum = 0.0;
```

```
        for (j = (i+1); j <= n; j++) {
            sum = sum + a[i][j]*x[j];
        }
        x[i] = (a[i][n+1]-sum)/a[i][i];
    }
    cout << "\n\nRequired solution is\n";
    cout.setf(ios::showpoint);
    cout << setprecision(4);
    for (i = 1; i <= n; i++)
        cout << "\nx[" << i << "] = " << setw(8) << x[i];
    cout << endl;
}
```

Gauss-Jordon Method . . .

```
//  Program to find solution of system of linear equation
//  using Gauss-Jordon technique
#include<iostream.h>
#include<iomanip.h>
#include<math.h>
#define MAX  20
void main() {
    float a[MAX][MAX], temp, comm;
    int i, j, k, m, n, p, q;
    cout << "Enter number of equations : ";
    cin >> n;
    cout << "\nEnter coefficients equation wise\n";
    for (i = 1; i <= n; i++ ) {
        for (j = 1; j <= (n+1); j++ ) {
            cin >> a[i][j];
        }
    }
    for (k = 1; k <= n; k++ ) {
        temp = a[k][k];
        for (j = k; j <= (n+1); j++ ) {
            a[k][j] = a[k][j]/temp;
        }
        for (i = 1; i <= n; i++ ) {
            if ( i != k ) {
                comm = a[i][k];
                for (j = k; j <= (n+1); j++ ) {
                    a[i][j] = a[i][j] - comm * a[k][j];
                }
            }
        }
    }
    cout << "\n\nRequired solution is\n";
    cout.setf(ios::showpoint);
    cout << setprecision(4);
    for (i = 1; i <= n; i++)
```

```
        cout << "\nx[" << i << "] = " << setw(8) << a[i][n+1]);
    cout << endl;
}
```

Matrix Inversion Method . . .

```
// Program to find solution of system of linear equation
// using Matrix Inversion technique
#include<iostream.h>
#include<iomanip.h>
#include<math.h>
#define MAX 20
void main() {
    float a[MAX][MAX], b[MAX], x[MAX], temp, comm;
    int i, j, k, m, n, p, q, r, s;
    cout << "Enter number of equations : ";
    cin >> n;
    cout >> "\nEnter coefficients equation wise\n";
    for (i = 1; i <= n; i++ ) {
        for (j = 1; j <= n; j++ ) {
            cin >> a[i][j];
        }
        cin >> b[i];
    }
    for (i = 1; i <= n; i++ ) {
        for (j = (n+1); j <= (n+n); j++ ) {
            if ((j - i) == n) {
                a[i][j]= 1.0;
            } else {
                a[i][j]= 0.0;
            }
        }
    }
    for (k = 1; k <= n; k++ ) {
        temp = a[k][k];
        for (j = k; j <= (n+n); j++ ) {
            a[k][j] = a[k][j]/temp;
        }
        for (i = 1; i <= n; i++ ) {
            if ( i != k ) {
                comm = a[i][k];
                for (j = k; j <= (n+n); j++ ) {
                    a[i][j] = a[i][j] - comm * a[k][j];
                }
            }
        }
    }
    for (i = 1; i <= n; i++ ) {
        for (j = (n+1); j <= (n+n); j++ ) {
            a[i][j-n] = a[i][j];
```

```
        }
    }
    for (i = 1; i <= n; i++ ) {
        x[i] = 0.0;
        for (j = 1; j <= n; j++ ) {
            x[i] = x[i] + a[i][j] * b[j];
        }
    }
    cout << "\n\nRequired solution is\n";
    cout.setf(ios::showpoint);
    cout << setprecision(4);
    for (i = 1; i <= n; i++)
        cout << "\nx[" << i << "] = " << setw(8) << x[i];
    cout << endl;
}
```

Modified Crout's Method . . .

```
// Program to find solution of system of linear equation
// using modified Crout's method
#include<iostream.h>
#include<iomanip.h>
#include<math.h>
#define MAX  20
void main() {
    float a[MAX][MAX], x[MAX], temp, sum;
    int i, j, jj, k, n;
    cout << "enter number of equations : ";
    cin >> n;
    cout << "\nenter coefficients equation wise\n";
    for (i = 1; i <= n; i++ ) {
        for (j = 1; j <= (n+1); j++ ) {
            cin >> a[i][j];
        }
    }
    temp = a[1][1];
    for ( j = 2; j <= n+1; j++ ) {
        a[1][j] = a[1][j] / temp;
    }
    for ( j = 2; j <= n; j++ ) {
        for ( i = j; i <= n; i++ ) {
            sum = 0.0;
            for ( k = 1; k <= j-1; k++ ) {
                sum = sum + a[i][k] * a[k][j];
            }
            a[i][j] = a[i][j] - sum;
        }
        i = j;
        for ( jj = i+1; jj <= n+1; jj++ ) {
            sum = 0.0;
```

```
        for ( k = 1; k <= i-1; k++ ) {
            sum = sum + a[i][k] * a[k][jj];
        }
        a[i][jj] = ( a[i][jj] - sum ) / a[i][i];
    }
  }
  x[n] = a[n][n+1];
  for (i = n-1; i >= 1; i-- ) {
    sum = 0.0;
    for ( j = i+1; j <= n; j++ ) {
        sum = sum + a[i][j] * x[j];
    }
    x[i] = a[i][n+1] - sum;
  }
  cout << "\n\nrequired solution is\n";
  cout.setf(ios::showpoint);
  cout << setprecision(4);
  for (i = 1; i <= n; i++)
    cout << "\nx[" << i << "] = " << setw(8) << x[i];
  cout << endl;
}
```

Jacobi's Method ...

```
//  Program to find solution of system of linear equation
//  using Jacobi's method
#include<iostream.h>
#include<iomanip.h>
#include<stdlib.h>
#include<math.h>
#define MAX  20
void main() {
    float a[MAX][MAX], old_x[MAX], new_x[MAX], sum, epsilon;
    float relerror, bigrelerror;
    int i, j, k, n, maxiter;
    cout << "Enter number of equations : ";
    cin >> n;
    cout >> "\nEnter coefficients equation wise\n";
    for (i = 1; i <= n; i++ ) {
        for (j = 1; j <= (n+1); j++ ) {
            cin >> a[i][j];
        }
    }
    cout << "Enter maximum number of iterations permited : ";
    cin >> maxiter;
    cout << "\nEnter acceptable precision : ";
    cin >> epsilon;
    for (i = 1; i <= n; i++ ) {
        old_x[i] = 0.0;
```

```
    }
    for (k = 1; k <= maxiter; k++ ) {
        bigrelerror = 0.0;
        for (i = 1; i <= n; i++ ) {
            sum = 0.0;
            for (j = 1; j <= n; j++ ) {
                if ( i != j ) {
                    sum = sum + a[i][j] * old_x[j];
                }
            }
            new_x[i] = (a[i][n+1]-sum)/a[i][i];
            relerror = fabs ( (new_x[i]-old_x[i])/new_x[i]);
            if ( relerror > bigrelerror ) {
                bigrelerror = relerror;
            }
        }
        if ( bigrelerror < epsilon ) {
            cout << "\n\nSolution converges in "
                 << k << " iterations\n";
            cout.setf(ios::showpoint);
            cout << setprecision(4);
            for (i = 1; i <= n; i++)
                cout << "\nx[" << i << "] = " << setw(8) << x[i];
            cout << endl;
            exit(1);
        }
        for ( i = 1; i <= n; i++ ) {
            old_x[i] = new_x[i];
        }
    }
    cout << "\n\nSolution does not converges in "
         << maxiter << " iterations\n";
}
```

Gauss-Seidel Method ...

```
//  Program to find solution of system of linear equation
//  using Gauss-Seidel method
#include<iostream.h>
#include<iomanip.h>
#include<stdlib.h>
#include<math.h>
#define MAX  20
void main() {
    float a[MAX][MAX], x[MAX], sum, epsilon;
    float relerror, bigrelerror, temp;
    int i, j, k, n, maxiter;
    cout << "Enter number of equations : ";
    cin >> n;
    cout >> "\nEnter coefficients equation wise\n";
```

```
    for (i = 1; i <= n; i++ ) {
        for (j = 1; j <= (n+1); j++ ) {
            cin >> a[i][j];
        }
    }
    cout << "Enter maximum number of iterations permitted : ";
    cin >> maxiter;
    cout << "\nEnter acceptable precision : ";
    cin >> epsilon;
    for (i = 1; i <= n; i++ ) {
        x[i] = 0.0;
    }
    for (k = 1; k <= maxiter; k++ ) {
        bigrelerror = 0.0;
        for (i = 1; i <= n; i++ ) {
            sum = 0.0;
            for (j = 1; j <= n; j++ ) {
                if ( i != j ) {
                    sum = sum + a[i][j] * x[j];
                }
            }
            temp = ( a[i][n+1]-sum)/a[i][i];
            relerror = fabs((temp-x[i])/temp);
            x[i] = temp;
            if ( relerror > bigrelerror ) {
                bigrelerror = relerror;
            }
        }
        if ( bigrelerror < epsilon ) {
            cout << "\n\nSolution converges in "
                 << k << " iterations\n";
            cout.setf(ios::showpoint);
            cout << setprecision(4);
            for (i = 1; i <= n; i++)
                cout << "\nx[" << i << "] = " << setw(8) << x[i];
            cout << endl;
            exit(1);
        }
    }
    cout << "\n\nSolution does not converges in "
        << maxiter << " iterations\n";
}
```

Power Method . . .

```
//  Program to find the largest eigen value using Power method
#include<iostream.h>
#include<iomanip.h>
#include<math.h>
#define MAX  20
```

```
void main() {
    float a[MAX][MAX], x[MAX], s[MAX], epsilon;
    float error, big_error, temp, lamda;
    int i, j, k, n;
    cout << "Enter number of equations : ";
    cin >> n;
    cout >> "\nEnter coefficients equation wise\n";
    for (i = 1; i <= n; i++ ) {
        for (j = 1; j <= n; j++ ) {
            cin >> a[i][j];
        }
    }
    cout >> "\nEnter required precision : ";
    cin >> epsilon;
    for (i = 1; i <= n; i++ ) {
        x[i] = 1.0;
    }
    do {
        big_error = 0.0;
        for (i = 1; i <= n; i++ ) {
            s[i] = 0.0;
            for (j = 1; j <= n; j++ ) {
                s[i] = s[i] + a[i][j] * x[j];
            }
        }
        lamda = fabs (s[1]);
        for (i = 1; i <= n; i++ ) {
            temp = s[i]/lamda;
            error = fabs((temp-x[i])/temp);
            x[i] = temp;
            if ( error > big_error ) {
                big_error = error;
            }
        }
    } while ( big_error > epsilon );
    cout.setf(ios::showpoint);
    cout << setprecision(4);
    cout << "\n\nLargest eigen value = " << lamda;
    cout << "\nAssociated eigen vector is\n";
    for (i = 1; i <= n; i++) {
        cout << "\nx[" << i << "] = " << setw(8) << x[i];
    }
    cout << endl;
}
```

Lagrangian Interpolation . . .

```
//  Program to implement the Lagrangian method of interpolation to
//  interpolate the value of variable 'y' for given value of
//  variable 'x'
```

```
#include<iostream.h>
#include<iomanip.h>
#define MAXPOINTS   20
void main()
{
    float x[MAXPOINTS], y[MAXPOINTS], sum, prod, a;
    int i, j, n;
    cout << "Enter number of points : ";
    cin >> n;
    for ( i = 1; i <= n; i++ ) {
       cout << "Enter x[" << i << "] : ";
       cin >> x[i];
       cout << "Enter y[" << i << "] : ";
       cin >> y[i];
    }
    cout << "Enter value for interpolation : ";
    cin >> a;
    if ( ( a < x[1] ) && ( a > x[n] ) ) {
       cout << "\nValue lies outside range\n";
       exit(1);
    }
    sum = 0;
    for ( i = 1; i <= n; ++i ) {
       prod = 1;
       for ( j = 1; j <= n; ++j ) {
          if ( i != j )
              prod *= (a-x[j])/(x[i]-x[j]);
       }
       sum += y[i] * prod;
    }
   cout.setf(ios::showpoint);
   cout << setprecision(4)
       << "\nInterpolated value = " << sum << endl;
}
```

Newton's Forward Difference Interpolation . . .

```
//  Program to implement Newton's forward difference interpolation
formula
#include<iostream.h>
#include<iomanip.h>
#include<stdlib.h>
void main() {
    int fact(int m);
    float x[10], y[10], d[10][10], a, u, sum, prod;
    int i, j, k, n;
    cout << "Enter number of points : ";
    cin >> n;
    for ( i = 1; i <= n; i++ ) {
       cout << "Enter x[" << i << "] : ";
```

```
        cin >> x[i];
        cout << "Enter y[" << i << "] : ";
        cin >> y[i];
    }
    cout << "Enter value for interpolation : ";
    cin >> a;
    if ( ( a < x[1] ) && ( a > x[n] ) ) {
        cout << "\nValue lies outside range\n";
        exit(1);
    }
    i = 2;
    while ( a > x[i] ) {
        i++;
    }
    k = i-1;
    u = (a - x[k])/(x[k+1] - x[k]);
    for ( j = 1; j <= (n-1); j++ ) {
        for ( i = 1; i <= (n-j); i++ ) {
            if ( j == 1 ) {
                d[i][j] = y[i+1] - y[i];
            } else {
                d[i][j] = d[i+1][j-1] - d[i][j-1];
            }
        }
    }
    sum = y[k];
    for ( i = 1; i <= (n-k); i++ ) {
        prod = 1.0;
        for ( j = 0; j <= (i-1); j++ ) {
            prod = prod *(u-j);
        }
        sum = sum + d[k][i] * prod / fact(i);
    }
    cout.setf(ios::showpoint);
    cout << setprecision(4)
        << "\nInterpolated value = " << sum << endl;
}
//   function to compute factorial of number 'm'
int fact(int m)
{
    int i, prod = 1;
    for ( i = 2; i <= m; i++ ) {
        prod = prod * i;
    }
    return prod;
}
```

Newton's Backward Difference Interpolation . . .

```
//  Program to implement Newton's backward difference interpolation
formula
#include<iostream.h>
#include<iomanip.h>
#include<stdlib.h>
void main() {
    int fact(int m);
    float x[10], y[10], d[10][10], a, u, sum, prod;
    int i, j, n, k;
    cout << "Enter number of points : ";
    cin >> n;
    for ( i = 1; i <= n; i++ ) {
       cout << "Enter x[" << i << "] : ";
       cin >> x[i];
       cout << "Enter y[" << i << "] : ";
       cin >> y[i];
    }
    cout << "Enter value for interpolation : ";
    cin >> a;
    if ( ( a < x[1] ) && ( a > x[n] ) ) {
       cout << "\nValue lies outside range\n";
       exit(1);
    }
    i = 2;
    while ( a > x[i] ) {
       i++;
    }
    k = i;
    u = (a - x[k])/(x[k] - x[k-1]);
    for ( j = 1; j < n; j++ ) {
        for ( i = j+1; i <= n; i++ ) {
            if ( j == 1 ) {
               d[i][j] = y[i] - y[i-1];
            } else {
               d[i][j] = d[i][j-1] - d[i-1][j-1];
            }
        }
    }
    sum = y[k];
    for ( i = 1; i <= (k-1); i++ ) {
        prod = 1.0;
        for ( j = 0; j <= (i-1); j++ ) {
            prod = prod * (u+j);
        }
        sum = sum + d[k][i] * prod / fact(i);
    }
    cout.setf(ios::showpoint);
    cout << setprecision(4)
```

```
        << "\nInterpolated value = " << sum << endl;
}
//   function to compute factorial of number 'm'
int fact(int m) {
   int i, prod = 1;
   for ( i = 2; i <= m; i++ ) {
       prod = prod * i;
   }
   return prod;
}
```

Newton's Divided Difference Interpolation . . .

```
//  Program to implement Newton's divided difference interpolation
formula
#include<iostream.h>
#include<iomanip.h>
#include<stdlib.h>
void main() {
    float x[10], y[10], d[10][10], a, sum, prod;
    int i, j, k, n;
    cout << "Enter number of points : ";
    cin >> n;
    for ( i = 1; i <= n; i++ ) {
       cout << "Enter x[" << i << "] : ";
       cin >> x[i];
       cout << "Enter y[" << i << "] : ";
       cin >> y[i];
    }
    cout << "Enter value for interpolation : ";
    cin >> a;
    if ( ( a < x[1] ) && ( a > x[n] ) ) {
       cout << "\nValue lies outside range\n";
       exit(1);
    }
    i = 2;
    while ( a > x[i] ) {
      i++;
    }
    k = i-1;
    for ( j = 1; j < n; j++ ) {
        for ( i = 1; i <= (n-j); i++ ) {
            if ( j == 1 ) {
               d[i][j] = (y[i+1] - y[i])/(x[i+1] - x[i]);
            } else {
               d[i][j] = (d[i+1][j-1]-d[i][j-1])/(x[i+j]-x[i]);
            }
        }
    }
    sum = y[k];
```

```
    for ( i = 1; i <= (n-k); i++ ) {
        prod = 1.0;
        for ( j = 0; j <= (i-1); j++ ) {
            prod = prod * (a-x[k+j]);
        }
        sum = sum + d[k][i] * prod;
    }
    cout.setf(ios::showpoint);
    cout << setprecision(4)
         << "\nInterpolated value = " << sum << endl;
}
```

Fitting a Straight Line . . .

```
//  Program developed to fit a straight line of Y on X
//  through a set of points using method of least squares
#include<iostream.h>
#include<iomanip.h>

void main() {
   float x[20], y[20];
   int i, n;
   float sx, sx2, sy, sxy, denom, a1, a2;
   cout << "Enter number of points : ";
   cin >> n;
   for ( i = 1; i <= n; i++ ) {
      cout << "Enter x[" << i << "] : ";
      cin >> x[i];
      cout << "Enter y[" << i << "] : ";
      cin >> y[i];
   }
   sx = 0.0;
   sx2 = 0.0;
   sy = 0.0;
   sxy = 0.0;
   for ( i = 1; i <= n; i++ ) {
        sx  = sx + x[i];
        sx2 = sx2 + x[i] * x[i];
        sy  = sy + y[i];
        sxy = sxy + x[i] * y[i];
   }
   denom = n * sx2 - sx * sx;
   a2 = ( n * sxy - sx * sy ) / denom;
   a1 = ( sy * sx2 - sx * sxy ) / denom;
   cout << "\nRegression coefficients are\n";
   cout.setf(ios::showpoint);
   cout << setprecision(4);
   cout << "\nSlope of the regression line = " << setw(8) << a2;
   cout << "\nY-intercept for the line = " << setw(8) << a1 << endl;
}
```

Fitting a Polynomial . . .

```
// Program to fit a polynomial of degree 'm' through a set
// of points using method of least squares
#include<iostream.h>
#include<iomanip.h>
#include<math.h>
void main() {
   float x[20], y[20], a[20], c[20][20];
   int i, j, n, m, k, m1, m2, l;
   float p, q, max, temp, sum;
   cout << "Enter degree of polynomial : ";
   cin >> m;
   cout << "Enter number of points : ";
   cin >> n;
   for ( i = 1; i <= n; i++ ) {
      cout << "Enter x[" << i << "] : ";
      cin >> x[i];
      cout << "Enter y[" << i << "] : ";
      cin >> y[i];
   }
   m1 = m + 1;
   m2 = m + 2;
   for ( i = 1; i <= m1; i++ ) {
      for ( j = 1; j <= m1; j++ ) {
          l = i + j - 2;
          c[i][j] = 0.0;
          for ( k = 1; k <= n; k++ ) {
             c[i][j] = c[i][j] + pow((double)x[k], (double)l);
          }
      }
      l = i - 1;
      c[i][m2] = 0.0;
      for ( k = 1; k <= n; k++ ) {
          c[i][m2] = c[i][m2]+ y[k] * pow((double)x[k], (double)l);
      }
   }
   for ( k = 1; k <= (m1-1); k++ ) {
       max = abs(c[k][k]);
       p = k;
       for ( q = k+1; q <= m1; q++ ) {
           if ( abs( c[q][k] ) > max ) {
               max = abs ( c[q][k] );
               p = q;
           }
       }
       if ( p != k ) {
           for ( q = k; q <= m2; q++ ) {
               temp = c[k][q];
               c[k][q] = c[p][q];
```

```
                c[p][q] = temp;
            }
        }
        for ( i = k+1; i <= m1; i++ ) {
            temp = c[i][k]/c[k][k];
            for ( j = k; j <= m2; j++ ) {
                c[i][j] = c[i][j] - temp * c[k][j];
            }
        }
    }
    a[m1] = c[m1][m2]/c[m1][m1];
    for ( i = (m1-1); i >= 1; i-- ) {
        sum = 0.0;
        for ( j = (i+1); j <= m1; j++ ) {
            sum = sum + c[i][j] * a[j];
        }
        a[i] = ( c[i][m2] - sum ) / c[i][i];
    }
    cout<< "\nRegression coefficients\n\n";
    cout.setf(ios::showpoint);
    cout << setprecision(4);
    for ( i = 1; i <= m1; i++ ) {
        cout << "\na[" << i << "] = " << setw(8) << a[i];
    }
    cout << endl;
}
```

Fitting an Exponential Curve . . .

```
//  Program to fit an exponential curve of type  y = ae^bx
#include<iostream.h>
#include<iomanip.h>
#include<math.h>
void main() {
   float x[20], y[20];
   int i, n;
   float xx, yy, sx, sx2, slogy, sxlogy, denom, a, b, c;
   cout << "Enter number of points : ";
   cin >> n;
   for ( i = 1; i <= n; i++ ) {
      cout << "Enter x[" << i << "] : ";
      cin >> x[i];
      cout << "Enter y[" << i << "] : ";
      cin >> y[i];
   }
   sx = 0.0;
   sx2 = 0.0;
   slogy = 0.0;
   sxlogy = 0.0;
   for ( i = 1; i <= n; i++ ) {
```

```
      xx = x[i];
      yy = log10(double(y[i]));
      sx  = sx + xx;
      sx2 = sx2 + xx * xx;
      slogy  = slogy + yy;
      sxlogy = sxlogy + xx * yy;
   }
   denom = n * sx2 - sx * sx;
   b = (n * sxlogy - sx * slogy) / denom;
   c = (slogy * sx2 - sx * sxlogy) / denom;
   a = exp(double(c));
   cout << "\nRegression coefficients are\n";
   cout.setf(ios::showpoint);
   cout << setprecision(5);
   cout << "\na = " << setw(8) << a;
   cout << "\nb = " << setw(8) << b << endl;
}
```

Differentiating a Function . . .

```
//  Program to compute first derivative at tabulated point
//  when the function is tabulated at equal intervals
#include<iostream.h>
#include<iomanip.h>
#include<stdlib.h>
void main() {
    float x[10], y[10], d[10][10], a, sum, derivative, h, term;
    int i, j, k, n, sign;
    cout << "Enter number of table points : ";
    cin >> n;
    cout >> "Enter interval size : ";
    cin("%f", &h);
    for ( i = 1; i <= n; i++ ) {
        cout << "Enter x[" << i << "] : ";
        cin >> x[i];
        cout << "Enter y[" << i << "] : ";
        cin >> y[i];
    }
    cout << "Enter tabulated point : ";
    cin >> a;
    if ((a < x[1]) && (a > x[n])) {
       cout << "\nValue lies outside range\n";
       exit(1);
    }
    i = 1;
    while ( a != x[i] )  {
       i++;
    }
    k = i;
```

```
    for ( j = 1; j <= (n-1); j++ ) {
        for ( i = 1; i <= (n-j); i++ ) {
            if ( j == 1 ) {
               d[i][j] = y[i+1] - y[i];
            } else {
               d[i][j] = d[i+1][j-1] - d[i][j-1];
            }
        }
    }
    sum = 0.0;
    sign = 1;
    for ( j = 1; j <= (n-k); j++ ) {
        term =  sign * d[k][j] / j;
        sum = sum + term;
        sign = - sign;
    }
    derivative = (1.0/h) * sum;
    cout.setf(ios::showpoint);
    cout << setprecision(4)
         << "\nValue of the derivative = " << derivative << endl;
}

// Program to compute first derivative at when the function
// is tabulated at unequal intervals
#include<iostream.h>
#include<iomanip.h>
#include<stdlib.h>
void main() {
    float x[10], y[10], d[10][10], a, sum1, sum2, prod;
    int i, j, k, n, m, p;
    cout << "Enter number of table points : ";
    cin >> n;
    for ( i = 1; i <= n; i++ ) {
        cout << "Enter x[" << i << "] : ";
        cin >> x[i];
        cout << "Enter y[" << i << "] : ";
        cin >> y[i];
    }
    cout << "Enter tabulated point : ";
    cin >> a;
    if ((a < x[1]) && (a > x[n])) {
       cout << "\nValue lies outside range\n";
       exit(1);
    }
    i = 1;
    while ( a > x[i] ) {
       i++;
    }
    k = i-1;
    for ( j = 1; j <= (n-1); j++ ) {
        for ( i = 1; i <= (n-j); i++ ) {
```

```
            if ( j == 1 ) {
                d[i][j] = (y[i+1]-y[i])/(x[i+1]-x[i]);
            } else {
                d[i][j] = (d[i+1][j-1]-d[i][j-1])/(x[i+j]-x[i]);
            }
        }
    }
    sum1 = d[k][1];
    for ( j = 2; j <= (n-k); j++ ) {
        sum2 = 0;
        for ( m = 0; m <= (j-1); m++ ) {
            prod = 1.0;
            for ( p = 0; p <= (j-1); p++ ) {
                if ( p != m ) {
                    prod = prod * ( a - x[k+p] );
                }
            }
            sum2 = sum2 + prod;
        }
       sum1 = sum1 + d[k][j] * sum2;
    }
    cout.setf(ios::showpoint);
    cout << setprecision(4)
         << "\nValue of the derivative = " << sum1 << endl;
}
```

Trapezoidal Rule . . .

```
//  Program to implement Trapezoidal rule for tabulated function
#include<iostream.h>
#include<iomanip.h>
void main() {
    int i, n;
    float x[20], y[20], h, sum, integral;
    cout << "Enter number of intervals : ";
    cin >> n;
    cout << "Enter size of interval : ";
    cin >> h;
    for ( i = 1; i <= (n+1); i++ ) {
        cout << "Enter x[" << i << "] : ";
        cin >> x[i];
        cout << "Enter y[" << i << "] : ";
        cin >> y[i];
    }
    sum = ( y[1] + y[n+1] )/2;
    for ( i = 2; i <= n; i++ ) {
        sum = sum + y[i];
    }
    integral = h * sum;
    cout.setf(ios::showpoint);
```

```
    cout << setprecision(4)
         << "\nValue of the integral = " << integral << endl;
}

//  Program to implement Trapezoidal rule for known function
#include<iostream.h>
#include<iomanip.h>
#include<math.h>
float f(float x) {
    return exp(-0.5*x);
}
void main() {
    int i, n;
    float a, b, h, sum, integral;
    cout << "Enter lower limit of integration : ";
    cin >> a;
    cout << "Enter upper limit of integration : ";
    cin >> b;
    cout << "Enter number of intervals : ";
    cin >> n;
    cout << "Enter number of intervals : ";
    cin >> n;
    h = (b - a)/n;
    sum = (f(a)+f(b))/2.0;
    for ( i = 1; i <= (n-1); i++ ) {
        sum = sum + f(a+i*h);
    }
    integral = h * sum;
    cout.setf(ios::showpoint);
    cout << setprecision(4)
         << "\nValue of the integral = " << integral << endl;
}

// Program to implement Modified Trapezoidal rule for known function
#include<iostream.h>
#include<iomanip.h>
#include<math.h>
float f(float x) {
    return exp(-0.5*x);
}
void main() {
    int i, k;
    float a, b, h, sum, integral, epsilon, i_old, i_new;
    cout << "Enter lower limit of integration : ";
    cin >> a;
    cout << "Enter upper limit of integration : ";
    cin >> b;
    cout << "Enter acceptable precision : ";
    cin >> epsilon;
    h = b - a;
    sum = (f(a)+f(b))/2.0;
```

```
    i_old = 0.0;
    i_new = h * sum;
    k = 1;
    while ( fabs((i_new - i_old)/i_new) > epsilon ) {
        k = k * 2;
        h = h / 2;
        for ( i = 1; i <= k; i += 2 ) {
            sum = sum + f(a+i*h);
        }
        i_old = i_new;
        i_new = h * sum;
    }
    cout.setf(ios::showpoint);
    cout << setprecision(4)
        << "\nValue of the integral = " << i_new << endl;
}
```

Simpson 1/3rd Rule . . .

```
//   Program to implement Simpson's (1/3)rd rule for tabulated function
#include<iostream.h>
#include<iomanip.h>
void main() {
    int i, n;
    float x[20], y[20], h, sum, integral, s2, s4;
    cout << "Enter number of intervals : ";
    cin >> n;
    cout << "Enter size of interval : ";
    cin >> h;
    for ( i = 1; i <= (n+1); i++ ) {
        cout << "Enter x[" << i << "] : ";
        cin >> x[i];
        cout << "Enter y[" << i << "] : ";
        cin >> y[i];
    }
    sum = y[1] + y[n+1];
    s2 = s4 = 0.0;
    for ( i = 2; i <= n; i += 2 ) {
        s4 = s4 + y[i];
    }
    for ( i = 3; i <= (n-1); i += 2 ) {
        s2 = s2 + y[i];
    }
    integral = (h/3.0) * ( sum + 2 * s2 + 4 * s4 );
    cout.setf(ios::showpoint);
    cout << setprecision(4)
        << "\nValue of the integral = " << integral << endl;
}
```

```
//   Program to implement Simpson's (1/3)rd rule for known function
#include<iostream.h>
#include<iomanip.h>
#include<math.h>
float f(float x) {
    return exp(-0.5*x);
}
void main() {
    int i, n;
    float a, b, h, x, s2, s4, sum, integral;
    cout << "Enter lower limit of integration : ";
    cin >> a;
    cout << "Enter upper limit of integration : ";
    cin >> b;
    cout << "Enter number of intervals : ";
    cin >> n;
    h = (b-a)/n;
    sum = f(a)+f(b);
    s2 = s4 = 0.0;
    for ( i = 1; i < n; i+=2 ) {
        x = a + i * h;
        s4 = s4 + f(x);
    }
    for ( i = 2; i < n; i+=2 ) {
        x = a + i * h;
        s2 = s2 + f(x);
    }
    integral = (h/3.0) * ( sum + 2 * s2 + 4 * s4 );
    cout.setf(ios::showpoint);
    cout << setprecision(4)
         << "\nValue of the integral = " << integral << endl;
}
```

```
//  Program to implement Modified Simpson (1/3)rd rule
//  for known function
#include<iostream.h>
#include<iomanip.h>
#include<stdlib.h>
#include<math.h>
inline float f(float x)
{
  return (1+3*x*x);
}
void main(void)
{
   float a, b, i_old, i_new, epsilon, sum, h, s1, s2, s4, x;
   int i, k;
   cout << "Enter lower limit of integration : ";
   cin >> a;
   cout << "Enter upper limit of integration : ";
   cin >> b;
```

```
    cout << "Enter acceptable precision : ";
    cin >> epsilon;
    h = (b - a)/2;
    s1 = f(a)+f(b);
    s2 = 0;
    s4 = f(a+h);
    i_old = 0;
    i_new = (h/3)* (s1 + 4 * s4);
    k = 2;
    while ( fabs((i_new - i_old)/i_new) > epsilon ) {
        k = k * 2;
        h = h / 2;
        s2 = s2 + s4;
        s4 = 0;
        for ( i = 1; i <= k; i += 2 ) {
            x = a + i * h;
            s4 = s4 + f(x);
        }
        i_old = i_new;
        i_new = (h/3) * (s1 + 4 * s4 + 2 * s2);
    }
    cout.setf(ios::showpoint);
    cout << setprecision(4)
        << "\nValue of the integral = " << i_new << endl;
}
```

Simpson 3/8th Rule . . .

```
//  Program to implement Simpson's (3/8)th rule for tabulated function
#include<iostream.h>
#include<iomanip.h>

void main() {
    int i, n;
    float x[20], y[20], h, sum, integral, s2, s3;
    cout << "Enter number of intervals : ";
    cin >> n;
    cout << "Enter size of interval : ";
    cin >> h;
    for ( i = 1; i <= (n+1); i++ ) {
        cout << "Enter x[" << i << "] : ";
        cin >> x[i];
        cout << "Enter y[" << i << "] : ";
        cin >> y[i];
    }
    sum = y[1] + y[n+1];
    s2 = s3 = 0.0;
    for ( i = 2; i <= (n-1); i += 3 ) {
        s3 = s3 + y[i] + y[i+1];
    }
```

```
    for ( i = 4; i <= (n-2); i += 3 ) {
        s2 = s2 + y[i];
    }
    integral = (3.0*h/8.0) * ( sum + 2 * s2 + 3 * s3 );
    cout.setf(ios::showpoint);
    cout << setprecision(4)
         << "\nValue of the integral = " << integral << endl;
}
//   Program to implement Simpson's (3/8)th rule for known function
#include<iostream.h>
#include<iomanip.h>
#include<math.h>
float f(float x) {
   return  1.0+x*x*x;
}
void main() {
    int i, n;
    float a, b, h, x, s2, s3, sum, integral;
    cout << "Enter lower limit of integration : ";
    cin >> a;
    cout << "Enter upper limit of integration : ";
    cin >> b;
    cout << "Enter number of intervals : ";
    cin >> n;
    h = (b-a)/n;
    sum = f(a) + f(b);
    s2 = s3 = 0.0;
    for ( i = 1; i < n; i+=3 ) {
        x = a + i * h;
        s3 = s3 + f(x)+f(x+h);
    }
    for ( i = 3; i < n; i+=3 ) {
        x = a + i * h;
        s2 = s2 + f(x);
    }
    integral = (3.0*h/8.0) * ( sum + 2 * s2 + 3 * s3 );
    cout.setf(ios::showpoint);
    cout << setprecision(4)
         << "\nValue of the integral = " << integral << endl;
}
```

Romberg Integration . . .

```
//  Program to implement Romberg integration for known function
#include<iostream.h>
#include<iomanip.h>
#include<math.h>
enum boolean { false, true };
float f ( float x ) {
  return (x*x+2.0*x);
```

```
}
void main() {
      float r[20][20], a, b, x, h, epsilon, sum, temp;
      int i, j, k, n;
      boolean flag;
      cout << "Enter lower limit for integration : ";
      cin >> a;
      cout << "Enter lower upper for integration : ";
      cin >> b;
      cout << "Enter allowed precision : ";
      cin >> epsilon;
      i = 1;
      n = 1;
      h = b - a;
      sum = ( f(a) + f(b) )/2;
      r[i][1] = h * sum;
      flag = false;
      while ( flag == false ) {
       n *= 2;
       h /= 2;
       i++;
       for ( j = 1; j <=n; j +=2 ) {
          x = a + j * h;
          sum += f(x);
       }
       r[i][1] = h * sum;
       for ( k = 2; k <= i; k++ ) {
           temp = pow(double(4), double(k-1));
           r[i][k] = (temp * r[i][k-1] - r[i-1][k-1])/(temp - 1);
       }
       if ( fabs(r[i][i] - r[i][i-1]) <= epsilon ) {
           cout << "\nElements of Romberg table are\n\n";
           cout.setf(ios::showpoint);
           cout.precision(4);
           for ( k = 1; k <= i; k++ ) {
             for ( j = 1; j <= k; j++ )
                 cout << setw(10) << r[k][j];
             cout << endl;
           }
           flag = true;
       }
      }
      cout << endl << "value of integral = "
           << setw(10) << r[i][i] << endl;

}
```

Gauss's Quadrature Formula . . .

```
//  Program to implement Gauss's Quadrature formula
#include<iostream.h>
#include<iomanip.h>
float g(float y) {
    return y*y + 8.0*y + 15;
}
void main() {
    int i, n;
    float a, b, r[20], y[20], sum, integral;
    cout << "Enter number of points : ";  cin >> n;
    cout << "Enter lower limit of integration : "; cin >> a;
    cout << "Enter upper limit of integration : ";    cin >> b;
    for ( i = 1; i <= n; i++ ) {
        cout << "Enter value of r[" << i << "] : ";
        cin >> r[i];
        cout << "Enter value of y[" << i << "] : ";
        cin >> y[i];
    }
    sum = 0;
    for ( i = 1; i <= n; i++ ) {
        sum = sum + r[i] * g(y[i]);
    }
    integral = ((b-a)/2.0) * sum;
    cout.setf(ios::showpoint);
    cout << setprecision(4);
    cout << "\nValue of the integral = " << integral << endl;
}
```

Euler's Method . . .

```
//  Program to solve ODE of type dy/dx = xy using Euler's method
#include<iostream.h>
#include<iomanip.h>
#include<conio.h>
float f(float x, float y) {
    return (x*y);
}
void main() {
    int i;
    float x, y, x1, y1, xf, h;
    cout << "Enter value of x1 : ";   cin >> x1;
    cout << "Enter value of y1 : ";   cin >> y1;
    cout << "Enter value of xf : ";   cin >> xf;
    cout << "Enter size of interval : ";  cin >> h;
    x = x1; y = y1;  i = 1;
    cout << "\n\nPoints constituting solution curve are\n";
    cout << "\n Point       x             y";
```

```
    cout << "\n ================================";
    cout.setf(ios::showpoint);
    cout << endl << setw(3) << i
       << "\t" << setw(8) << setprecision(2) << x
       << "\t" << setw(8) << setprecision(4) << y;
    while ( x < xf ) {
       y = y + h * f(x, y);
       x = x + h;
       i = i + 1;
       cout << endl << setw(3) << i
            << "\t" << setw(8) << setprecision(2) << x
            << "\t" << setw(8) << setprecision(4) << y;
    }
    cout << "\n ================================\n";
}
```

Runge-Kutta Second Order Method . . .

```
//  Program to solve ODE of type dy/dx = xy using
//  second order Runge-Kutta method (Heun's method)
#include<iostream.h>
#include<iomanip.h>
float f(float x, float y) {
    return (x*y);
}
void main() {
    int i;
    float x, y, x1, y1, xf, h, s1, s2;
    cout << "Enter value of x1 : ";  cin >> x1;
    cout << "Enter value of y1 : ";  cin >> y1;
    cout << "Enter value of xf : ";  cin >> xf;
    cout << "Enter size of interval : ";  cin >> h;
    x = x1;
    y = y1;
    i = 1;
    cout << "\n\nPoints constituting solution curve are\n";
    cout << "\n Point          x              y";
    cout << "\n ================================";
    cout.setf(ios::showpoint);
    cout << endl << setw(3) << i
       << "\t" << setw(8) << setprecision(2) << x
       << "\t" << setw(8) << setprecision(4) << y;
    while ( x < xf ) {
       s1 = f(x, y);
       x = x + h;
       s2 = f(x, y+h*s1);
       y = y + h * (s1+s2)/2;
       i = i + 1;
       cout << endl << setw(3) << i
            << "\t" << setw(8) << setprecision(2) << x
```

```
            << "\t" << setw(8) << setprecision(4) << y;
    }
    cout << "\n =================================\n";
}
```

Runge-Kutta Fourth Order Method . . .

```
//  Program to solve ODE of type dy/dx = xy
//  using fourth order Runge-Kutta method
#include<iostream.h>
#include<iomanip.h>
float f(float x, float y) {
    return (x*y);
}
void main() {
    int i;
    float x, y, x1, y1, xf, h, s, s1, s2, s3, s4;
    cout << "Enter value of x1 : ";   cin >> x1;
    cout << "Enter value of y1 : ";   cin >> y1;
    cout << "Enter value of xf : ";   cin >> xf;
    cout << "Enter size of interval : ";  cin >> h;
    x = x1;    y = y1;
    i = 1;
    cout << "\n\nPoints constituting solution curve are\n";
    cout << "\n Point       x              y";
    cout << "\n =================================";
    cout.setf(ios::showpoint);
    cout << endl << setw(3) << i
       << "\t" << setw(8) << setprecision(2) << x
       << "\t" << setw(8) << setprecision(4) << y;
    while ( x < xf ) {
       s1 = f(x, y);
       s2 = f(x+h/2, y+h*s1/2);
       s3 = f(x+h/2, y+h*s2/2);
       s4 = f(x+h, y+h*s3);
       s = (s1+2*s2+2*s3+s4)/6;
       y = y + h*s;
       x = x + h;
       i = i + 1;
       cout << endl << setw(3) << i
            << "\t" << setw(8) << setprecision(2) << x
            << "\t" << setw(8) << setprecision(4) << y;
    }
    cout << "\n =================================\n";
}
```

Euler's Predictor Corrector Method . . .

```
//  Program to solve ODE of type dy/dx = xy
//  using Euler's predictor corrector method
#include<iostream.h>
#include<iomanip.h>
#include <math.h>
enum boolean { false, true };
float f(float x, float y) {
    return (x*y);
}
void main() {
    int i;
    float x, y, x1, y1, y2, xf, h, epsilon;
    boolean flag;
    cout << "Enter value of x1 : ";     cin >> x1;
    cout << "Enter value of y1 : ";     cin >> y1;
    cout << "Enter value of xf : ";     cin >> xf;
    cout << "Enter size of interval : ";     cin >> h;
    cout << "Enter allowed precision : ";    cin >> epsilon;
    x = x1;
    y = y1;
    i = 1;
    cout << "\n\nPoints constituting solution curve are\n";
    cout << "\n Point          x                y";
    cout << "\n =================================";
    cout.setf(ios::showpoint);
    cout << endl << setw(3) << i
       << "\t" << setw(8) << setprecision(2) << x
       << "\t" << setw(8) << setprecision(4) << y;
    while ( x < xf ) {
       y2 = y + h * f(x, y);
       flag = true;
       while ( flag ) {
       y1 = y2;
       y2 = y + (h / 2) * (f(x, y) + f(x+h, y2));
       if ( fabs((y2-y1)/y2) <= epsilon )
          flag = false;
       }
       y = y2;
       x = x + h;
       i = i + 1;
       cout << endl << setw(3) << i
            << "\t" << setw(8) << setprecision(2) << x
            << "\t" << setw(8) << setprecision(4) << y;
    }
    cout << "\n =================================\n";
}
```

❑ ❑ ❑

Appendix C

Listing of Python Programs

Bisection Method . . .

```
#   program for Bisection method to find the root of equation
#     x^3-4x-9 = 0
# function to compute expression x^3-4x-9
def f(x):
    return (x**3-4*x-9)
# driver code
print("\nProgram to find solution of ' x^3-4x-9 = 0' \
using Bisection method\n")

x1 = float(input("\nEnter first point of the search interval : "))
x2 = float (input("Enter second point of the search interval : "))
epsilon = float( input("Enter permitted error : "))
while ( abs(x1-x2) > epsilon ):
    x3 = ( x1+x2)/2             # compute mid point of the interval
    if ( f(x1)*f(x3) < 0 ):     # select the appropriate subinterval
        x2 = x3
    else:
        x1 = x3
print( "\nApproximate root = %9.4f" %(x3) )
```

False Position Method . . .

```
#   program for False position method to find the root of equation
#     x^3-2x-5 = 0
import math
import sys
# function to compute expression x^3-2x-5
def f(x):
    return (x**3-2*x-5)
```

```
# driver code
print("\nProgram to find solution of ' x^3-2x-5 = 0' \
using False Position method\n")

x1 = float(input("\nEnter first point of the search interval : "))
x2 = float(input("Enter second point of the search interval : "))
epsilon = float(input("Enter permitted error : "))
delta = float(input("Enter lower bound on slope : "))

f1 = f(x1)
f2 = f(x2)
while (math.fabs((x1-x2)/x2) > epsilon):   # check relative error
    if (math.fabs(f2-f1) <= delta ):
        print("\nSlope of the function becomes too small")
        sys.exit(1)
    x3 = (x1*f2-x2*f1)/(f2-f1)
    f3 = f(x3)
    if (f1*f3 < 0 ):
        x2 = x3
        f2 = f3
    else:
        x1 = x3
        f1 = f3
# end of while loop
print( "\nApproximate root = %9.4f" %(x3) )
```

Secant Method . . .

```
#   program for Secant method to find the root of equation
#     cos(x)- xe^x = 0

import math
import sys

# function to compute expression cos(x)- xe^x
def f(x):
    return (math.cos(x)-x*math.exp(x))

# driver code
print("\nProgram to find solution of 'cos(x)- xe^x = 0' \
using secant method\n")
print("\nProgram to find solution of 'cos(x)- xe^x = 0' using secant
method\n")
x1 = float(input("\nEnter first point of the search interval : "))
x2 = float(input("Enter second point of the search interval : "))
epsilon = float(input("Enter permitted error : "))
delta = float(input("Enter lower bound on slope : "))
n = int(input("Enter maximum iterations permitted : "))
```

```
f1 = f(x1)
f2 = f(x2)
for i in range(1,n+1):
    if (math.fabs(f2-f1) <= delta ):
        print("\nSlope of the function becomes too small")
        sys.exit(1)

    x3 = (x1*f2-x2*f1)/(f2-f1)
    f3 = f(x3)
    x1 = x2
    f1 = f2
    x2 = x3
    f2 = f3

    if (math.fabs((x1-x2)/x2) <= epsilon):   # check relative error
        print("\nSolution converges in",i,"iterations")
        print( "\nApproximate root =%8.4f" %(x3) )
        sys.exit(1)
# end of while loop
print("\nSolution does nt converges in",n,"iterations\n")
```

Newton–Raphson Method . . .

```
#   Program for Newton-Raphson method to find the root of equation
#     x^3-4x-9 = 0

import math
import sys

# function to compute expression x^3-4x-9
def f(x):
    return (x**3-4*x-9)

# function to compute expression 3x^2-4
def df(x):
    return (3*x**2 - 4)

# driver code
print("\nProgram to find solution of 'x^3-4x-9=0' \
using Newton-Raphson method\n")
x0 = float(input("\nstarting approximation : "))
epsilon = float(input("Enter permitted error : "))
delta = float(input("Enter lower bound on slope : "))
n = int(input("Enter maximum iterations permitted : "))

for i in range(1,n+1):
    f0 = f(x0)
    df0 = df(x0)
    if (math.fabs(df0) <= delta ):
        print("\nSlope of the function becomes too small")
```

```
        sys.exit(1)

    x1 = x0 - f0 / df0
    relerror = math.fabs( ( x1 - x0) / x1 )
    x0 = x1

    if (relerror <= epsilon):   # check relative error
        print("\nSolution converges in",i,"iterations")
        print( "\nApproximate root =%8.4f" %(x1) )
        sys.exit(1)
# end of while loop
print("\nSolution does not converges in",n,"iterations\n")
# end of program
```

Successive Approximation Method . . .

```
#  Program to implement method of successive approximations to find the
#  root of equation    2^x - x - 3 = 0

import math
import sys

# function to compute expression 2^x - 3
def g(x):
    return (2**x - 3)

# driver code
print("\nProgram to find solution of '2^x - x - 3 = 0' \
using successive approximations method\n")
x1 = float(input("\nstarting approximation : "))
epsilon = float(input("Enter prescribed tolerance : "))
n = int(input("Enter maximum iterations permitted : "))
for i in range(1,n+1):
    x2 = g(x1)
    relerror = math.fabs( ( x2 - x1) / x2 )
    x1 = x2
    if (relerror <= epsilon):   # check relative error
        print("\nSolution converges in",i,"iterations")
        print( "\nApproximate root =%8.4f" %(x2) )
        sys.exit(1)
print("\nSolution does not converges in",n,"iterations\n")
```

Muller's Method . . .

```
#  Program to implement Muller's method to find the
#  root of equation 3x + sin(x) - e^x = 0

# import appropriate modules
import math
```

```
import sys

# function to compute expression 3x + sin(x) - e^x
def f(x):
    return (3*x + math.sin(x)-math.exp(x))

# driver code
a1 = float(input("\nEnter first point of the search interval : "))
b1 = float(input("Enter second point of the search interval : "))
epsilon = float(input("Enter prescribed tolerance : "))
x2 = a1
x1 = b1
x0 = (a1+b1)/2
xrp = 0.0
while (True):
    h1 = x1 - x0
    h2 = x0 - x2
    gama = h2/h1
    f0 = f(x0)
    f1 = f(x1)
    f2 = f(x2)
    c = f0
    a = (gama*f1-f0*(1.0+gama)+f2)/(gama*h1*h1*(1.0+gama))
    b = (f1-f0-a*h1*h1)/h1
    disc = math.sqrt(b*b-4*a*c)
    if ( b > 0 ):
        xrn = x0-2.0*c/(b+disc)
    else:
        xrn = x0-2.0*c/(b-disc)
    relativeerror = math.fabs((xrp-xrn)/xrn)
    xrp = xrn
    if ( xrn > x0 ):
        x2 = x0
        x0 = xrn
    else:
        x1 = x0
        x0 = xrn
    if ( relativeerror <= epsilon ):     # check relative error
        print( "\nApproximate root = %9.4f" %(xrn) )
        break
```

Quotient-Difference Method . . .

```
#  Program to implement Quotient-Difference method to find real
#  roots of polynomial of type
#     a(n+1)*x^n+a(n)*x^(n-1)+a(n-1)*x^(n-2)+...+a(1)

# import appropriate modules
import math
```

```
n = int(input("Enter order of polynomial : "))

# Initialize list
a = [0]*20
e = [0]*20
q = [0]*20

# For user input
for i in range(1,n+2):
    print("Enter coefficient of x^",i-1," : ", sep = '', end = '')
    a[i] = float(input())

epsilon = float(input("Enter prescribed tolerance : "))

q[1] = -a[n]/a[n+1]
for i in range (2,n+1):
    q[i] = 0.0
e[1] = 0.0
e[n+1] = 0.0
for i in range (2,n+1):
   e[i] = a[n-i+1]/a[n-i+2]
while (True):
    for i in range (1,n+1):
        q[i] = e[i+1]-e[i]+q[i]
    for i in range (2,n+1):
        e[i] = (q[i]/q[i-1])*e[i]
    maxx = math.fabs(e[1])
    for i in range (2,n+2):
       if ( math.fabs(e[i]) > maxx ):
            maxx = math.fabs(e[i])
    if ( maxx <= epsilon ):
        print("\nRoots of polynomial are")
        print("  s#\troot")
        print("-------------------")
        for i in range (1,n+1):
            print(" %2d\t%8.3f" %(i,q[i]))
        break
```

Inverse of a Matrix . . .

```
#   Program to find inverse of a matrix using Gauss Jordon method

import sys
n = int(input("Enter order of matrix : "))
# Initialize matrix
a = []
print("\nEnter elements row-wise\n")
# For user input
for i in range(n):              # for loop for row entries
    col_entries =[]
```

```
    for j in range(n):       # for loop for column entries
        print("Enter Element a[",i,"][",j,"] : ", sep='', end='')
        t = float(input())
        col_entries.append(t)
    a.append(col_entries)
#  append identity matrix
for i in range(n):
    for j in range(n,n+n):
        if ((j-i) == n):
            a[i].append(1)
        else:
            a[i].append(0)
for k in range(n):
    temp = a[k][k]
    if (temp == 0):     #Pivot element becomes zero
        print("\nPivot element becomes zero\n")
        sys.exit(1)
    for j in range(k,n+n):
        a[k][j] = a[k][j]/temp
    for i in range(n):
        if ( i != k ):
            temp = a[i][k]
            for j in range(k,n+n):
                a[i][j] = a[i][j] - temp * a[k][j]
#  copy inverse on original matrix place
for i in range(n):
    for j in range(n,n+n):
        a[i][j-n] = a[i][j]
# For printing the matrix
print("\nInverse of given matrix...\n")
for i in range(n):
    for j in range(n):
        print("%6.2f" %(a[i][j]), end = " ")
    print()
```

Expanding a Determinant . . .

```
# Program to expand determinant using Gauss-elimination technique

import sys
n = int(input("Enter order of square matrix : "))
# Initialize matrix
a = []
print("\nEnter the entries row-wise\n")
# For user input
for i in range(n):               # for loop for row entries
    col_entries =[]
    for j in range(n):       # for loop for column entries
        print("Enter Element a[",i,"][",j,"] : ", sep='', end='')
        x = float(input())
```

```
            col_entries.append(x)
        a.append(col_entries)
    for k in range(0,n-1):
        temp1 = a[k][k]
        if (temp1 == 0):    #Pivot element becomes zero
            print("\nPivot element becomes zero\n")
            sys.exit(1)
        for i in range(k+1,n):
            temp2 = a[i][k]/temp1
            for j in range(k,n):
                a[i][j] = a[i][j] - temp2 * a[k][j]
    prod = 1.0
    for i in range(n):
        prod = prod * a[i][i]
    print("\nValue of determinant = ", prod)
```

Gauss Elimination Method . . .

```
#  Program to find solution of linear system of equations
#  by Gauss elimination method

import sys
n = int(input("Enter number of equations : "))
# Initialize augment matrix
a = []
# Initialize solution vectro
x = [0]*(n+1)
print("\nEnter the entries row-wise\n")
# For user input
for i in range(n):                # for loop for row entries
    col_entries =[]
    for j in range(n+1):    # for loop for column entries
        print("Enter Element a[",i,"][",j,"] : ", sep='', end='')
        t = float(input())
        col_entries.append(t)
    a.append(col_entries)
for k in range(n-1):
    temp1= a[k][k]
    if (temp1 == 0):    #Pivot element becomes zero
        print("\nPivot element becomes zero\n")
        sys.exit(1)
    for i in range(k+1,n):
        temp2 = a[i][k]/temp1
        for j in range(k,n+1):
            a[i][j] = a[i][j] - temp2 * a[k][j]
x[n-1] = a[n-1][n]/a[n-1][n-1]
for i in range(n-2,0,-1):
    s = 0
    for j in range(i+1,n):
        s = s + a[i][j] * x[j]
```

```
    x[i] = (a[i][n]-s)/a[i][i]
print("\nSolution of system of equations is\n")
for i in range(n):
    print("x[",i,"] = ", x[i], sep = "")
```

Gauss–Jordon Method . . .

```
#  program to find solution of linear system of equations
#  by Gauss-Jordon method

import sys
n = int(input("Enter number of equations : "))
# Initialize augment matrix
a = []
# Initialize solution vectro

x = [0]*(n+1)

print("\nEnter the entries row-wise\n")
for i in range(n):              # for loop for row entries
    col_entries =[]
    for j in range(n+1):        # for loop for column entries
        print("Enter Element a[",i,"][",j,"] : ", sep='', end='')
        t = float(input())
        col_entries.append(t)
    a.append(col_entries)
for k in range(n):
    temp = a[k][k]
    if (temp == 0):      #Pivot element becomes zero
        print("\nPivot element becomes zero\n")
        sys.exit(1)
    for j in range(k,n+1):
        a[k][j] = a[k][j]/temp
    for i in range(n):
        if ( i != k ):
            temp = a[i][k]
            for j in range(k,n+1):
                a[i][j] = a[i][j] - temp * a[k][j]
print("\nSolution of system of equations is\n")
for i in range(n):
    print("x[",i,"] = %5.2f" %a[i][n], sep = "")
```

Matrix Inversion Method . . .

```
#   Program to find solution of linear system of equations
#   by Matrix Inversion method

import sys
n = int(input("Enter number of equations : "))
```

```
# Initialize coefficient matrix
a = []
# Initialize constant
b = [0]*(n+1)
# Initialize solution vector
x = [0]*(n+1)

print("\nEnter the entries row-wise\n")
for i in range(n):                    # for loop for row entries
    col_entries =[]
    for j in range(n):        # for loop for column entries
        print("Enter Element a[",i,"][",j,"] : ", sep='', end='')
        t = float(input())
        col_entries.append(t)
    a.append(col_entries)
    print("Enter Element b[",i,"] : ", sep='', end='')
    b[i] = float(input())
#  append identity matrix
for i in range(n):
    for j in range(n,n+n):
        if ((j-i) == n):
            a[i].append(1)
        else:
            a[i].append(0)
for k in range(n):
    temp = a[k][k]
    if (temp == 0):    #Pivot element becomes zero
        print("\nPivot element becomes zero\n")
        sys.exit(1)
    for j in range(k,n+n):
        a[k][j] = a[k][j]/temp
    for i in range(n):
        if ( i != k ):
            temp = a[i][k]
            for j in range(k,n+n):
                a[i][j] = a[i][j] - temp * a[k][j]
#  copy inverse on original matrix place
for i in range(n):
    for j in range(n,n+n):
        a[i][j-n] = a[i][j]
#  multiply inverse with rhs
for i in range(n):
    for j in range(n):
        x[i] = x[i] + a[i][j] * b[j]

print("\nSolution of system of equations is\n")
for i in range(n):
    print("x[",i,"] = %6.2f" %x[i], sep = "")
```

Jacobi's Method . . .

```
#  Program to find solution of system of linear equation
#  using Jacobi's method

import sys
import math
n = int(input("Enter number of equations : "))
# Initialize augment matrix
a = []
# Initialize solution vectro
old_x = []*(n+1)
new_x = []*(n+1)

print("\nEnter the entries row-wise\n")
for i in range(n):              # for loop for row entries
    col_entries =[]
    for j in range(n+1):        # for loop for column entries
         print("Enter Element a[",i,"][",j,"] : ", sep='', end='')
         t = float(input())
         col_entries.append(t)
    a.append(col_entries)
maxiter = int(input("Enter maximum iterations permitted : "))
epsilon = float(input("Enter permitted error : "))
for i in range(n):
    old_x[i] = 0.0
for k in range(1,maxiter+1):
    bigrelerror = 0.0
    for i in range(n):
        sum = 0.0
        for j in range(n):
            if ( i != j ):
                sum = sum + a[i][j] * old_x[j]
        new_x[i] = (a[i][n]-sum)/a[i][i]
        relerror = math.fabs( (new_x[i]-old_x[i])/new_x[i])
        if ( relerror > bigrelerror ):
            bigrelerror = relerror
    if ( bigrelerror < epsilon ):
        print("\n\nSolution converges in", k, "iterations\n")
        for i in range(n):
            print("x[",i,"] = %8.4f" %new_x[i], sep = "")
        sys.exit(0)
    for i in range(n):
        old_x[i] = new_x[i]
print("\n\nSolution does not converges in", maxiter, "iterations\n")
```

Gauss–Seidel Method . . .

```
#  Program to find solution of system of linear equation
#  by Gauss(Seidel method

import sys
import math
n = int(input("Enter number of equations : "))
# Initialize augment matrix
a = []
# Initialize solution vector
x = []*(n+1)
print("\nEnter the entries row-wise\n")
for i in range(n):            # for loop for row entries
    col_entries =[]
    for j in range(n+1):      # for loop for column entries
         print("Enter Element a[",i,"][",j,"] : ", sep='', end='')
         t = float(input())
         col_entries.append(t)
    a.append(col_entries)
maxiter = int(input("Enter maximum iterations permitted : "))
epsilon = float(input("Enter permitted error : "))
for k in range(1,maxiter+1):
    bigrelerror = 0.0
    for i in range(n):
        sum = 0.0
        for j in range(n):
            if ( i != j ):
                sum = sum + a[i][j] * x[j]
        temp = (a[i][n]-sum)/a[i][i]
        relerror = math.fabs( (x[i]-temp)/temp)
        if ( relerror > bigrelerror ):
            bigrelerror = relerror
        x[i] = temp
    if ( bigrelerror < epsilon ):
        print("\n\nSolution converges in", k, "iterations\n")
        for i in range(n):
            print("x[",i,"] = %8.4f" %x[i], sep = "")
        sys.exit(0)
print("\n\nSolution does not converges in", maxiter, "iterations\n")
```

Power Method . . .

```
#  Program to solve a system of homogeneous linear equations using
#  Power method for largest eigen value and its associated eigen vector

import sys
import math
```

```
n = int(input("Enter number of equations : "))
# Initialize augment matrix
a = []
# Initialize solution vector
x = []*(n+1)
s = []*(n+1)
print("\nEnter the entries row-wise\n")
for i in range(n):          # for loop for row entries
    col_entries =[]
    for j in range(n):      # for loop for column entries
        print("Enter Element a[",i,"][",j,"] : ", sep='', end='')
        t = float(input())
        col_entries.append(t)
    a.append(col_entries)
epsilon = float(input("Enter required accuracy : "))
for i in range(n):              # for loop for solution vector
    x[i] = 1.0
while True:
    bigerror = 0.0
    for i in range(n):
        s[i] = 0.0
        for j in range(n):
            s[i] = s[i] + a[i][j] * x[j]
    lamda = math.fabs(s[1])
    for i in range(n):
        temp = s[i] / lamda
        error = math.fabs ( temp - x[i] )
    if ( error > bigerror ):
        bigerror = error
    x[i] = temp
    if ( bigerror <= epsilon ):
        break

print("\nLargest eigen value = %8.4f" %lamda)
print("\nand associated eigen vector is\n")
for i in range(n):
    print("x[",i,"] = %8.4f" %x[i], sep = "")
```

Lagrangian Interpolation Method . . .

```
#  program to interpolate a value of dependent variable 'y'
#  for given value of independent variable 'x' using
#  Lagrangian Interpolation Method

import sys
n = int(input("Enter number of tabulated points (n) : "))
# Initialize vectors
x = []
y = []
print("\nEnter", n, " pairs of (x,y)\n")
```

```
# For user input
for i in range(n):
    print("Enter x[",i,"] : ", sep='', end='')
    x.append(float(input()))
    print("Enter y[",i,"] : ", sep='', end='')
    y.append(float(input()))
a = float(input("Enter value of x for interpolation : "))
if ( (a < x[0]) and (a > x[n-1]) ):
    print("\nValue outside tabulated range\n")
    sys.exit(0)
sum = 0.0
for i in range(n):
    prod = 1.0
    for j in range(n):
        if ( i != j ):
            prod = prod * ((a-x[j])/(x[i]-x[j]))
    sum = sum + y[i] * prod
print("\nInterpolated value of y = %8.4f\n" %sum)
```

Newton's Forward Difference Interpolation Formula . . .

```
#  Program to interpolate a value of dependent variable 'y'
#  for given value of independent variable 'x' using
#  Newton's Forward Difference Interpolation Method

import sys
# function to compute factorial of 'm'
def factorial(m):
    p = 1
    for i in (2,m+1):
        p *= i
    return p

n = int(input("Enter number of tabulated points (n) : "))
# Initialize vectors
x = [0]
y = [0]
# initialize difference 'd' matrix
d = []
for i in range(1,n+1):
    t = []
    for j in range(1,n+1):
        t.append(0)
    d.append(t)
print("\nEnter", n, " pairs of (x,y)\n")
for i in range(1,n+1):
    print("Enter x[",i,"] : ", sep='', end='')
    x.append(float(input()))
    print("Enter y[",i,"] : ", sep='', end='')
```

```
    y.append(float(input()))
a = float(input("Enter value of x for interpolation : "))
if ( (a < x[1]) and (a > x[n]) ):
    print("\nValue outside tabulated range\n")
    sys.exit(0)
i = 2;
while ( a > x[i] ):
    i += 1
k = i-1;
u = (a - x[k])/(x[k+1] - x[k])
for j in range(1,n):
    for i in range(1,n-j+1):
        if ( j == 1 ):
            d[i][j] = y[i+1] - y[i]
        else:
            d[i][j] = d[i+1][j-1] - d[i][j-1]
sum = y[k];
for i in range(1,n-k+1):
    prod = 1.0;
    for j in range(i):
        prod = prod *(u-j);
    sum = sum + d[k][i] * prod / factorial(i)
print("\nInterpolated value of y = %8.4f\n" %sum)
```

Newton's Backward Difference Interpolation Formula . . .

```
#  Program to interpolate a value of dependent variable 'y'
#  for given value of independent variable 'x' using
#  Newton's Backward Difference Interpolation Method

import sys
# function to compute factorial of 'm'
def factorial(m):
    p = 1
    for i in (2,m+1):
        p *= i
    return p
n = int(input("Enter number of tabulated points (n) : "))
# Initialize vectors
x = [0]
y = [0]
# initialize difference 'd' matrix
d = [[0,0]]
for i in range(1,n+1):
    t = []
    for j in range(1,n+1):
        t.append(0)
    d.append(t)
print("\nEnter", n, " pairs of (x,y)\n")
```

```
for i in range(1,n+1):
    print("Enter x[",i,"] : ", sep='', end='')
    x.append(float(input()))
    print("Enter y[",i,"] : ", sep='', end='')
    y.append(float(input()))
a = float(input("Enter value of x for interpolation : "))
if ( (a < x[1]) and (a > x[n]) ):
    print("\nValue outside tabulated range\n")
    sys.exit(0)
i = 2;
while ( a > x[i] ):
    i += 1
k = i;
u = (a - x[k])/(x[k] - x[k-1])
for j in range(1,n):
    for i in range(j+1,n+1):
        if ( j == 1 ):
            print("i=",i,"j=",j)
            d[i][j] = y[i] - y[i-1]
        else:
            print("i=",i,"j=",j)
            d[i][j] = d[i][j-1] - d[i-1][j-1]
sum = y[k];
for i in range(1,k):
    prod = 1.0
    for j in range(i):
        prod = prod *(u+j)
    sum = sum + d[k][i] * prod / factorial(i)
print("\nInterpolated value of y = %8.4f\n" %sum)
```

Newton's Divided Difference Interpolation Formula . . .

```
#  Program to interpolate a value of dependent variable 'y'
#  for given value of independent variable 'x' using
#  Newton's Divided Difference Interpolation Method

import sys
n = int(input("Enter number of tabulated points (n) : "))
# Initialize vectors
x = [0]
y = [0]
# initialize difference 'd' matrix
d = [[0,0]]
for i in range(1,n+1):
    t = []
    for j in range(1,n+1):
        t.append(0)
    d.append(t)
print("\nEnter", n, " pairs of (x,y)\n")
```

```
for i in range(1,n+1):
    print("Enter x[",i,"] : ", sep='', end='')
    x.append(float(input()))
    print("Enter y[",i,"] : ", sep='', end='')
    y.append(float(input()))
a = float(input("Enter value of x for interpolation : "))
if ( (a < x[1]) and (a > x[n]) ):
    print("\nValue outside tabulated range\n")
    sys.exit(0)
i = 2;
while ( a > x[i] ):
    i += 1
k = i-1;
u = (a - x[k])/(x[k+1] - x[k])
for j in range(1,n):
    for i in range(1,n-j+1):
        if ( j == 1 ):
            d[i][j] = (y[i+1] - y[i])/(x[i+1] - x[i])
        else:
            d[i][j] = (d[i+1][j-1] - d[i][j-1])/(x[i+j] - x[i])
sum = y[k];
for i in range(1,n-k+1):
    prod = 1.0;
    for j in range(i):
        prod = prod *(a - x[k+j]);
    sum = sum + d[k][i] * prod
print("\nInterpolated value of y = %8.4f\n" %sum)
```

Fitting a Straight Line . . .

```
#  Program to fit a straight line of Y on X

import sys
n = int(input("Enter number of tabulated points (n) : "))
# Initialize vectors
x = [0]
y = [0]
print("\nEnter", n, " pairs of (x,y)\n")
for i in range(1,n+1):
    print("Enter x[",i,"] : ", sep='', end='')
    x.append(float(input()))
    print("Enter y[",i,"] : ", sep='', end='')
    y.append(float(input()))
sx = sx2 = sy = sxy = 0.0
for i in range(1,n+1):
    sx  = sx + x[i]
    sx2 = sx2 + x[i] * x[i]
    sy  = sy + y[i]
    sxy = sxy + x[i] * y[i]
```

```
denom = n * sx2 - sx * sx
a2 = ( n * sxy - sx * sy ) / denom
a1 = ( sy * sx2 - sx * sxy ) / denom
print("\nRegression coefficients are\n")
print("\nSlope of the regression line = %6.3f" %a2)
print("\nY-intercept for the line = %6.3f\n" %a1)
```

Fitting a Polynomial . . .

```
#  Program to fit a polynomial of degree 'm' through a set
#  of points using method of least squares

import math
m = int(input("Enter degree of polynomial(m) : "))
n = int(input("Enter number cf observations(n) : "))
m1 = m + 1
m2 = m + 2
# Initialize vectors
x = [0]
y = [0]
a = [0]
c = [ [0,0] ]
print("\nEnter", n, " pairs of (x,y)\n")
for i in range(1,m+2):
    a.append(0)
for i in range(1,n+1):
    print("Enter x[",i,"] : ", sep='', end='')
    x.append(float(input()))
    print("Enter y[",i,"] : ", sep='', end='')
    y.append(float(input()))
for i in range(1,m+5):
    t = []
    for j in range(1,m+5):
        t.append(0)
    c.append(t)
for i in range (1, m+2):
    for j in range (1, m+3):
        p = i+j-2;
        for k in range (1, n+1):
            c[i][j] = c[i][j] + math.pow(x[k],p)
    q=i-1
    for k in range (1, m+1):
        print("c[i][m2]=%8.4f" %c[i][m2])
        kk=input()
        c[i][m2] = c[i][m2]+ y[k] * math.pow(x[k],q)
for k in range (1,m1):
    max = math.fabs(c[k][k])
    p = k;
    for q in range (k+1,m1+1):
        if ( math.fabs(c[q][k]) > max ):
```

```
                max = math.fabs( c[q][k] )
                p = q
        if ( p != k ):
            for q in range (k,m2+1):
                temp = c[k][q]
                c[k][q] = c[p][q]
                c[p][q] = temp

        for i in range (k+1,m1+1):
            temp = c[i][k]/c[k][k]
            for j in range (k,m2+1):
                c[i][j] = c[i][j]-temp*c[k][j]
    a[m1] = c[m1][m2]/c[m1][m1]
    for i in range (m1-1,1,-1):
        sum = 0.0
        for j in range (i+-1,m1+1):
            sum = sum + c[i][j] * a[j]
        a[i] = ( c[i][m2]-sum ) / c[i][i]
    print("\nRegression coefficients\n\n");
    for i in range (1,m1+1):
        print("a[%d]= %8.4f\n" %(i,a[i]))
```

Fitting an Exponential Curve . . .

```
#  Program to fit an exponential curve of type  y = ae^bx

import math
n = int(input("Enter number of tabulated points (n) : "))
# Initialize vectors
x = [0]*(n+1)
y = [0]*(n+1)
print("\nEnter", n, " pairs of (x,y)\n")
for i in range(1,n+1):
    print("Enter x[",i,"] : ", sep='', end='')
    x[i]= float(input())
    print("Enter y[",i,"] : ", sep='', end='')
    y[i]= float(input())
sx = sx2 = slogy = sxlogy = 0.0
for i in range(1,n+1):
    xx = x[i];
    yy = math.log10(y[i])
    sx  = sx + xx
    sx2 = sx2 + xx * xx
    slogy  = slogy + yy
    sxlogy = sxlogy + xx * yy
denom = n * sx2 - sx * sx
b = (n * sxlogy - sx * slogy) / denom
c = (slogy * sx2 - sx * sxlogy) / denom
a = math.exp(c)
print("\nRegression coefficients are\n")
```

```
print("\na = %.5f" %a)
print("\nb = %.5f" %b)
```

Trapezoidal Rule . . .

```
#  Program to implement Trapezoidal rule for tabulated function

import math
n = int(input("Enter number of intervals (n) : "))
h = float(input("Enter size of interval (h) : "))
# Initialize vectors
x = [0]*(n+2)
y = [0]*(n+2)
print("\nEnter", n+1, " pairs of (x,y)\n")
for i in range(1,n+2):
    print("Enter x[",i,"] : ", sep='', end='')
    x[i]= float(input())
    print("Enter y[",i,"] : ", sep='', end='')
    y[i]= float(input())
sum = ( y[1] + y[n+1] )/2
for i in range(2,n+1):
    sum = sum + y[i]
integral = h * sum
print("\nValue of the integral = %.4f\n" %integral)
```

```
#  Program to implement Trapezoidal rule for tabulated function

import math
# definition of function to compute expression e^(1/2)x
def f(x):
    return math.exp(-0.5*x)
a = float(input("Enter lower limit of integration (a) : "))
b = float(input("Enter upper limit of integration (b) : "))
n = int(input("Enter number of interval (n) : "))
h = (b - a)/n
sum = (f(a)+f(b))/2.0
for i in range(1,n):
    sum = sum + f(a+i*h)
integral = h * sum
print("\nValue of the integral = %.4f\n" %integral)
```

Simpson's 1/3rd Rule . . .

```
#  Program to implement Trapezoidal rule for tabulated function

import math
n = int(input("Enter number of intervals (n) : "))
h = float(input("Enter size of interval (h) : "))
```

```
# Initialize vectors
x = [0]*(n+2)
y = [0]*(n+2)
print("\nEnter", n+1, " pairs of (x,y)\n")
# For user input
for i in range(1,n+2):
    print("Enter x[",i,"] : ", sep='', end='')
    x[i]= float(input())
    print("Enter y[",i,"] : ", sep='', end='')
    y[i]= float(input())
sum = y[1] + y[n+1]
s2 = s4 = 0.0
for i in range(2,n+1,2):
    s4 = s4 + y[i]
for i in range(3,n,2):
    s2 = s2 + y[i]
integral = (h/3.0) * ( sum + 2 * s2 + 4 * s4 )
print("\nValue of the integral = %.4f\n" %integral)
```

```
#   Program to implement Simpson's 1/3rd rule for known function

import math
# definition of function to compute expression e^(1/2)x
def f(x):
    return math.exp(-0.5*x)
a = float(input("Enter lower limit of integration (a) : "))
b = float(input("Enter upper limit of integration (b) : "))
n = int(input("Enter number of interval (n) : "))
h = (b - a)/n
sum = f(a)+f(b)
s2 = s4 = 0.0;
for i in range(1,n,2):
    x = a + i * h
    s4 = s4 + f(x)
for i in range(2,n,2):
    x = a + i * h;
    s2 = s2 + f(x);
integral = (h/3.0) * ( sum + 2 * s2 + 4 * s4 )
print("\nValue of the integral = %.4f\n" %integral)
```

Romberg Integration . . .

```
#   Program to implement Simpson's 1/3rd rule for known function

import math
# definition of function to compute expression e^(1/2)x
def f(x):
    return (math.exp(-0.5*x))
```

```
e = [0]*10
r = [e]*10
a = float(input("Enter lower limit of integration (a) : "))
b = float(input("Enter upper limit of integration (b) : "))
epsilon = float(input("Enter allowed precision : "))
i = 1
n = 1
h = b - a
sum = ( f(a) + f(b) )/2
r[i][1] = h * sum
print(r[i][1])
flag = True
while ( flag ):
    n *= 2
    h /= 2
    i += 1
    for j in range(1,n+1,2):
        x = a + j * h
        sum += f(x)
    r[i][1] = h * sum
    for k in range(2,i+1):
        temp = 4 ** (k-1)
        r[i][k] = (temp * r[i][k-1] - r[i-1][k-1])/(temp - 1)
    if ( math.fabs(r[i][i]-r[i][i-1]) <= epsilon ):
        print("\nElements of Romberg table are\n\n")
        for k in range(1,i+1):
            for j in range(1,k+1):
                print("%.4f\t" %r[k][j],end="")
            print()
            flag = False
print("\nValue of integral = %.4f\n" %r[i][i])
```

Gauss's Quadrature Formula . . .

```
#   Program to implement Simpson's 1/3rd rule for known function

import math
# definition of function to compute expression x*x + 8.0*x + 15
def g(y):
    return (y*y + 8*y + 15)
n = int(input("\nEnter number of points  : "))
r = [0]*(n+1)
y = [0]*(n+1)
a = float(input("Enter lower limit of integration (a) : "))
b = float(input("Enter upper limit of integration (b) : "))
for i in range(1,n+1):
    print("Enter r[",i,"] : ", sep='', end='')
    r[i] = float(input())
    print("Enter y[",i,"] : ", sep='', end='')
    y[i] = float(input())
```

```
print(r)
print(y)
sum = 0.0
for i in range(1,n+1):
    print(r[i])
    sum += r[i] * g(y[i])
integral = ((b-a)/2.0)*sum
print("\nValue of integral = %.4f\n" %integral)
```

Euler's Method . . .

```
#  program to solve a first order differential equation using
#  Euler's method

# definition of function to compute expression x*y
def f(x,y):
    return (x*y)
print("\nEnter starting solution point (x1,y1)\n")
x1 = float(input("\nEnter vlue for x1 : "))
y1 = float(input("Enter vlue for y1 : "))
xf = float(input("\nEnter last point of the interval (xf) : "))
h = float(input("\nEnter step size to be used (h) : "))
x = x1
y = y1
i = 1
print("\nSolution points\n")
print("\t i\tx\ty")
print("\t%2d\t%.3f\t%.3f" %(i,x,y))
while ( x  < xf ):
    y = y + h * f(x, y)
    x = x + h
    i = i + 1
    print("\t%2d\t%.3f\t%.3f" %(i,x,y))
```

Runge–Kutta Second Order Method . . .

```
#  program to solve a first order differential equation using
#  Runge-Kutta (Heun's) second order method

# definition of function to compute expression x*y
def f(x,y):
    return (x*y)
print("\nEnter starting solution point (x1,y1)\n")
x1 = float(input("\nEnter vlue for x1 : "))
y1 = float(input("Enter vlue for y1 : "))
xf = float(input("\nEnter last point of the interval (xf) : "))
h = float(input("\nEnter step size to be used (h) : "))
```

```
x = x1
y = y1
i = 1
print("\nSolution points\n")
print("\t i\tx\ty")
print("\t%2d\t%.3f\t%.3f" %(i,x,y))
while ( x  < xf ):
    s1 = f(x,y)
    x = x + h
    s2 = f(x,y+h*s1)
    y = y+h*(s1+s2)/2
    i = i + 1
    print("\t%2d\t%.3f\t%.3f" %(i,x,y))
```

Runge–Kutta Fourth Order Method . . .

```
#  program to solve a first order differential equation using
#  Runge-Kutta fourth order method

# definition of function to compute expression x*y
def f(x,y):
    return (x*y)
print("\nEnter starting solution point (x1,y1)\n")
x1 = float(input("\nEnter vlue for x1 : "))
y1 = float(input("Enter vlue for y1 : "))

xf = float(input("\nEnter last point of the interval (xf) : "))

h = float(input("\nEnter step size to be used (h) : "))

x = x1
y = y1
i = 1

print("\nSolution points\n")
print("\t i\tx\ty")
print("\t%2d\t%.3f\t%.3f" %(i,x,y))
while ( x  < xf ):
    s1 = f(x,y)
    s2 = f(x+h/2,y+h*s1/2)
    s3 = f(x+h/2,y+h*s2/2)
    s4 = f(x+h,y+h*s3)
    s = (s1+2*s2+2*s3+s4)/6
    y = y+h*s
    x = x + h
    i = i + 1
    print("\t%2d\t%.3f\t%.3f" %(i,x,y))
```

Euler's Predictor Corrector Method . . .

```
#  Program to solve a first order differential equation using
#  predictor corrector  (Modified Euler's) method s

import math

# definition of function to compute expression x*y
def f(x,y):
    return (x*y)

print("\nEnter starting solution point (x1,y1)\n")
x1 = float(input("\nEnter vlue for x1 : "))
y1 = float(input("Enter vlue for y1 : "))

xf = float(input("\nEnter last point of the interval (xf) : "))

h = float(input("\nEnter step size to be used (h) : "))

epsilon = float(input("Enter prescribed tolerance : "))

x = x1
y = y1
i = 1

print("\nSolution points\n")
print("\t i\tx\ty")
print("\t%2d\t%.3f\t%.3f" %(i,x,y))
while ( x  < xf ):
    y2 = y + h * f(x, y)
    flag = True
    while ( flag ):
      y1 = y2
      y2 = y + (h / 2) * (f(x, y) + f(x+h, y2))
      if ( math.fabs(y2-y1) <= epsilon ):
            flag = False
    y = y2
    x = x + h
    i = i + 1
    print("\t%2d\t%.3f\t%.3f" %(i,x,y))
```

❑ ❑ ❑

Appendix D

Proficiency Test

Multiple Choice Questions

Each of the following multiple choice question has only one correct response. You are to select the most appropriate correct response among the options given.

1. What is the number of significant digits in number 0.002063

(*a*) 6
(*b*) 3
(*c*) 4
(*d*) None of above

2. The sum of numbers 0.5467E5 and 0.7253E2 on a computer system that represent real number as normalized floating-point numbers with four digit mantissa and two digit exponent is

(*a*) 0.5474E5
(*b*) 0.5470E2
(*c*) 0.4327E5
(*d*) None of above

3. If the number 0.005998 is truncated to three digits, the relative error due to truncation will approximately be of the order of

(*a*) 0.133×10^{2}
(*b*) 0.133×10^{-2}
(*c*) 0.144×10^{-2}
(*d*) 0.153×10^{-2}

4. If the number 0.005998 is rounded-off to three digits, the relative error due to rounding will approximately be of the order of

   (*a*) 0.321×10^{-4}
   (*b*) 0.300×10^{-4}
   (*c*) 0.333×10^{3}
   (*d*) 0.333×10^{-3}

5. If the number 0.00458529 is rounded-off to three digits, the absolute error due to rounding will approximately be of the order of

   (*a*) 0.471×10^{-3}
   (*b*) 0.475×10^{-5}
   (*c*) 0.471×10^{-5}
   (*d*) 0.521×10^{-5}

6. If the number 0.005998 is truncated to three digits, the relative error due to truncation will approximately be of the order of

   (*a*) 0.521×10^{-5}
   (*b*) 0.521×10^{-4}
   (*c*) 0.550×10^{-5}
   (*d*) 0.529×10^{-5}

7. In using an iterative method, if the solution is desired to be accurate to four significant digits, the algorithm must ensure that the relative error is

   (*a*) Less than or equal to 0.0001
   (*b*) Less than or equal to 1.0×10^{-4}
   (*c*) Less than or equal to 10.0×10^{-5}
   (*d*) All of above

8. Which of the following is a transcendental equation

   (*a*) $2^x - x - 3 = 0$
   (*b*) $xe^x - x\sin x = 0$
   (*c*) $e^{3x} - 0.5x = 0$
   (*d*) All of above

9. Which of the following is not a polynomial equation

   (*a*) $x^3 - 2x + 5 = 0$
   (*b*) $2^x - 5x + 10 = 0$
   (*c*) $0.5x^3 - 5x^2 - 20 = 0$
   (*d*) None of above

10. Which of the following statement is not true about iterative methods?

(*a*) Iterative methods can be concisely expressed as computational algorithms.
(*b*) Iterative method give desired solution in fixed number of steps.
(*c*) Algorithms for iterative methods are capable of handling class of similar problems.
(*d*) None of above

11. Which of the following statement is true about roots of an nth degree polynomial?

(*a*) It has exactly n roots.
(*b*) Roots can be real and repeated.
(*c*) Roots can be complex.
(*d*) All of above
(*e*) None of above

12. Suppose a root of a non-linear equation lies in the interval (1,2). If bisection method is used to find the root of the equation, accurate to four significant digits, how many iterations will be required?

(*a*) 15
(*b*) 14
(*c*) 16
(*d*) 32

13. ____________ method has a fastest rate of convergence.

(*a*) Regula-falsi method
(*b*) Bisection method
(*c*) Secant method
(*d*) Method of Successive Approximations

14. Rate of convergence of secant method is

(*a*) 2
(*b*) 1
(*c*) 1.618
(*d*) 1.5

15. Consider the equation $x^3 - 4x - 9 = 0$. If starting approximation is 2.95, the next approximation by Newton-Raphson method will be

(*a*) 2.7068
(*b*) 2.7501
(*c*) 2.7296
(*d*) 2.7198

16. In which of the following methods convergence is guaranteed?

(a) Newton-Raphson method
(b) Secant method
(c) Method of Successive Approximations
(d) None of above

17. Which of the following statement is not true about determinants?

(a) If even number of rows are interchanged, the value of the determinant remain unchanged.
(b) If any two rows or columns are identical, the determinant will vanish.
(c) If each element of a given row or column is multiplied by 5, then the resultant value of determinant will be 5 time that of original value.
(d) All of above

18. The objective of pivoting is

(a) Reduce the number of operations to be performed to find the desired solution.
(b) Improve the accuracy of the solution.
(c) Both (a) & (b)
(d) None of above

19. Consider the following system of linear equations

$$2x_1 + 8x_2 + 2x_3 = 14$$
$$x_1 + 6x_2 - x_3 = 13$$
$$2x_1 - x_2 + 2x_3 = 5$$

If current approximation is $x_1 = 4.5, x_2 = 0.5, x_3 = -1.5$, then the Jacobi's iterative method will give next approximation as

(a) $x_1 = 4.5, x_2 = 0.5, x_3 = -1.5$
(b) $x_1 = 4.5, x_2 = 0.5, x_3 = -1.5$
(c) $x_1 = 4.5, x_2 = 0.5, x_3 = -1.5$
(d) $x_1 = 4.5, x_2 = 0.5, x_3 = -1.5$

20. Consider the following system of linear equations

$$2x_1 + 8x_2 + 2x_3 = 14$$
$$x_1 + 6x_2 - x_3 = 13$$
$$2x_1 - x_2 + 2x_3 = 5$$

If current approximation is $x_1 = 4.5, x_2 = 0.5, x_3 = -1.5$, then the Gauss-Seidel iterative method will give next approximation as

(a) $x_1 = 4.5, x_2 = 0.5, x_3 = -1.5$

(*b*) $x_1 = 4.5, x_2 = 0.5, x_3 = -1.5$
(*c*) $x_1 = 4.5, x_2 = 0.5, x_3 = -1.5$
(*d*) $x_1 = 4.5, x_2 = 0.5, x_3 = -1.5$

21. If a function is given in the tabulated form, then the process of approximating the function on one of non-tabulated point within the tabulation interval is known as

(*a*) Extrapolation
(*b*) Interpolation
(*c*) Regression analysis
(*d*) Correlation analysis

22. If a function is given in the tabulated form, then the process of approximating the function on a point outside the tabulated interval is known as

(*a*) Extrapolation
(*b*) Interpolation
(*c*) Regression analysis
(*d*) Correlation analysis

23. Suppose a function is given in the tabulated form at unequal intervals, which of the following methods cannot be used for interpolation?

(*a*) Lagrangian method
(*b*) Newton's forward difference method
(*c*) Newton's divided difference method
(*d*) None of above

24. In fitting a fourth degree polynomial using method of least squares, the number of normal equations obtained is

(*a*) 4
(*b*) 5
(*c*) 6
(*d*) 8

25. The method of least squares minimizes the

(*a*) Sum of deviations
(*b*) Sum of squares of deviations
(*c*) Either (*a*) or (*b*) depending on the degree of polynomial
(*d*) None of above

26. Regression line of y on x and x on y passing through following points

x:	0.1	0.2	0.3	0.4	0.5	0.6
y:	5.1	5.3	5.6	5.7	5.9	6.1

intersects at point

(*a*) (0.3, 5.6)
(*b*) (0.4, 5.7)
(*c*) (0.35, 5.62)
(*d*) (0.45, 5.60)

27. Consider the following first order ordinary differential equation

$dy/dx = xy$ with $y(1) = 5$

If interval of size = 0.1 is used, then what will be the next solution point by Euler's method?

(*a*) (1.1, 5.2)
(*b*) (1.1, 5.5)
(*c*) (1.1, 6.0)
(*d*) (1.1, 5.3)

28. Which of the following is a multi-step method?

(*a*) Euler's method
(*b*) Runge-Kutta fourth order method
(*c*) Taylor's method
(*d*) Adams-Bashforth method

Answers

1.	c	**2.**	a	**3.**	b	**4.**	d
5.	c	**6.**	d	**7.**	d	**8.**	d
9.	b	**10.**	b	**11.**	d	**12.**	b
13.	c	**14.**	c	**15.**	c	**16.**	d
17.	d	**18.**	b	**19.**	b	**20.**	b
21.	b	**22.**	a	**23.**	b	**24.**	b
25.	b	**26.**	c	**27.**	b	**28.**	d

❑ ❑ ❑

Appendix E

Numerical and Statistical Software Libraries

Many excellent software libraries are available for desktop computers and workstations. Some of these are

1. **DERIVE** is one the computer algebra systems. In addition to the PC version, it offers a ROM-card version for the HP 95LX Palmtop computer.

   Source: Soft Warehouse, Inc., 3660 Waialae Avenue, Suite 304, Honolulu, HI 96816-3236.

2. **ESSL** (Engineering and Scientific Subroutine Library) version 2, developed by IBM, consists of more than 400 subroutines callable from C and FORTRAN. The package includes topics in numerical quadrature, interpolation, random number generation, FFTs, linear systems, and eigenvalue problems. This package is available for workstation.

   Source: (914)385-9829.

3. **IMSL** (International Mathematical and Statistical Library) is a library of over 900 useful FORTRAN subroutines that cover both mathematical and statistical problems. There is a comparable C/Base/Library of numerical functions as well. Moreover, this library can be installed on a variety of machines, including CRAY Y-MP.

   Source: IMSL, Inc. Houston, TX (713)242-6776.

4. **LAPACK** is a library of FORTRAN 77 suboutines for solving problems in numerical linear algebra, including eigenvalue problems, and singular value problems. This library supersedes LINPACK and EISPACK. Source: Individual routines available through *netlib*. Currently, the email addresses are *netlib@ornl.gov* and *netlib@research.att.com*. For just general information send mail to one of these two addresses with message "*send index from lapack.*". The complete package, consisting of about 6,00,000 lines of FORTRAN code, can be obtained on magnetic media from NAG.

5. **MACSYMA** is a software package that has been around for a long time. Some of its features include simplification of trigonometric functions, differential-equation packages, and color graphics. It is now available for the PC.

   Source: MACSYMA, Inc., 20 Academy Street, Arlington, MA 02174.

6. **MAPLE V** is very powerful software package that performs both symbolic and numeric computations. it provides both two-dimensional and three-dimensional color graphics capabilities. This software runs on almost all computers including PC's, Macintoshes, Workstations, and Mainframes.

   Source: Waterloo Maple Software, 160 Columbia Street West, Dept. 507, Waterloo, Ontario, Canada N2L 3L3. Email: *wmsi@daisy.waterloo.edu*.

7. **MATHEMATICA** is one of the best-known software packages for computing a large variety of mathematical problems. Moreover, it has excellent two-dimensional and three-dimensional graphics capabilities and does both symbolic and numeric computations. This system is available on a very large variety of computers.

   Source: Wolfram research, Inc., P.O. Box 6059, Champaign, IL 61821-9902.

8. **MATLAB** covers several topics in mathematics, including linear algebra, digital signal processing, graphing, and numerical methods. It is available both for the MS-DOS machines as well as the Macintosh computers.

   Source: The Math Works Inc., Cochituate Place, 24 Prime Park Way, Natick, MA 01760. Email: *info@mathworks.com*.

9. **NAG** (Numerical Algorithms Group) has provided a large library of FORTRAN subprograms. In addition, this group has now come out with a symbolic solver software package, called AXIOM. This package also produces two-dimensional and three-dimensional graphics.

   Source: Numerical Algorithms Group, Inc. 1400 Opus Place, Suite 200, Downers Grove, IL 60515-5702.

10. **NETLIB**, mentioned with regard to the LAPACK software, is a tremendous resource for numerical libraries. Rather than deal with a long list, one could send an email message containing the line "help" to one of the Internet addresses: *netlib@research.att.com*, *netlib@ornl.gov*, *netlib@nec.no* for information on how to use *netlib* and for an overview of the mathematical software.

11. **NUMERICAL RECIPES** is a well-known set of 300-plus numerical routines for a variety of programming languages, including C, FORTRAN, Pascal, and BASIC. In addition to the books, the source code for each version is available on diskette.

    Source: Cambridge University Press, 40 West 20th Street, New York, NY 10011-4211.

❑ ❑ ❑

Index

F

G

H

I

J

L

M

N

O

P

Q

R

S

T

U

V

W

Z

❑ ❑ ❑